U0259444

ARCHITECTURE
–STUDIO

法国AS建筑工作室

Philip Jodidio / 菲利普 · 朱迪狄欧
Born in New Jersey, Philip Jodidio studied art history and economics at Harvard. In 1979, he became the Editor in Chief of the French monthly *Connaissance des Arts,* a position he held until 2002. A specialist in contemporary Architecture, Philip Jodidio has published over 100 books including the Architecture Now! series (Taschen), and monographs on such influential architects as Tadao Ando, Santiago Calatrava, Norman Foster, Zaha Hadid, Richard Meier, Álvaro Siza and Renzo Piano. His body of work has served to bring contemporary architecture to the fuller attention of the general public in numerous countries.
菲利普·朱迪狄欧出生于美国新泽西州，1979年曾在哈佛大学研读艺术史和经济学。随后，他成为法国著名艺术杂志《艺术知识（Connaissance des arts）》的主编，并任职二十余年。作为当代建筑领域的专家，菲利普·朱迪狄欧先后出版了上百本著作，其中包括由Taschen出版社出版的《Architecture Now!》系列书籍，以及安藤忠雄、圣地亚哥·卡拉特拉瓦、诺曼·福斯特、扎哈·哈迪德、查理·麦尔、阿尔瓦罗·西扎以及伦佐·皮亚诺等著名建筑师的专著。他的这些文学作品让不同国家的众多读者更加关注并了解当代建筑。

Architecture-Studio / 法国AS建筑工作室
Founded in Paris in 1973, Architecture-Studio is based in Paris, Shanghai and Venice. The firm has a team of 150 architects, planners and interior designers, representing 25 different nationalities, who support the 12 partners.
法国AS建筑工作室在1973年创立于巴黎，在上海和威尼斯均设有分支机构。如今，工作室十二位合伙建筑师身边集结了约一百五十名由25个不同国籍成员组成的专业团队，包括建筑师、规划师和室内设计师等。
Martin Robain, Rodo Tisnado, Jean-François Bonne, Alain Bretagnolle, René-Henri Arnaud, Laurent-Marc Fischer, Marc Lehmann, Roueïda Ayache, Gaspard Joly, Marie-Caroline Piot, Mariano Efron, Amar Sabeh el Leil.
马丁·罗班、罗多·蒂斯纳多、让-弗朗索瓦·博内、阿兰·布勒塔尼奥勒和勒内-亨利·阿诺、洛朗-马克·费希尔、马克·莱曼、罗伊达·阿亚斯、以及贾斯帕·朱利、玛丽卡·碧欧、马里亚诺·艾翁和艾马·萨布埃雷。
The international presence of Architecture-Studio is particularly strong in China. Architecture-Studio China, which managed by Vincent Hannotin, Pan Li and Ying Chaojun, pursues Architecture-Studio's philosophy. It combines the ideas and characteristics of the French and Chinese cultures.
法国AS建筑工作室尤其注重在中国的发展，其中国团队由文森·汉诺丁、潘莉和应朝君联合负责。在中国的事务所延续了法国AS建筑工作室的理念及工作方式，并与巴黎总部紧密联合将中法两国的精神理念和文化特点融合在一起。

Note / 说明
The project descriptions in this book are preceded by a code, which is the internal classification system of Architecture-Studio. The letters refer to the location of the project and this is followed by a number which indicates the specific project at this site. If the letter code is followed by a dash this indicates there was more than one project. An index at the end of the book lists the 1 200 project codes which sum up the work of Architecture-Studio since its inception.
本书中介绍的建筑项目均配有对应的编号，此编号由字母和数字组成，为法国AS建筑工作室工作时内部使用的项目编号。字母为项目所在地地名的缩写，跟随的数字代表的是所在城市所开展的项目序号，若字母后跟随的是小横线，则代表的是同一个地区几个不同项目的合并。列于本书最后索引部分的1200个编号是法国AS建筑工作室从建立起至今所有项目列表。

Editorial coordination / 英法双语原版出版协作
Olivia Barbet-Massin
Vanessa Clairet & Tiphaine Riva (Architecture-Studio)

Chinese version coordination / 中文版协作
Wu Qiong, Hao Yanjie, Shen Dan(Architecture-Studio), Liu Zhanhui /
吴琼、郝艳杰、沈丹（法国AS建筑工作室）、柳战辉

Layout and Graphic Design / 排版设计
Change is good (Paris)
YIWEI culture / 乙未文化

Project descriptions / 项目介绍
Hugo Lacroix / 雨果 · 拉克鲁瓦

Translations / 翻译
Tianjin Boven Translation Co.,Ltd. / 天津市博文翻译有限公司

Copy editing / 审稿
Jeanne El Ayeb
Wang YunShi / 王云石

isbn978-2-37368-026-3
Dominique Carré éditeur / 出版人
Is an imprint of Éditions La Découverte
本书英法双语原始版本为 Éditions La Découverte出版
www.editionscarre.com

ARCHITECTURE —STUDIO

法国AS建筑工作室

PHILIP JODIDIO
菲利普·朱迪狄欧 编著

METEO
Canal couvert

NATURAL ARCHITECTURE THE WORK OF ARCHITECTURE STUDIO

ARCHITECTURE NATURELLE LE TRAVAIL D'ARCHITECTURE STUDIO

浑然天成的建筑——法国AS建筑工作室的作品

Contemporary architecture has long since ceded to the cult of the "genius" – the creator whose every gesture speaks of art and originality – a man with a hat like Frank Lloyd Wright, a man with little round glasses like Corbu. The very presence of the "starchitect" brings media attention and political acceptability, at least until budgets get out of hand. Even beyond the rarified world of such famous figures, the commonest organization of architectural offices revolves around an individual whose name is put forward at the front, even if that person is not always involved in the projects concerned. This is not the case with Architecture-Studio, which was founded in 1973 in Paris. Its twelve partners do not put themselves forward as individuals and in the very real equality that they have established lies the first reason for the longevity and continued, or rather increasing, presence of the office[1].

Starting from a conceptual approach, projects are discussed in weekly meetings in Paris, and each partner is free to comment, to make proposals, and thus to participate in the final result. Speaking of the early days of Architecture-Studio, Martin Robain says, "Our ideas had to do with equality and sharing." To this day, the structure of the firm is not pyramidal, but consists of a more horizontal layering of responsibilities, with the partners, associates and other architects who work for Architecture-Studio all participating in what might be best called an eco-system, rather than an ego-driven quest for fame. No single architect working for the firm can be identified as the "inventor" of the concept or the form of a building; rather theirs is a participatory process that ultimately resembles that of a natural system, where each person or entity plays a role in creating the final result. The work of the office, and indeed its list of partners, speaks to what was defined from the outset as an open, international approach, one that retains a "militant" view about the role and usefulness of architecture. The point has never been to change the world, but to create buildings that are an integral part of their site and their neighborhood, somehow not only rooted in the earth they are built on, but also contributing to their environment, making it better in ways that are not only aesthetic, but also social. This approach inevitably means that the office eschews the stylistic foibles that have so marked the evolution of contemporary architecture from Post-Modernism to Minimalism and beyond. If the partners reject the idea of the architect as artist, they do all agree that time and its users may find that a well-conceived building is indeed a work of art. Each of their works is based on the programmatic requirements, ecology, economics and innovation as those terms are defined by the context concerned. Context for them is not just a matter of aligning rooflines or window patterns; it is also about achieving a deeper appreciation of each project, an appreciation that can also delve into the historical background.

For Architecture-Studio, being "international" has more than one meaning. Members of the team, including partners, are from a number of different countries, ranging from Peru to Syria. Additionally, from the time that their European Parliament building made it clear that they could handle very large-scale projects, the firm has been building extensively outside of France, including China, the Middle East and, more recently, Africa. A sense of openness, which indeed characterizes the office itself, is an essential element of their approach to work in other countries. With their base in Venice at the Ca'asi, they have organized numerous exhibitions that have focused on developing parts of the world, from China to Africa and on to the Middle East. Not only do they acknowledge that each of these areas has its own characteristics, but they have actively sought not to impose their own point of view but to learn from the contexts concerned. As they say, in Africa ecology is not so much a matter of avoiding excess energy use or pollution as it is an issue of pure survival. In China or the Middle East, they have combined their sensitivity towards the historic context with the needs, or even desires, of emerging countries to be fully modern.

当代建筑一直以来都是“天才”的狂热追求——这里的“天才”是指举手投足都彰显艺术气息和创意的创造者——他们头戴一顶帽子，看起来像弗兰克·劳埃德·赖特；戴着一副小圆框眼镜，看起来像科尔布。唯独“明星建筑师”的出现能够吸引媒体的关注，同时在政治上得到反响，这种现象至少持续至预算失控之前。即使抛开这些建筑界稀有的著名代表，最常见的建筑事务所也时常以个人名字命名，并且所有工作都围绕着此人展开，即便该人不一定真正参与到某些项目中。然而，1973年成立于巴黎的法国AS建筑工作室并不会出现此种情况。它的十二名合伙人不把自己视为单独的个体，而是做到真正的平等，他们的运作首先考虑持久性和可持续性，更确切地说，是增加事务所的稳定性[1]。

每个进行中的项目都会在巴黎周会上被讨论，从概念构思开始，每位合伙人都可自由发表意见，献计献策，从而分享最终成果。在说起法国AS建筑工作室的初期时，马丁·罗班说：“我们的想法必须与平等和共享有关”，至今，该公司的架构仍然不是金字塔形的，而是由更多同等级别的人组成的，合伙人、合作伙伴以及其他建筑师都参与到工作中，而不是单纯追求所谓的自我名声，称该运作系统为生态体系最合适不过了。没有哪个为公司效力的建筑师可以被定义为概念构思或建筑形式上的“创建者”；在该体系中每个人或实体都对最终结果作出贡献。工作室及其许多合伙人的工作，从一开始就被定义为具有开放性和国际化的方式，其中，对关于建筑的角色和作用保留“激进”的看法。该观点不是为了改变世界，而是将建筑物设计成为其周边环境和附近区域的组成部分，在某种程度上，建筑不仅根植于它们的建造位置，还应对其环境发展作出贡献，使其以更好的形式呈现，不仅美观而且社会化。这种方法无疑意味着事务所可以避免追随从后现代主义到极简主义以及随之而来的当代建筑演变中在风格上的潮流。如果说合伙人否定了建筑师即是艺术家的想法，他们都承认时间会让其用户发现一座构思巧妙的建筑其实就是一件艺术品。每个项目都有生态性、经济性和创新性等不同的需求，而这些条条框框与建筑的内涵密切相关。建筑设计不仅仅关注齐整的屋顶轮廓线与窗户的样式和排列，更是要实现建筑更深沉的意义，其中包括促进建筑与其历史背景的和谐。

对于法国AS建筑工作室而言，“国际化”具有多种含义。团队成员包括合伙人来自不同国家（从秘鲁到叙利亚）。此外，从欧洲议会大厦的顺利竣工来看，很明显，该工作室有能力处理国际大型项目。该工作室一直在法国以外的地区进行设计，包括中国、中东，最近还扩展到非洲。工作室本身就具有开放和包容的特征，这为设计师们在其他国家开展设计工作提供了基本的方法。事务所在威尼斯建立的CA'ASI艺术展览馆已经成功举办众多的展览，展览主题主要集中在世界各地的发展中地区，从中国到中东以及非洲。他们承认每一个地区都有各自的特点，他们也不寻求强加自己的观点，而是从相关环境中学习。正如他们所说，在非洲，生态与其说是一种避免过多能源使用或污染的问题，不如说它是一种纯粹的生存问题。在中国或中东地区，工作室依靠对历史的认识和感知，尽量了解当地对实现国家全面现代化的渴望和需求。

References to the economic consciousness of Architecture-Studio have several implications. Working within an established budget, and indeed offering a client more than they expect, is a part of their understanding of economics. So, too, is the related idea of an economy of means, whereby superfluous additions are consciously rejected. Interestingly, this includes some aspects of high technology that are often added to buildings, perhaps in an effort to demonstrate how up-to-date they are. Architecture-Studio takes the position that technology must be part and parcel of the architecture, of the project, not something that is tacked on for the sake of proving an interest in contemporary means and methods. It might be said that the firm's sense of "economy" also takes technology into account – what is truly useful and integrated into the architecture is surely present, whether the means employed are of recent invention or of a more established variety. For example, for them, correct orientation and shading are more important than a field of solar panels added to a roof, though they never exclude the use of technological means when they are justified.

Architecture-Studio has had a number of projects where existing buildings have been retained as the basis for new uses, or as a transition into modernity. In contrast with some of their well-known colleagues, their goal has never been to erase the past, but rather they seek to embrace what is clearly another element of context. Where architectural quality exists they acknowledge and amplify this, reworking circulation patterns, making use of interstitial spaces to stimulate the activity and flow of users. Here again, the image of the creation of an eco-system, most frequently in a dense urban environment, is one that is appropriate. Their sense of rootedness and movement in the city has also led them to be active in the area of urban planning, in the sense of enlarging their vision of an individual building in order to take into account broader swathes of future communities.

Shared responsibility, openness, a sense that architecture still has a social meaning – these guiding ideas are combined at Architecture-Studio with a broad sense of the economics of buildings and of context, whether the context is historical (defined in terms of an urban environment) or even beyond into the realm of city planning. Imagine architecture that is not the fruit of one ego, but of openness and sharing. Imagine a building that is fully rooted in its environment, not only echoing its real context but also defining it, designed and built with a sense of respect for all concerned. Perhaps this is a way forward for contemporary architecture, which critical references to style and "isms" have ignored for far too long. And what if ego and fame were less important than the quality of the architecture? Rather than being a kind of foreign body imposed on a city, what if a new building was instead to be an integral part of an eco-system that the architecture itself helps to define and invent? This architecture is not "organic" in the sense of imitating nature; it is natural in the sense that it seeks to an active part of what already exists and what is to come.

Philip Jodidio
May 2016

1. Architecture-Studio was awarded the Chaptal Construction Committee and Fine Arts Prize in 2016.

法国AS建筑工作室的经济意识体现在不同方面。在既定的预算内工作，为客户提供高于他们期望值的产品，这是工作室理解经济学的一部分。此外，工作室的经济意识还体现在对使用材料的节约上，即有意识地拒绝多余的附加物。有趣的是，当今的建筑设计中，某些高科技时常被生硬地植入建筑物中，也许是为了彰显这些建筑是何等的现代化。法国AS建筑工作室认为高科技须是建筑项目的必要部分，而不是单纯为了证明现代手段和方法而附加的事物。可以说工作室的经济意识也包含对高新科技的考量——对于建筑具有真正价值的技术是一定会被考虑的，无论是最新的科研成果还是已被广泛运用的技术。例如，对于他们来说，相比在屋顶上安装太阳能电池板，正确的朝向和合理的遮蔽才更为重要，但他们从来没有排斥合理使用高新科技。

法国AS建筑工作室曾负责许多整修项目，目的在于保留现有的建筑物，使其作为新用途的基础，或作为向现代化的过渡。与他们的一些知名同行相比，他们的目标从不是抹掉过去的痕迹，而是设法找出环境中其他清晰而可利用的元素。他们承认并将优化其已存在的建筑优点，通过改进环流模式，比如利用间隙空间来促进使用者的活动和流动。需要强调的是，建立一个生态系统，尤其是在密集的城市环境下的生态系统是他们所追求的。深厚的生态意识和在城市的活跃程度让他们在城市规划领域有所建树，他们的视野往往超越单一的建筑体，而是考虑整个社区未来的发展。

建筑仍然具有分担责任和对公开放的社会意义——法国AS建筑工作室将这些建筑理念结合到建筑和环境的广义经济学中，无论该环境是具有城市环境规划历史性又或者是超越城市当前规划的领域。建筑构想并不是一个自我意识的果实，而是开放和共享的成果。一个建筑物的构想需要完全根植于其环境，不仅应当呼应其真实环境，也应在尊重与其有关因素的基础上对环境进行再定义、设计和建造。也许这是当代建筑的发展方向，因为这是对于风格和“主义”的过度吹捧而被长时间忽视的重要参考。如果将自我和名声看得比建筑的质量轻，那会如何？如果一个新建筑成为生态体系的组成部分并起到建立和优化生态系统的作用，而不是强加于城市的一种外来事物，那又会如何？从效仿自然的意义上讲，建筑并不是“有机的”；只有寻求在已存在和未来的环境中发挥积极作用，建筑才能与环境融为一体。

菲利普·朱迪狄欧
2016年5月

1.2016年法国AS建筑工作室荣获Chaptal建筑与艺术奖

01

ARCHITECTURE STUDIO
STIMULI

ARCHITECTURE-STUDIO

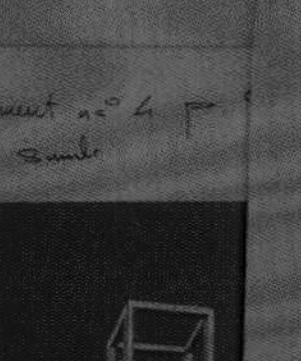

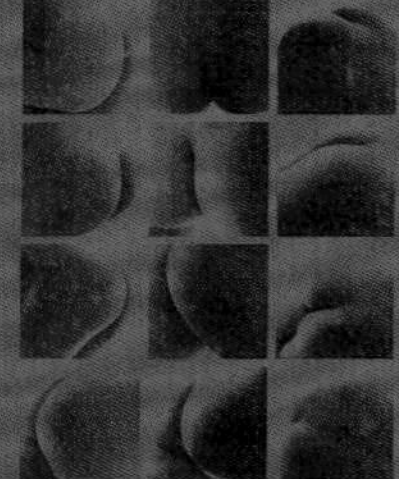

ARCHITECTURE · STUDIO
"Nous voilà."

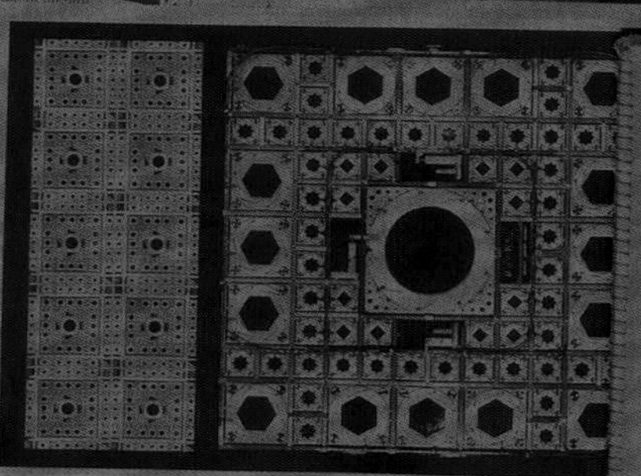

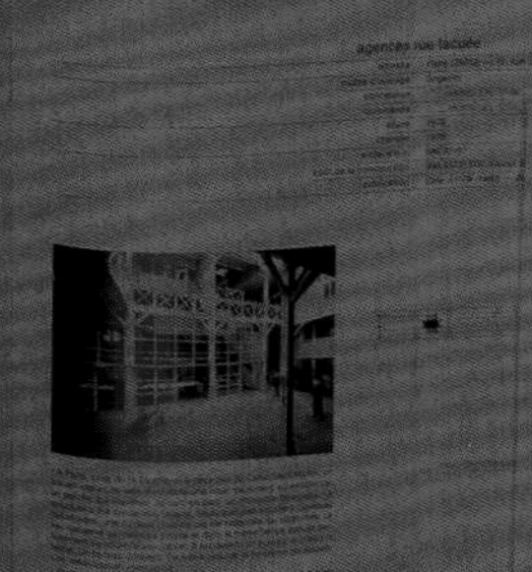

The Aga Khan Award for Architecture

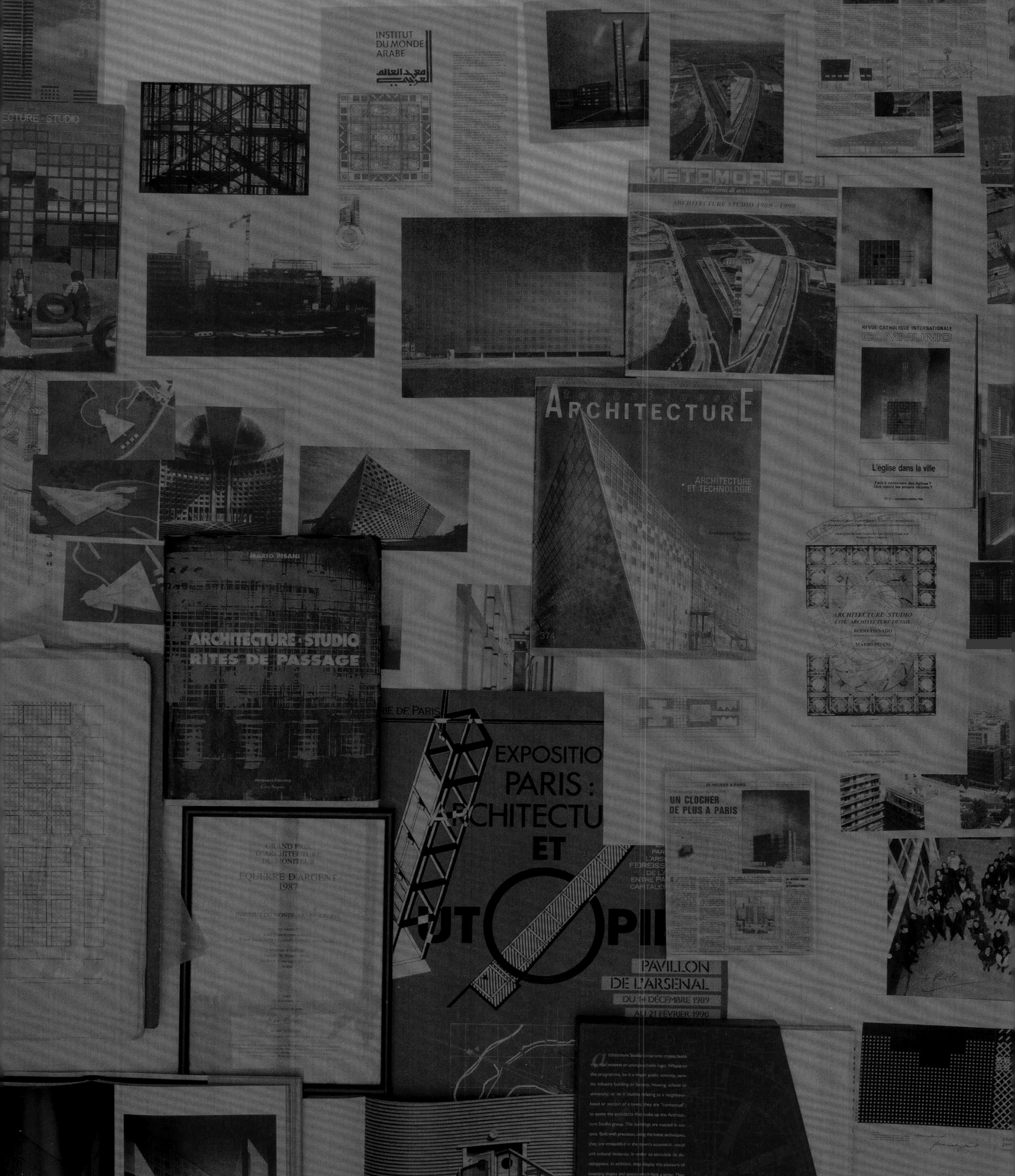
RESTORING ITS SPIRIT
INSTITUT DU MONDE ARABE
METAMORFOSI
ARCHITECTURE
ARCHITECTURE ET TECHNOLOGIE
COMMUNIO
L'église dans la ville
MARIO PISANI
ARCHITECTURE-STUDIO RITES DE PASSAGE
GRAND PRIX D'ARCHITECTURE DU MONITEUR
EQUERRE D'ARGENT 1987
EXPOSITIO
PARIS :
ARCHITECTU
ET
UT
PI
PAVILLON DE L'ARSENAL
DU 14 DECEMBRE 1989
AU 21 FEVRIER 1990
UN CLOCHER DE PLUS A PARIS

01 THE DNA OF ARCHITECTURE-STUDIO

01 L'ADN D'ARCHITECTURE STUDIO

法国AS建筑工作室的基因

The story of Architecture-Studio began in Paris in 1973. Created by Martin Robain, the Atelier de l'Arbre Sec was named after the street where Robain had offices at the time. He was later joined by Jean-François Galmiche (1974–1989), and Rodo Tisnado (1976). This was the founding team that assumed the name Architecture-Studio. Jean-François Bonne arrived in 1979. Since that time, Alain Bretagnolle and René-Henri Arnaud became partners in 1989, Laurent-Marc Fischer in 1993, and Marc Lehmann in 1998. Confirming the international direction of the office, Roueïda Ayache joined this group in 2001, while Gaspard Joly, Marie-Caroline Piot, Mariano Efron and Amar Sabeh el Leil became partners in 2009. The studio now employs approximately 150 people, directed by the 12 partners. The team of 25 different nationalities is composed of architects, urban planners, designers and interior designers. The Arab World Institute in Paris, the European Parliament in Strasbourg, the master plan of Expo 2010 in Shanghai, Our Lady of the Ark of the Covenant Church in Paris, the expansion and renovation of Créteil Cathedral, and the Onassis Foundation in Athens are amongst the best-known projects of Architecture-Studio, which also recently had the responsibility of rehabilitating the French national radio headquarters (*Maison de la Radio*) and the Jussieu University Campus in Paris. Based in Paris near the Bastille, the firm has offices in Shanghai and Venice. In recent years, Architecture-Studio has increased its presence on the international stage, especially in China and in the Middle East. Jinan's Regional Cultural Center, the Rotana Hotel tower in Amman, the Muscat Cultural Center in the Sultanate of Oman, and Bahrain's National Theatre are amongst Architecture-Studio's latest major projects. The firm has also engaged in numerous urban projects such as the Fort of Issy-les-Moulineaux,[1] the planning of the Parc Marianne – Port Marianne in Montpellier[2] and the landscape approach for public spaces at Lusail (Doha, Qatar).

This brief introduction to the history of Architecture-Studio does not seem unusual, but as Alain Bretagnolle puts it, "The office was created at a certain time and with certain principles." Martin Robain goes directly to the point: "We were all children of the 1968 uprisings. Our ideas had to do with equality and sharing." Although Robain makes clear that he had no grand plan for the future of the office, his earliest work was clearly generated in a collaborative mode. "Our first project with Jean-François Galmiche concerned the so-called SAP (*Surfaces d'activité partagée),*

法国AS建筑工作室的故事始于1973年的巴黎。马丁・罗班为法国AS建筑工作室的创始人，他起初根据当时工作室所在的街道名称将事务所命名为Atelier de l'Arbre Sec。后来，让–弗朗索瓦・加米彻（1974—1989）和罗多・蒂斯纳多（1976）加入了团队，这便是将工作室更名为Architecture-Studio（法国AS建筑工作室）的创始团队。让–弗朗索瓦・博内于1979年加入。自此以后，阿兰・布勒塔尼奥勒和勒内–亨利・阿诺在1989年成为合伙人，洛朗–马克・费希尔于1993年加入，而马克・莱曼于1998年加入。确认国际发展方向以后，罗伊达・阿亚斯于2001年加入了该团队，随后贾斯帕・朱利、玛丽卡・碧欧、马里亚诺・艾翁和艾马・萨布埃雷于2009年成为了合伙人。法国AS建筑工作室在十二位建筑师合伙人的周围团结了一支由25个不同国籍的优秀建筑师、规划师、室内设计师及成本评估师等组成的专业团队，员工约150人。巴黎的阿拉伯世界文化中心、斯特拉斯堡的欧洲议会中心、2010年上海世界博览会会址总规划、巴黎约柜圣母教堂、克雷泰伊大教堂的扩展与改造，以及雅典奥纳西思基金会文化中心等为法国AS建筑工作室最为知名的项目代表。而最近，法国AS建筑工作室承担起了修复法国国家电台总部（Maison de la Radio）和巴黎第六大学Jussieu校区的任务。该工作室总部位于巴黎，毗邻巴士底广场，上海和威尼斯均设置了办事处。最近几年，法国AS建筑工作室在国际舞台上的地位不断提升，尤其在中国和中东地区。济南的山东省会文化艺术中心、安曼的罗塔娜酒店大厦、阿曼苏丹国的马斯喀特文化中心以及巴林的国家大剧院等都是法国AS建筑工作室近期的大型项目。该工作室也参与了很多城市规划项目，例如伊西莱穆林诺数码要塞[1]、蒙彼利埃玛丽安公园—港口区域的规划[2]以及卢赛尔的公共空间景观设计（位于卡塔尔多哈）。

法国AS建筑工作室的历史简介看起来并非不同寻常，但阿兰・布勒塔尼奥勒称："工作室在特定时间且基于特定原则创建。"马丁・罗班直言："我们都是1968年法国风暴运动的'后代'。我们的观念在于平等和分享。"虽然罗班曾清楚地表明对于工作室的未来，他并没有什么宏远规划，但他最早的工作理念明显来源于合作模式。"我们与让–弗朗索瓦・加米彻达成的第一个项目就与SAP（共享活动区）有关。"

The twelve partners of Architecture-Studio, from left to right: Alain Bretagnolle, Mariano Efron, Gaspard Joly, Jean-François Bonne, Martin Robain, Roueïda Ayache, Laurent-Marc Fischer, Marie-Caroline Piot, Rodo Tisnado, Amar Sabeh el Leil, René-Henri Arnaud, Marc Lehmann.

法国AS建筑工作室12位合伙人（左起）：阿兰・布勒塔尼奥勒、马里亚诺・艾翁、贾斯帕・朱利、让–弗朗索瓦・博内、马丁・罗班、罗伊达・阿亚斯、洛朗–马克・费希尔、玛丽卡・碧欧、罗多・蒂斯纳多、艾马・萨布埃雷、勒内–亨利・阿诺、马克・莱曼。

SYNDICAT DE L'ARCHITECTURE

JOURNAL

NUMERO 1 NOV 1978

- Mais Papa, Il n'y a pas d'eau!
- Ca ne fait rien... Tais-toi et nage!

C'est un peu le discours que nous tient l'Ordre des Architectes et son UNSFA qui s'acharnent à protéger un titre qui n'a même plus de support dans notre société.

Car enfin, il n'y a plus d'Architecture, ou si peu...

Le Syndicat de l'Architecture émet un préalable à son action. Il ne peut défendre les "intérêts" d'une profession que dans la mesure où il démontre préalablement que cette profession présente un intérêt.

L'Architecture ne se décrète pas "d'intérêt public", elle n'a jamais cessé d'intéresser les hommes.

- Papa, c'est loin l'Amérique?
- Tais-toi, fils... et nage.

CITATION

Chers amis de l'Oeuvre!

Je voudrais faire la suggestion suivante: aujourd'hui il n'y a pas de commandes d'architecture et si nous arrivons à construire quand même quelque part, nous travaillons pour vivre. Où avez-vous eu la chance d'exécuter une belle commande? la pratique m'écoeure et vous êtes en principe dans la même situation. Soyons francs, de nos jours, il vaut peut-être mieux ne pas construire. Ainsi les choses peuvent-elles mûrir, et nous prenons des forces et lorsque de nouvelles possibilités se présenteront nous connaîtrons notre but, et serons assez forts pour protéger notre mouvement de la dégénération. Soyons volontairement des architectes imaginaires!!! Nous croyons que seule une révolution complète peut nous conduire à l'oeuvre. Les bourgeois et même les collègues flairent avec raison le côté révolutionnaire de nos tendances, la désorganisation et la destruction des notions et des principes traditionnels. C'est stupide! Et nous sommes un germe dans la nouvelle terre végétale. Renoncement à la personnalité qui sera absorbée par un idéal plus grand.

Si l'architecte réapparaît, le maître devient anonyme.

...ma suggestion ayant pour but de fortifier cette unité est la suivante: chacun de nous dessinera ou notera librement.. les idées qu'il voudrait communiquer à notre cercle et il enverra à chacun une photocopie. Ainsi naîtra un échange, la question, la réponse, la critique...

Bruno Taut
Lettre circulaire du 19.12.1919

POURQUOI UN SYNDICAT

En désaccord avec l'Ordre des Architectes et son syndicat potiche l'UNSFA, déçus de ne pas voir se réaliser dans des pratiques concrètes les velléités des associations, un petit groupe d'architectes a pris le risque de créer le Syndicat de l'Architecture. Depuis les premiers contacts pris entre nous il y a tout juste un an, nous pouvons nous estimer satisfaits du chemin parcouru malgré les obstacles rencontrés, malgré notre inexpérience dans cette forme de combat nouvelle pour beaucoup d'entre nous, et malgré la léthargie du militantisme qui a suivi les événements politiques du printemps dernier.

Parmi les difficultés que nous avons à surmonter, il en est une qui tient à la fois à la nature même du Syndicat et des différentes personnalités qui le composent. Je veux évoquer la nécessité de maintenir l'équilibre entre les revendications purement corporatistes et le penchant que nous éprouvons tous à développer un discours idéologique éloigné de la réalité quotidienne.

Cette réalité quotidienne, quelle est-elle pour plus de 10 000 architectes en France?

- pour 50% d'entre eux, le chômage; pour la grande majorité des autres, un taux d'activité en baisse constante et de plus en plus d'incertitude pour l'avenir;
- des conditions d'exercice et de rémunération qui les placent plus au service des entreprises capitalistes qu'à celui de la population;
- un carcan réglementaire et normatif laissant de moins en moins de liberté de création;
- des interventions politiques désordonnées, allant jusqu'à nier la nécessité de l'Architecture.

Du point de vue de la population, la situation est tout aussi alarmante, mais ne débouche pas toujours sur une véritable prise de conscience qui dénoncerait:

- l'urbanisme ségrégatif et déshumanisé,
- l'accablante médiocrité d'une large proportion de la production architecturale,
- le faux choix "collectif - individuel",
- un marché du logement quasi saturé dans le secteur libre, et un déficit insoluble en logements sociaux,

tous ces éléments concourant à renforcer les inégalités sociales et a conforter les positions du pouvoir en place.

Face à cette situation, le Syndicat de l'Architecture a jugé que la défense des intérêts professionnels et corporatistes, pour indispensables qu'en soient certains aspects, devait s'effacer devant la défense de l'*Architecture*, et que l'existence même du concept est lié à des formes différentes d'exercice professionnel, l'exercice libéral n'étant qu'une forme particulière liée à un instant historique déterminé.

Nos publications antérieures, et notamment le n° 0 du journal, ont déjà popularisé nos thèmes d'action et de réflexion, qui se retrouvent aujourd'hui diffusés à travers toute la profession par la convocation au premier congrès.

Il n'est sans doute pas nécessaire de les rappeler ici in extenso. Nous devons prendre conscience que notre projet est ambitieux, et que pour obtenir l'abolition de la loi du 3 Janvier 1977 et de l'Ordre, le contingentement de la commande, la mise en place des modalités d'urgence, la création des 1000 chantiers libres par an, l'abolition du décret d'Ornano sur l'enseignement, etc..., nous devrons représenter un véritable rapport de force par rapport aux pouvoirs publics, à l'Ordre et à l'UNSFA. Ce rapport de force, nous ne l'obtiendrons que si nous sommes *nombreux*, *unis* et *organisés* .

Nombreux: le recrutement doit être notre souci permanent. Nous devons diffuser nos thèses auprès de tous ceux qui sont susceptibles d'adhérer au Syndicat.

Unis: au delà de la diversité et de la richesse des échanges dues aux multiples personnalités qui se rencontrent au Syndicat, doit apparaître clairement une ligne d'action cohérente et unique.

Organisés: le réglement intérieur doit fixer les rôles et les responsabilités de chaque organe du Syndicat. Mais au-delà d'un texte, toujours perfectible, chaque militant, et en particulier chaque responsable, doit prendre conscience de l'importance de la tâche et de son engagement.

En créant le Syndicat de l'Architecture, nous avons pris, vis à vis des professionnels qui refusent l'Ordre et le pouvoir actuels, et aussi vis à vis de la population, une lourde responsabilité.

Nous sommes l'équipage d'un navire qui va devoir affronter les éléments avant d'aborder sur de nouveaux rivages, et d'y découvrir de nouvelles architectures.

Jean-Claude Boussat

or Shared Activity Areas. We sent a poster to residents of low-cost housing in Poitiers as part of a competition, and we won the competition to design nearly 300 apartments in that city. The residents had a budget to fit out the shared spaces that amounted to an extra 20 percent of floor area, which is to say quite a bit. Prior to building, we organized a North African style *méchoui* so that future residents could meet the persons who were going to become their neighbors. The idea was to create groups of about 12 people who wanted to live together with private and shared spaces. We were in the logic of participation and what the French call '*autoconstruction*' or self-construction. Architects usually build housing and know nothing about the people who are going to live there – in this case, we knew everybody. I would say that our approach amounted to a vision of a different way of living."

Alain Bretagnolle, who received his own diploma as an architect in 1985, reflects on events that occurred before his arrival and explains: "What is interesting is that the principles that were behind the creation of the office in the early 1970s remained, whereas other architectural firms founded at the time bit by bit assumed a more 'normal' pyramidal structure, often with one leading 'star' architect. The name Architecture-Studio, which is also legible in English, is an indication of this philosophy – there are no individuals named, it is an open structure. Though it may be a coincidence, the early arrival of Rodo Tisnado, who is Peruvian, began to define the international nature of the group. The open, international approach with a somewhat 'militant' view of the role of architecture has remained the basis for what we do."

The commitment to an open structure was indeed one of the elements that Martin Robain insisted on. "The lack of a personal identification was a purposeful gesture. The idea was that there should be no visible named architect to lead the office," he says. "It made it easier to integrate new partners, and using a name makes you seem more like an artist. Our colleagues were convinced that our form of office organization would fall apart. We never set out to create a larger office, it was a matter of circumstances that led us in that direction." Looking back, Alain Bretagnolle sees the seeds of the present success of Architecture-Studio in its original structure. "The horizontal, open organization of Architecture-Studio logically was integrated into our way of working and our rapport with the outside world. We refer to a quote of Søren Kirkegaard who said, we must 'leave open the wounds of possibility'. This is a commitment

"作为竞赛的组成部分，我们向普瓦捷廉住房居民发送了一张海报，并最终赢得竞赛，获得为该市设计近300套公寓的资格。居民获得为共享空间建设提供的预算，而共享空间约是建筑底层面积20%的新增空间，可以说是较大面积了。在开始建设前，我们组织了一次北非烤羊肉聚餐，让未来的居住者能够有机会接触以后会成为他们邻居的人。理念在于拥有私人空间的前提下，愿意一同共享空间的12个人创建一个团体。这是共同参与逻辑，法国人也称它为'自主建设'或自我建设概念。一般情况下，建筑师只建设房屋，而对将在房屋内生活的人知之甚少——在这个设计案例中我们了解了所有人。可以说我们的方法带来了对不同的生活方式的观察视角。"

阿兰·布勒塔尼奥勒于1985年获得了建筑师学位，他回顾了来工作室之前发生的事件并解释道："有趣的是，20世纪70年代早期工作室创建时的最初原则得以保留，而同时期兴起的其他建筑事务所则逐渐转向更为'普遍'的金字塔结构，通常由一位'明星'建筑师领导。Architecture-Studio（法国AS建筑工作室）这个名称用法语和英语都很易理解，该名称体现了这种哲学——并没有根据个人进行命名，这是一个开放的结构，或许是一种巧合，秘鲁人罗多·蒂斯纳多在工作室创建初期的加入定义了该团队的国际性质。这种带着对建筑某些'激进'视角的开放的、国际化的理念始终是我们工作的依据。"

对开放型结构的承诺确实是马丁·罗班一贯的坚持。"个人辨识度的缺失意味着我们是一个整体，呼应了工作室不是由某一位知名的建筑

Cover of the first issue of the *Journal du Syndicat de l'architecture*, founded in 1978, as a follow up to the Mars 76 movement.

成立于1978年的建筑工会期刊首个发行物封面，追踪Mars 76运动。

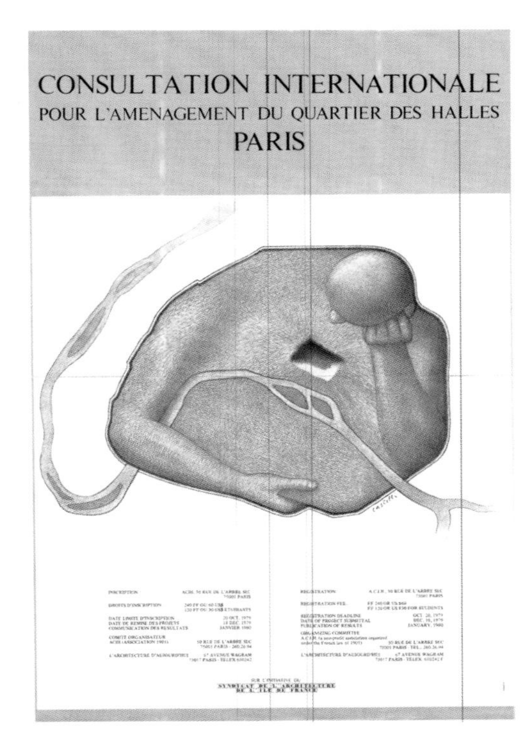

Poster of Les Halles District Redevelopment International Consultation, launched in 1979 by the ACIH, to which 600 teams, from all over the world, responded. Drawing by Castelli.

巴黎Les Halles区再发展规划国际咨询海报，由ACIH于1979年发布，最终，来自世界各地的600个团队参与。由Castelli绘图。

to openness, whether it be in our own mode of work, or vis-à-vis the outside world."

The early history of the office is closely associated with the strong currents that shook the architecture world in France in the 1970s. The founding partners participated with Jean Nouvel in the Mars 76 movement, which was opposed to the corporatist rigidity of architecture and urban planning, seen as being a result of Le Corbusier's 1933 Athens Charter. This movement also aimed at promoting the participation of inhabitants in the conception of their architectural and living environment. Martin Robain explains: "What we did as the follow up to the Mars 76 movement was the creation of the *Syndicat de l'architecture* (Architectural Union) in 1978, which was based on a cultural and philosophical approach. We wanted architecture to be seen as a significant part of culture, which was not obvious at that time. The *Syndicat* didn't really go very far, but it did launch an ideas competition for the use of the Halles market area of Paris (transferred to Rungis 1969). We worked only on that project for six months." Co-founded by Martin Robain and Rodo Tisnado in 1979, the ACIH (*Association pour la Consultation Internationale pour l'Aménagement des Halles*) presented more than 600 projects from all over the world in opposition to the scheme selected by the Paris City Council for the former market area located in the heart of the city. Figures like Henri Ciriani and Francois Barré who was later Director of Architecture at the Ministry of Culture (1996–2000) were all involved in this process as well.

The collective method employed in the Halles consultation and the earlier movements was, in some sense, a rehearsal for a major design by Architecture-Studio, associated with Pierre Soria and Jean Nouvel, who worked with Gilbert Lézènès for the *Institut du Monde Arabe* or IMA (Arab World Institute) in Paris. Martin Robain recalls: "We did the IMA competition in three weeks. Things went well, there were no conflicts. At the time Jean Nouvel blended in very well with our collective approach. Rodo Tisnado worked on the diaphragms that had to do with controlling natural light but also with the cultural reference to the geometric patterns of the Arab world. The morphology of the building was clearly the result of its context, on the river, near the Jussieu campus of the university. We took a considerable risk by asking to do away with a roundabout at the tip of the structure, in order to recreate the former alignment of the Boulevard Saint-Germain. We defined the technological appearance of

师所领导的创建理念。”他说道，“这种方式更方便新合伙人融入，而用名字命名更像是艺术家的做法。我们的同行认为工作室的这种运作形式容易导致分裂。我们绝不会去创建更大的办公室，环境会为我们提供方向指引。”回顾过去，阿兰·布勒塔尼奥勒认为法国AS建筑工作室成功的根源在于其原始结构。“法国AS建筑工作室的平行开放结构已像逻辑般融入我们的工作方式和我们与外界世界的关系中。我们引用了索伦·克尔凯郭尔的话语，我们必须‘使充满可能性的开口始终保持开放’，这是对于开放的承诺，无论是我们自己的工作方式，或者对待外界的理念。”

工作室的早期历史与20世纪70年代法国建筑界的强大潮流紧密相关。创始合伙人们与让·努维尔一同参加了Mars 76运动，该运动反对建筑和城市规划行业的僵化，这个问题被视为由勒·柯布西耶主导颁布的1933年《雅典宪章》所致。此运动也旨在促使居民参与对建筑和生活环境的构想。马丁·罗班表示：“Mars 76运动之后，以文化和哲学作为支撑，我们在1978年创建了建筑工会。我们希望让建筑成为文化的重要组成部分，而当时的社会缺少这种观点。工会的实际发展速度并不快，但曾提出组织关于重新规划原巴黎中央市场Les Halles区域的方案征集竞赛的主张（原巴黎中央市场于1969年转移至Rungis）。我们在这个项目上的工作仅用了六个月的时间。”马丁·罗班和罗多·蒂斯纳多于1979年联合创建了ACIH（Les Halles区域规划国际咨询协会），收集了来自世界各地超过600个规划方案，以反对巴黎市政厅针对巴黎市中心原中央市场区域改造的原始方案。还有亨利·西里安尼以及后来成为法国文化部建筑司负责人的弗朗索瓦·巴莱（1996年至2000年）也参与到整个项目流程中。

Les Halles的规划咨询工作和更早的运动中所展现的合作意识，在某种意义上可以说是法国AS建筑工作室与皮尔埃·索利亚、让·努维尔以及吉列博·雷泽能共同设计巴黎阿拉伯世界研究中心这一代表作的提前预演。马丁·罗班回忆道：“我们在三周时间内完成了巴黎阿拉伯世界研究中心的竞赛设计，合作很顺利，没有发生什么冲突。当时让·努维尔将其工作与我们的集体合作方法实现了充分融合。罗多·蒂斯纳多主要负责借鉴阿拉伯世界特色几何图样，设计控制自然光线的控光装置。建筑的外观形态明显受其位于塞纳河畔，且毗邻巴黎第

The architects based at Rue Lacuée, close to the Bastille Plaza, in 1980. In the beginning, the four practices (Architecture-Studio, Archigroup, Pierre Soria, Gilbert Lézènes et Jean Nouvel) shared these offices and sometimes collaborated together on projects.

1980年，在毗邻巴士底广场Lacuée街工作的建筑师们。起初，四家事务所（法国AS建筑工作室、Archigroup、皮尔埃·索利亚、吉列博·雷泽能和让·努维尔）共用这些办公场所，有时还在某些项目上进行合作。

the structure and its rapport with Paris." Defining another characteristic of the work of Architecture-Studio, Alain Bretagnolle comments, "This building marked the passage in France from the Post-Modern era to a period where *contextualism* was more important."

The office developed greatly through competitions and public projects. The surprising form of the *Lycée du Futur* (High School of the Future) in Poitiers earned the office a good deal of press coverage at the time. Martin Robain confirms this: "It was important for us to begin to become known under our own name and for our completed work. Our way of working may have made this more complicated with some clients who enjoy being in touch with a 'star' architect." Despite its rather radical forms, the *Lycée du Futur* provides an interesting insight into the modesty that is an element of the work of the office. Martin Robain explains: "We came to the idea of this form because of what we call the 'Theory of the Hippopotamus.' The site was near the Futuroscope theme park in Poitiers which was a project carried forward by Réné Monory who had been Minister of Economy in France (1978–1981). Ours was the first building to be completed in *Technopole* area associated with the Futuroscope, opening in September 1987. Our thought was to touch a watering point like a hippopotamus and to mark the smallest territory possible. The watering point here was the shared space of the *Technopole*, and it was necessary to leave space for other buildings to also reach that location." Again insisting on the importance of context for the work of Architecture-Studio, Bretagnolle says, "the only context in this instance was this concept of the organization of the entire complex and its focus on the future. The triangular volume was in rapport with this context and also had the advantage of creating a very visible structure."

Another important idea in the early development of the office was that of the so-called *Stimuli*. Martin Robain states, 'It was the time of the Grands Travaux of President François Mitterrand such as the Louvre Pyramid. So we thought the big projects were nice, but that there could also be small ones. For the price of one major government project, we said that it might be possible to carry forward 1,000 small ones. We did a broad analysis of the empty parcels in the middle of Paris and in the surrounding areas, to find to whom they belonged, and we made proposals for the use of those plots. We asked other architects like Massimiliano Fuksas to work on this with us. Almost all of the parcels of land that we

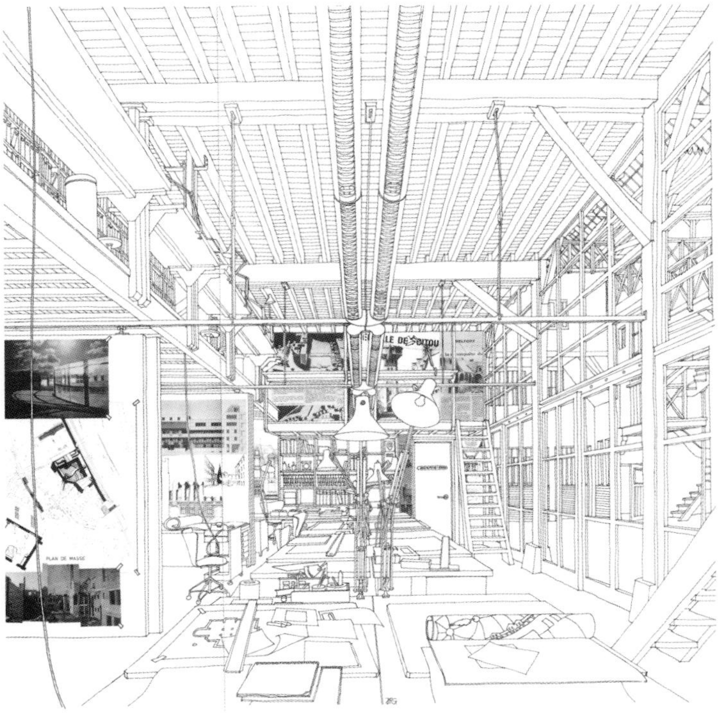

六大学Jussieu校区的环境所影响。对于修改阿拉伯世界研究中心前的道路，使之与圣日耳曼大道的走向一致，我们承担了相当大的风险。我们确定了建筑的技术外观，并让它融入巴黎的环境中。”这定义了法国AS建筑工作室作品的另一重要特征，阿兰·布勒塔尼奥勒称：“该建筑标志着法国从后现代时期进入了更加注重文化内涵的时期。”

工作室通过竞赛和公共项目迅速发展。普瓦捷的未来中学的特色造型在那时为法国AS建筑工作室赢得了大量的媒体报道机会。马丁·罗班承认：“我们因为自己完成的项目而闻名于世，这对我们来说非常重要。对于想要接触星级建筑师的客户来说，我们的工作方式可能使情况变得更加复杂。”尽管未来中学的建筑形式过于前卫，但它其实给朴实无华的设计理念带去了有趣的视角，而这种理念也是工作室的一个元素。”马丁·罗班解释道：“这个建筑造型的设计概念源自一种‘河马理论’。项目地址位于普瓦捷的由曾于1978年至1981年间任法国经济部长的勒内·莫诺利主导建造的Futuroscope未来主题公园附近。我们的建筑是Futuroscope未来主题公园周边的高新区域内的第一幢建筑，于1987年9月正式开幕。我们的思想旨在像河马一样接触水点并标出可能的最小区域，此处所指的水点为高新区的共享区域，必须为该地区内的其他建筑留出空间。”这个设计理念再次体现了法国AS建筑工作室在设计时注重环境因素。阿兰说：“这个案例中，只有环境为整个建筑的依托及其对于未来的关注。三角形体量与这样的背景相符合，也具有创建显著结构的优势。”

工作室早期发展的另一项重要项目是“激励（Stimuli）”。马丁·罗班说道：“当时弗朗索瓦·密特朗总统正着手统筹宏伟大工程，例如卢浮宫金字塔。我们认为大项目非常不错，但当时也可能存在许多值得关注的小项目。一个重要的政府大项目的资金可能可以分配给1 000个小项目。我们对巴黎中心和周围区域内的空白区域进行了大量研究，旨在确定这些空白地块所属对象并为这些土地的使用提供建议。我们邀请其他建筑师如马西米利亚诺·福克萨斯等与我们一同工作。”

“我们当时挑选的所有地块后来几乎都被规划建设，但这并不存在压倒一切的逻辑——我们的项目目的是要让土地被利用起来。”很显然，对城市中这些没有被利用起来的地块进行系统分析的行为，显示了工

Architecture-Studio's New Year's greetings card, 1983. Drawn by Jean-François Galmiche.

法国AS建筑工作室1983年的新年贺卡。绘图人：让-弗朗索瓦·加米彻

8 AVRIL 1987
PARIS·FRANCE
"LA VILLE EST UNE ILE ? DES PHARES DANS LA CITE"
ARCHITECTURE STUDIO (ROBAIN, BONNE, GALMICHE, TISNADO, architectes)
JEAN-CHRISTOPHE BAILLY, écrivain
29 AVRIL 1987
MARRAKECH·MAROC
"CONSTRUCTION D'UNE LEGENDE DE SABLE"
CHARLES BOCCARA, architecte
SERGE SANTELLI, architecte
6 MAI 1987
URBINO·ITALIE
"DES REMPARTS DE BRIQUE POUR LA CITTA IDEALE"
GIANCARLO DE CARLO, architecte
PAOLO FABBRI, sémiologue
13 MAI 1987
CORDOBA·ARGENTINE
"UN DAMIER : LA DECOLONISATION DES SOUVENIRS"
MIGUEL A. ROCA, architecte
ROSALBA CAMPRA, écrivain
20 MAI 1987
BARCELONE·ESPAGNE
"A LA RENCONTRE DE LA VOCATION PREMIERE"
ORIOL BOHIGAS, architecte
JOSE AUGUSTIN GOYTISOLO, écrivain
3 JUIN 1987
BERLIN·ALLEMAGNE (RFA)
"L'HISTOIRE AU FUTUR, RECONSTRUIRE LE PASSE"
JOSEF PAUL KLEIHUES, architecte
THORSTEN BECKER, écrivain
ARCHITECTURE ET CONSTRUCTION
LA VILLE DECLINE SES MEMOIRES
RENCONTRES-SESSION PRINTEMPS 87
Centre Georges Pompidou
Petite salle - 1er sous-sol, de 18 heures à 20 heures
Entrée libre dans la mesure des places disponibles
Renseignements : Architecture et Construction
136, bd Saint-Germain - 75006 Paris - Tél. 43 29 00 60
Centre de Création Industrielle
Centre Georges Pompidou

identified at the time have been built on since then, but without any overriding logic – our project was to link these plots and their use." Clearly the idea of making use of sites that had remained untouched in the city, and doing so in a systematic way was indicative of the participative and intentionally modest approach of the office. "We were imagining what might be called 'urban acupuncture'," says Alain Bretagnolle. "We actually built only one of these projects, public housing called the Château des Rentiers."

The partners have shared the principles laid down at the origins of Architecture-Studio but have also integrated these ideas into a unique work method over the years. A key element of the work method is the weekly meeting of partners where projects are discussed and were long sketched by those in attendance with what the firm calls *Tempos* which are drawings originally made with Papermate flair pens (Tempo). The tempo drawings are now superimposed using computer methods to create what is called the *tracé rouge* (red drawing) that serves as the matrix for the development of the full architectural project. Unlike many offices where a single architect, or partners working independently, may simply pass on their personal sketches to assistants who then execute them in computer form, Architecture-Studio intentionally avoids signature drawings.

The preparation of this book was approached a bit like an architectural project of the office in the sense that a number of partners sat with the author at a large table and expressed their ideas on what the publisher, Dominique Carré, has called the "DNA" of the office, its underlying concepts and methods. The method of the office, based on the equality of the partners, their sharing of information, their engagement in what must be called the art of architecture in the fullest sense, emerge in this chapter from an unscripted conversation between four of the 12 partners in the presence of the author in Paris on February 17, 2016, but these values are also the basis for the organization of this book – each of its chapters reflects a different aspect of the approach of Architecture-Studio, which, taken together make the firm quite unique.

Discourse on Method

The first task of the team is to analyze the program for a given project and then to reduce it to a number of concepts. They create a document that outlines the constraints. At the outset they talk about these constraints, and other elements, such as the social

作室的社会参与度和节制性。“我们设想了所谓的‘城市针灸疗法’”阿兰·布勒塔尼奥勒说道，“我们实际仅完成建设了一个项目，即被称为‘Château-des-Rentiers’的公共住房。”

合伙人分享了法国AS建筑工作室最初推行的几项重要原则，多年来工作室已将这些理念融入独特的工作方法中。主要工作形式是合伙人周会，合伙人通过周会讨论项目，参与工作室称作Tempos的草图绘制，Tempos该名字源于绘图使用的毛毡笔的名称（Tempo）。现在工作室已通过计算机叠加Tempo图纸并将之整合，被称作为Tracé Rouge（红线图），其可充当完整建筑设计项目发展的基础。与其他多数工作室建筑师或合作伙伴单独工作，然后将图纸交给助手，由助手将其转变为电子版本这种工作方式不同，法国AS建筑工作室会有意避免使用带签名的图纸。

本书的筹备过程在某些意义上有点像工作室处理建筑项目的方法，即合伙人与作者围坐在一张大桌子旁，对出版人Dominique Carré提出的工作室“DNA”及其暗含的概念和方法表达各自的看法。本章节的介绍源自2016年2月17日作者与12位合伙人中的4位在巴黎的无剧本对话，工作室立足于合伙人的平等地位、信息共享以及完整意义上的建筑艺术参与的工作理念，在本章节中得到充分的展现。同时，这些价值观念也是本书架构的基础，每章均体现了法国AS建筑工作室工作方法的各个方面，更综合体现了法国AS建筑工作室是独一无二的。

方法论述

团队的首要任务是分析给定项目的背景内容，并将其简化为一系列概念。随后编写一份有关各项限制的框架文件。对于一些限制条件和其他因素，例如所涉及的社会与政治背景等，在项目的最初就被讨论。建筑产生的社会后果从一开始便是他们思考的内容之一。构思具有吸

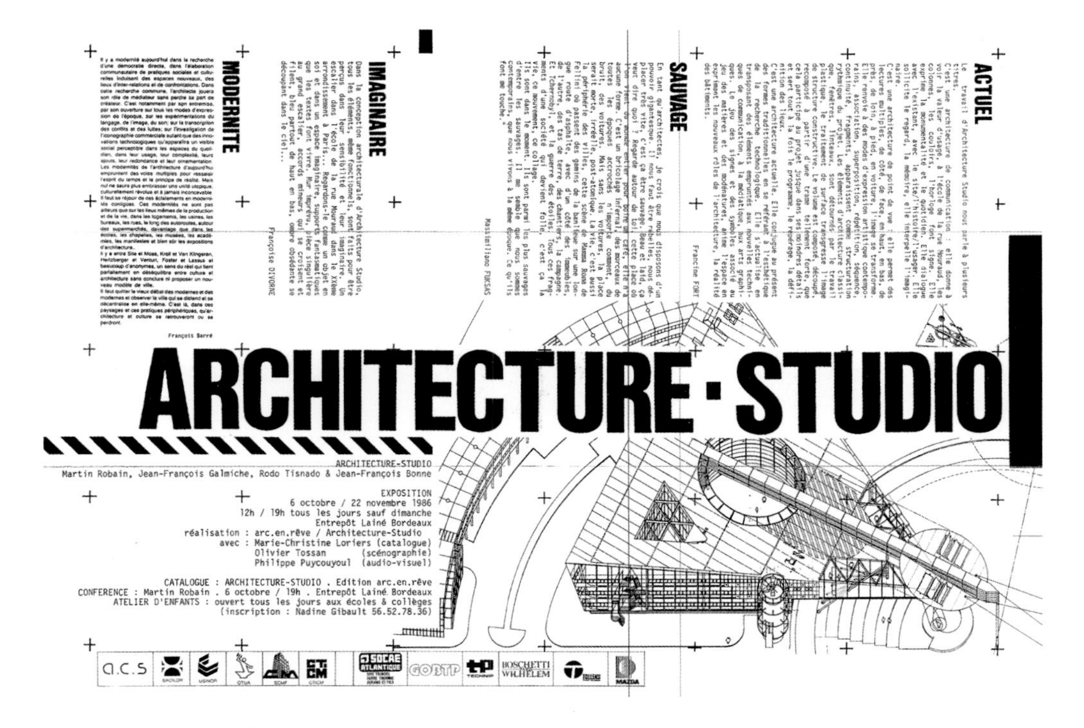

"Architecture and construction. The city states its memories" – a poster for the cycle of meetings organized by the Centre Georges Pompidou in 1987. Architecture-Studio presented the opening conference: "The city as an island? Lighthouses in the city".

Poster for the exhibition dedicated to Architecture-Studio in 1986, at arc en rêve in Bordeaux. The texts are by Francine Fort, Massimiliano Fuksas, Françoise Dirvone and François Barré.

“建筑与建设，城市与它的历史”——由蓬皮杜艺术中心于1987年组织的系列研讨会的海报。法国AS建筑工作室出席开幕研讨会，并以“城市是否是一座岛屿？灯塔位于城市中”为标题进行演讲。

波尔多建筑中心arc en rêve于1986年为法国AS建筑工作室举办的展览的海报。文本由Francine Fort、Massimiliano Fuksas、Françoise Dirvone与François Barré编写。

and political context involved. The social consequences of architecture are part of their thinking from the beginning. Imagining an attractive building is not the first priority, there are other things to take into account that matter more. First, the group has to understand the project. They then proceed to think about it, and then they test various options amongst themselves. Words come before drawings in this method, but there is always an interaction between the two. As one of the partners, Rodo Tisnado says, "Some other architects do a sketch and give it to their collaborators – I would call this the Walt Disney system – here things are not like that." The firm considers that an outline of the constraints is already a kind of virtual drawing. And when that exists, everyone participates. There is no lead architect, even on a case-by-case basis – they operate as a team. The office organizes itself around projects rather than around architects. The equality of the partners is highlighted by the fact that they all receive the same salary. Martin Robain further explains: "The partners are listed not in alphabetical order but in chronological order according to when they became partners. We try to encourage fidelity and longevity in the firm. We have minimal profits because we reinvest almost everything we earn in our office and in our work. There are also 16 partners who may eventually become partners." Finally, and importantly, the intellectual property of the office belongs not to individuals but to Architecture-Studio itself.

The method of Architecture-Studio starts from the realization that there is a particular strength in the action of a group that leaves little place for individual egos to emerge, but does allow for contradiction and the confrontation of different ideas. This is an element that each partner has accepted in joining the group. Sketching in the more traditional sense is still part of their method, but discussion is at the heart of the meetings between partners that occur in Paris every Monday and Friday. Interestingly, the partners never put decisions to a vote; their response to a given project emerges out of their conversations and, gradually, in their drawings.

Architecture/Art

The partners of Architecture-Studio have long maintained that they do not presume to be artists. "But we do believe that architecture is an art," states Martin Robain. "The two concepts are not the same. We do not want to claim that we are artists. We are involved in many aspects of design and construction, and that means that

引力的建筑并不是首先考虑的内容，还有许多至关重要的条件需要优先考虑。第一，团队必须理解项目。随后他们继续进行思考，尝试各种可能性。在这种情况下，文字比图纸更先在脑海中出现，但二者之间始终存在关联。作为合伙人之一的罗多 · 蒂斯纳多说道："有些建筑师会绘制指导草图，然后将草图交给他们的同事去完成，我将其称为沃尔特 · 迪斯尼系统，我们事务所绝不会这么运作。" 工作室认为限制框架本身已是一种虚拟图纸。当其被定义之后，所有人都能参与其中。工作室没有担任领导的建筑师，即使是某个单独的项目，他们仍然以团队形式进行工作。工作室围绕着项目开展工作，而不是围绕着建筑师开展工作。所有合伙人拥有相等的薪酬更进一步突显了合伙人之间的平等性。马丁 · 罗班进一步解释道："合伙人排序不按字母顺序，而是按照成为合伙人的时间。我们在工作室内部鼓励忠诚和长久工龄。我们会将工作室和工作的收入进行再投资，因此我们的利润并不高。目前还有16位合作伙伴可能最终成为合伙人。" 最后且最重要的是，工作室的知识产权并不属于个人，而是属于法国AS建筑工作室。

法国AS建筑工作室的工作方法来源于对于团体合作的强大力量的认知，在集体中允许个人主义涌现的空间很小，但允许不同想法和观念间的碰撞和对话。这是每位合伙人在加入之前必须接受的理念。使用传统的绘制草图方法仍是他们工作中的一部分，但每周一及周五在巴黎召开的合伙人周会更重要的是进行语言层次的讨论。有趣的是，合伙人从不通过投票的方法确定决议，会议中对于特定项目的反馈来自他们的对话，并逐渐体现在图纸上。

建筑/艺术

法国AS建筑工作室的合伙人一直声明他们并不是艺术家。"但是我们认为建筑是一种艺术，" 马丁 · 罗班说，"这两个概念并不相同、我们并不想说我们是艺术家、因为我们会处理设计和施工工作的很多事

Architecture-Studio is based in Paris, Shanghai (picture), and Venice.

法国AS建筑工作室总部设于巴黎，在上海（左图）与威尼斯均设有分支机构。

Overleaf: The library of A3 albums containing project sketches and drafts.

次页：工作室档案间保存的A3专集簿包含项目草图与草稿。

creating art is definitely not the initial goal." For Rodo Tisnado, it is clear that "a building can indeed be considered to be a work of art once it is completed, but architects are not artists." Indeed, the approach of the partners to the question of "beauty" or the artistic value of architecture allows a good deal of their individuality and originality to emerge. Mariano Efron states that "Our sharing of responsibility and creation is antithetical with what I would call the narcissistic approach that can often be associated with the statute of 'artist.' I do believe that our insistence on linking a project to its context is related to the concept of 'beauty,' but not to the mark of an artist. It is the integration of a project into its context that generates the aesthetic aspect of the built work. There is not an aesthetic signature that we intentionally carry forward from project to project so that we become better known. Our DNA is expressed in our method of collective work." As Alain Betagnolle concludes, "We might say that our sense of aesthetics is linked to the idea of our responsibility as architects."

The sharing of responsibility, and indeed the equality of the partners come forward in other ways that are further explained in this volume. These include a continual search for the correct economy of means and choice of materials for a given project, but also the idea of reaching out to younger architects around the world, as evidenced in the ongoing exhibitions at the CA'ASI, the showplace of the firm in Venice. Here, young architects from China, Africa or the Middle East have been gathered for shows that seek to reveal the underlying issues that are driving emerging architecture from other parts of the world. So, too, Architecture-Studio has reached out considerably in its own efforts to build in places as different as Amman and Shanghai. Wherever the firm works, context has been considered a guiding issue, in terms of bringing not only a building, but a site, a location, an area to life. The reference to DNA thus comes full circle and can be explained in terms of viewing architecture itself as a living discipline, with buildings forming vital elements of the larger, changing city.

1. Eco-quarter Award, 2011, in the Urban Redevelopment and Eco-Digital Innovation categories and winner of the Green City Solution Awards, 2016, in the Smart City category.
2. Eco-quarter Award, 2011, in the Global Ecological category and Eco-quarter Award, 2015

项，而不是单纯的艺术创造。”很明显，对罗多·蒂斯纳多来说，“建筑建成后的确可被看作是一件艺术品，但建筑师并不是艺术家。”以集体的方式共同审视“美”或艺术价值，能够促进每个人的个性和原创性。马里亚诺·艾翁说道，“我们所说的责任分享与共同创造和坚持自我是对立的，后者更像是‘艺术家’的态度。我确信我们坚持将建筑与环境紧密联系与‘美’的概念有关，但却不是艺术家的标志。将建筑项目融入环境中能够创造建筑的美学。我们没有为了提高知名度而在不同的项目上运用某一种特定的形态美学来营造风格。集体工作方法是我们DNA的一种体现。就像阿兰总结的一样，“可以说，我们对于美学的感知与我们作为建筑师的责任息息相关。”

责任共担与真实的平等精神在本书中将有更深入的说明。这包括对于正确的经济手段和针对不同项目的材料选择的持续探索；也包含希望发掘更多年轻一代杰出建筑师的意愿，这从CA'ASI艺术展览馆（法国AS建筑工作室在威尼斯的艺术展览馆）举办的多个展览可得知，组织来自中国、非洲和中东地区的青年杰出建筑师汇聚一堂，展示这些新兴地区的机遇与挑战。也是因为这些机遇，法国AS建筑工作室在世界高速发展的城市及地区如安曼和上海等地投入了大量精力。但无论该工作室在何处工作，环境始终都是最重要的指导因素，工作室希望自己不仅仅是建设一幢建筑，更是为周边带来一个充满活力的空间。关于DNA的参考因此绕了一个圈又回到原点，且可通过将建筑本身看作一个具有生命力的存在进行解释，因为建筑是不断扩大和变化的城市中的重要元素。

1.获2011年度生态街区：规划改造及生态智能创新类奖；获2016年绿色生态建筑探索奖：智慧城市大奖。
2.获2011年度生态街区：综合生态规划奖；获2015年生态街区称号。

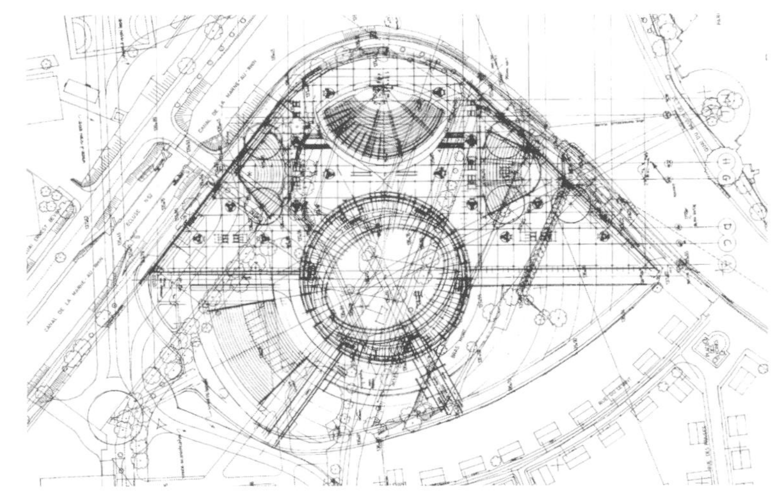

The *tracé rouge* is a significant work method for Architecture-Studio. It brings together all the features of a building in one drawing. Here, for example, is the drawing for the European Parliament (1991). It is both the result of, and the support for, the collective conception.

Tracé rouge（红线图）是法国AS建筑工作室的主要工作方法。它将建筑的所有特征融合到一张图纸上。上图的例子为欧洲议会中心的红线图纸（1991）。这是集体观念的结果也是集体观念的力证。

A preview of Architecture-Studio's project for the restructuring of the "racket" shape of La Défense, conducted in 1988 at the request of Prime Minister Michel Rocard.

1988年，应时任法国总理米歇尔·罗卡尔的要求，法国AS建筑工作室为整修拉德芳斯“racket”项目所做设计的预览图。

CROKO
CROKO
CROKO

P02

Shared Activity Areas
Poitiers, France
1975/1980
11 300 m²

共享活动空间公寓
法国，普瓦捷
1975/1980
11 300 m²

To plan housing projects with spaces shared by different activities would have been considered very avant-garde in the 1970s. However, Poitiers' public housing body gave it a try. In 1975, the year of the start of the design research, uniformity seemed to have been the solution for every question that architecture was posing itself as regards design for low-income households. But here the human scale was to be thought of differently and revisited above the low ceiling advocated by the Modernist movement. The rethink concerned shared spaces, i.e. the social distribution of spaces, and the division between private space and common space. Taking into account the residents' enjoyment of architecture and entrusting them with the functionality and use of semi-private spaces attached to the projects were quite new ideas. These extra rooms could be appropriated by four to six households: some dedicated to a common activity, others reserved for private use only. This innovative offer entailed new requirements, which involved the improvement of noise insulation – adapting it to the flexible use of space, the fluidity of neighbors' activity and a seamless circulation between private and public spheres, shared by every generation and open to the city. These 274 experimental apartments have been the harbinger of some of the most advanced projects of 21st century contemporary architecture, and are mostly to be found in Northern Europe. H.L.

在住宅项目中，规划不同活动的共享空间在20世纪70年代是非常前卫的做法。但是，普瓦捷的公共住房机构仍决定尝试一次。1975年是设计研究开始的一年，单一的设计似乎一直被视为解决廉住房建筑设计所有问题的方案。但是，在这个设计中，超越现代主义建筑设下的限制，人性的尺度被予以更多的考虑。新想法考虑到了共享空间，例如，通过私人空间和共享空间的结合以创造更好的社会邻里关系，通过附加的半私人共享空间和各项功能让居民体验建筑的乐趣在当时是一种新潮的观点。这些额外的房间可供4到6户家庭使用：一些用作共享活动区域，而一些仅作为私人空间。这个创新的想法对建筑提出了新的要求，保证共享活动空间的灵活使用、邻居活动的流畅性和私人范围与公共范围之间的无缝对接需要提升建筑的隔音效果。这274间试验性公寓是建筑界的思想先驱，这种共享精神在21世纪当代建筑，尤其是北欧建筑中得以发扬。H.L.

Poster of the SAP project launch in Poitiers.
The architectural utopia became a reality and the SAP project was displayed at the Paris Biennale in 1980.

发布于普瓦捷的SAP（共享活动空间公寓）项目海报。建筑乌托邦得以实现，SAP项目于1980年在巴黎双年展上展出。

HUM HUM...,
ET SI ON FAISAIT QUELQUE CHOSE ENSEMBLE ?
BEAULIEU
POITIERS
RALLYE
alors, ces brochettes !
cette fois, clou ! je t'aurai
... il était une fois,
Pépé racontes-nous une histoire !
et après l'épluche patates
le repasse-cravatte !
C'EST QUOI CET IMMEUBLE ?
C'EST UNE RÉALISATION EXPERIMENTALE DE L'OFFICE HLM DE POITIERS, SUBVENTIONNÉE PAR LE MINISTÈRE DE L'ÉQUIPEMENT.
L'IDÉE C'EST QU'IL YA DE GRANDES SALLES COMMUNES À DIFFÉRENTS NIVEAUX
IL YA DES ENDROITS AU REZ DE CHAUSSÉE OÙ LES HABITANTS PEUVENT SE CONSTRUIRE D'AUTRES LOCAUX COMMUNS
IL YA DES TERRASSES ACCESSIBLES
IL YA UNE SERRE IL YA UN BUDGET DISPONIBLE POUR QUE LES HABITANTS PUISSENT AMÉNAGER TOUT CELA
AH ! J'OUBLIAIS L'ESSENTIEL : IL YA DES APPARTEMENTS BIEN SÛR ! 274 HLM EN LOCATION
ET COMMENT FAIRE POUR Y HABITER ?
ON SE GROUPE À PLUSIEURS FAMILLES QUI SE CONNAISSENT ET S'ENTENDENT BIEN (DES VOISINS, DES AMIS...)
CHAQUE FAMILLE DISPOSE D'UN APPARTEMENT, TANDIS QUE LE GROUPE DES FAMILLES AMÉNAGE ET UTILISE LES LOCAUX COMMUNS, LES TERRASSES, LA SERRE ETC... EN PLUS IL FAUT AVOIR DROIT AUX HLM LOCATIVES BIEN SÛR.
DES RÉUNIONS SONT PRÉVUES À PARTIR DE DÉCEMBRE 1978 POUR DISCUTER AVEC LES GENS INTERESSÉS, LEUR PERMETTRE DE SE RENCONTRER — SI CELA VOUS TENTE, PASSEZ A L'OFFICE DEMANDER DES RENSEIGNEMENTS !
Imp. TYPOFSET

PACHR1

Housing, rue du Château-des-Rentiers
Paris, France
1984/1986
1 500 m²

Château-des-Rentiers街公寓
法国，巴黎
1984/1986
1 500 m²

The more difficult the context is, the more imagination and care are required. The idea underlying the "Stimuli' project was the use of the residual building plots still vacant in Paris and its neighboring suburbs. Architecture-Studio had an interest in regenerating "urban wasteland' and giving it a second chance. The ambition was to promote urban planning and development in excluded terrains, but which, once reconquered, could prove dynamic and materialize themselves into the new frontiers of development. The angled building of the rue du Château-des-Rentiers, in Paris's 13th arrondissement, is like an acupuncture needle stuck into the neighborhood's lively social fabric in order to revitalize the whole organism. The conditions of the project were strict: next to a 30-meter high wall, and to be built within the limited budget usually allocated to public housing, the surface had been reduced to 150 square meters. This resulted in the physical transformation of the plot's original triangular shape. Out-of-scale façades, due the circumstances of the constructible site, are constraints within which architecture likes to play and, from it, draw its own logic. This building's angled façade and the stilts raising it above the ground bring elegance to the project and give it some breathing space. For more dramatic effect, the north wall has been covered with a Parisian metro map with backlit stations, not unlike the ones Beatnik author Jack Kerouac saw on his visit to the French capital and loved so much as he was writing *Satori in Paris*. H.L.

环境越复杂，越需要想象力与关注。"Stimuli"项目所隐含的观点是将巴黎及其附近郊区的可用于建设的空白地块利用起来。法国AS建筑工作室对改造"城市废弃地块"及赋予它们新生和活力方面深感兴趣。工作室希望对空白地块进行的改造一旦实现，这些地块将重获活力并成为城市发展与规划的新前沿。位于巴黎十三区的Château-des-Rentiers街角的公寓就像插入周边环境中的针头，目的是使整个机制恢复活力。项目的条件非常苛刻：占地面积仅150平方米且位于高度为30米的墙体旁，只有公共廉价住房建设般有限的预算。这些导致建筑随地块的三角形状成为三角柱体。由地面条件所致的不规则外立面使建筑具有独特的材质特征和自身的逻辑。这个建筑的夹角和支撑它的支柱使项目变得高雅并赋予了一丝时尚气息。为赋予建筑更多的活力，建筑北墙覆盖着巴黎地铁图，这与"垮掉派"作者杰克·凯鲁亚克在拜访法国首都时看到并深深爱上，甚至写下《Satori in Paris》的建筑并没有什么不同。H.L.

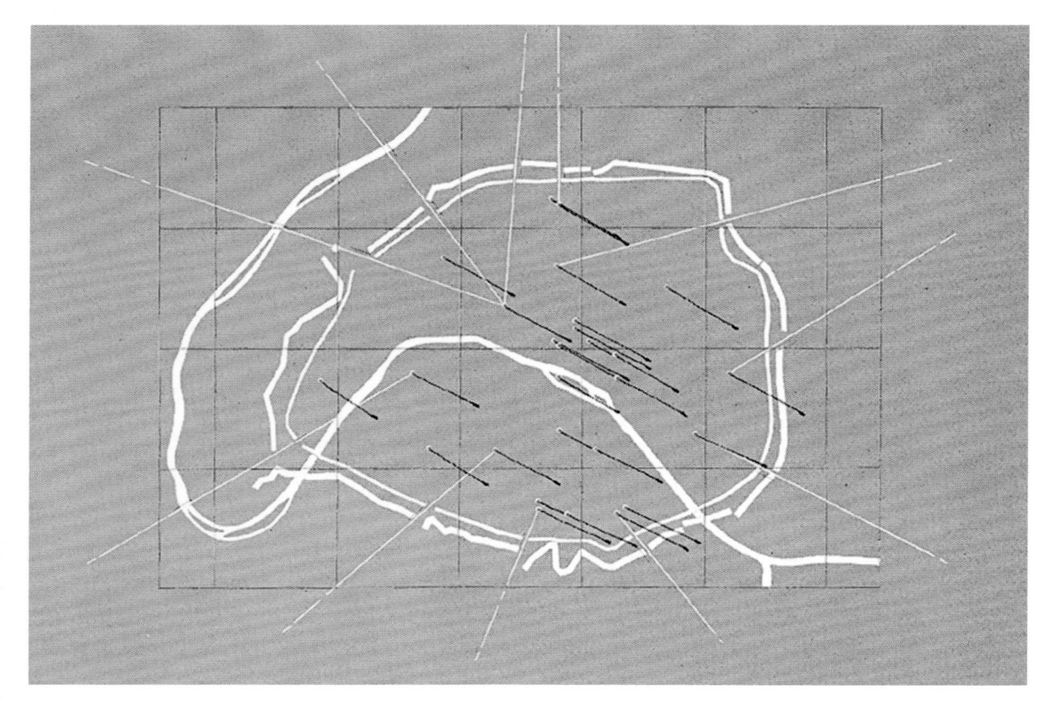

A drawing of the vacant plots identified in Paris by Architecture-Studio for the "Stimuli" project. Like hollow teeth in the fabric of Paris, these plots were able to join with others and form a connected network of projects, like an urban acupuncture map.

法国AS建筑工作室为"Stimuli"项目挑选的巴黎空白地块。这些地块能够相互呼应并且共同构成项目网，犹如城市针灸点地图。

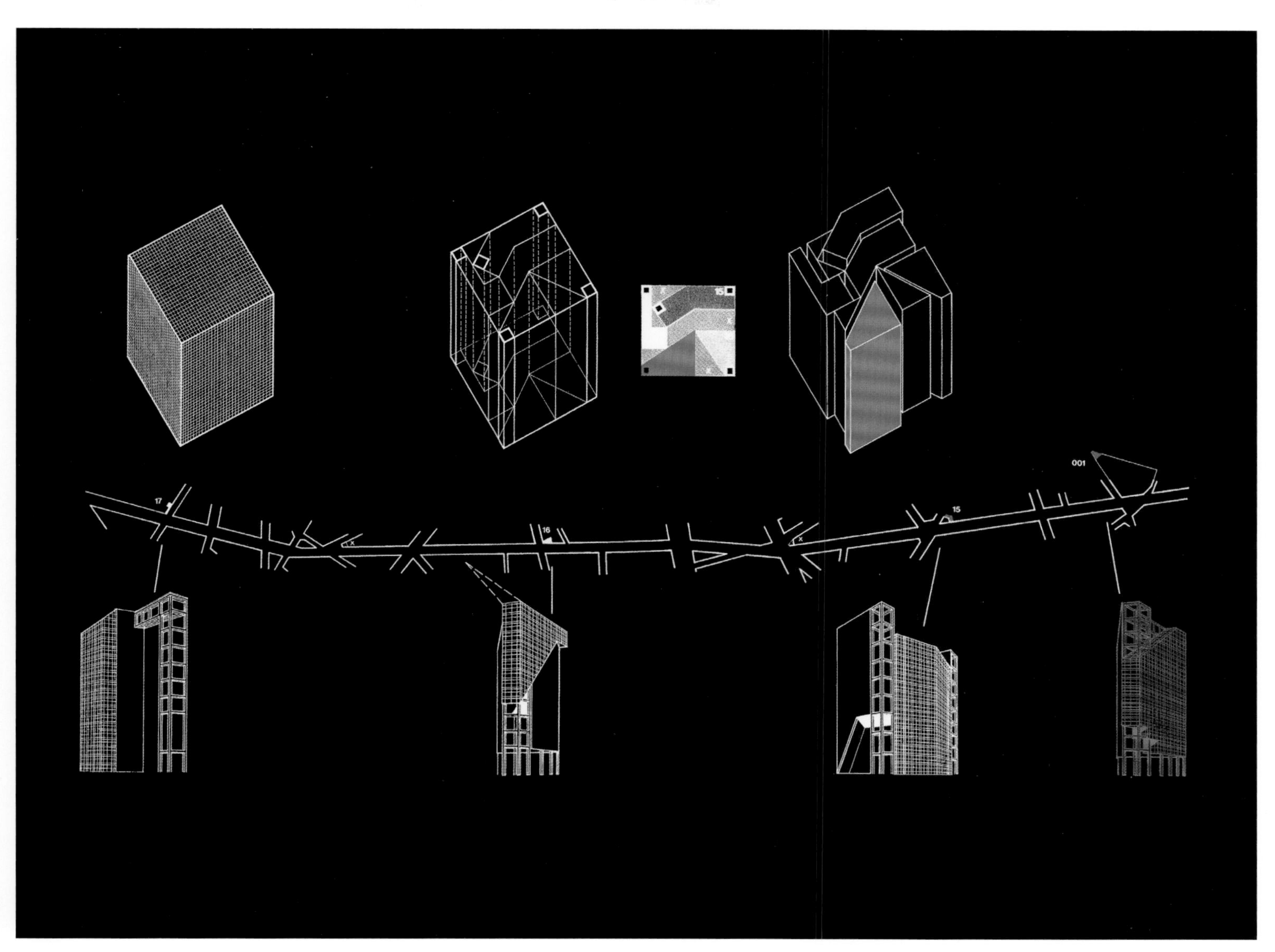

The urban dwellings of the Stimuli project appear as fragments of the city as a whole. Here are some research shapes and implantation studies for the project.

Stimuli项目的城市住宅是整座城市的片段。图为相关地块的形态与植入性研究。

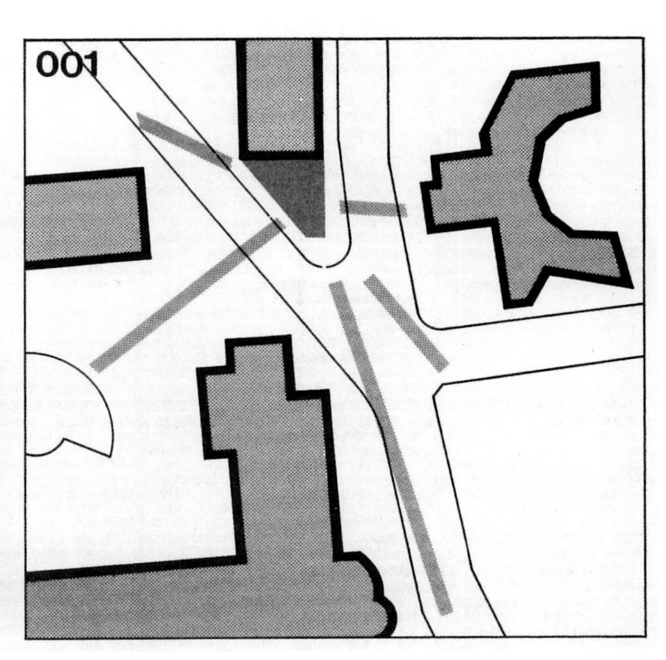

The plot before the intervention of Architecture-Studio. The small size of the aera reverses the ratio of the living space and the facades. Their functions are radically changed.

法国AS建筑工作室施工前的地块。狭小的地块改变了居住面积和建筑外墙之间的比例，项目建成后该地块的功能被彻底改变了。

The exoskeleton of the building – resembling a construction site tower crane – keeps the public spaces clear and open.

建筑外部构造，像建筑工地塔式起重机一样架起整个建筑，让底层成为开放式的公共空间。

Next page:
The building on rue du Château-des-Rentiers, 13th arrondissement, Paris, is the first application of the Stimuli approach.

下一页：
位于巴黎十三区Château-des-Rentiers街的公寓建筑是Stimuli研究项目的首次实现。

CAMPO FORMIO
CHEVALERET
10 MN
NATIONALE
PLACE D'ITALIE
27
TOLBIAC
62
5 MN
132
IVRY
MAISON BLANCHE
47

High School of the Future
Jaunay-Clan, France
1986/1987
19 000 m²

未来中学
法国，若奈克朗
1986/1987
19 000 m²

The triangle, the old masonic emblem, has been expanded into an arts and technological high-school situated in the vicinity of the Poitou-Charentes region's flagship scientific theme park, Futuroscope. A curvilinear building separates itself from the triangle-shaped main building in order to echo the nearby curving road, integrating with the environment, as if to embrace a series of future educational venues. Designing means opening up to new possibilities. The curve welcomes more space and "lost" functions are regained by the empty space in the middle of the main building's triangular form. A playground has been placed within these precincts – an elliptical form that enjoys the magic scale of the edifice while toying with the curvaceous swirl of the traditionally baroque figure. This courtyard has a sliding roof that opens and closes according to the changes in the weather. It is built in the same metal as the building shell and equipped with a fantastic opening mechanism. Jules Verne meets high-tech engineering... but with the ingenious simplicity of Jean Prouvé's famous sunroof. This humorous use of technical skill reminds young souls that innovation can be fun. H.L.

三角形是共济会旧式的标志，已被应用到位于普瓦图—夏朗德地区的Futuroscope未来主题公园附近的艺术及科技中学中。曲线建筑与三角形主建筑相分离，从而与附近弯曲的道路相呼应，实现与周围的环境相结合，就如同拥抱着这些未来将被建成的教育场所一样。设计象征着接受新的可能性。曲线建筑可获得更多的空间，三角形主建筑中的镂空区域所“失去的”功能由此得到填补。而镂空区域则设置为椭圆形的操场，既能够反衬出建筑物的规模，同时还能够体现传统巴洛克风格的曲线美。这个庭院有一个能够按照天气变化而开启或关闭的滑动式顶棚，使用与建筑外壳相同的金属制作并配有神奇的开关系统。在这里，作家儒勒·凡尔纳遇到了高科技结构——那便是Jean Prouvé著名的既简易又巧妙的顶棚。这种幽默地使用工艺技术的方式提醒了年轻人创新可以很有趣。H.L.

The heart of the building consists of an elliptical courtyard surrounded by a web of red concrete. A flying wing, a piece of the building, moves over the course of the day and accommodates cultural events.

建筑的中心是一个椭圆形的庭院，由红色混凝土墙包围着。滑动式顶棚如羽翼般的构造会根据阳光的变化而变化，并在举行文化活动时为场地提供遮蔽。

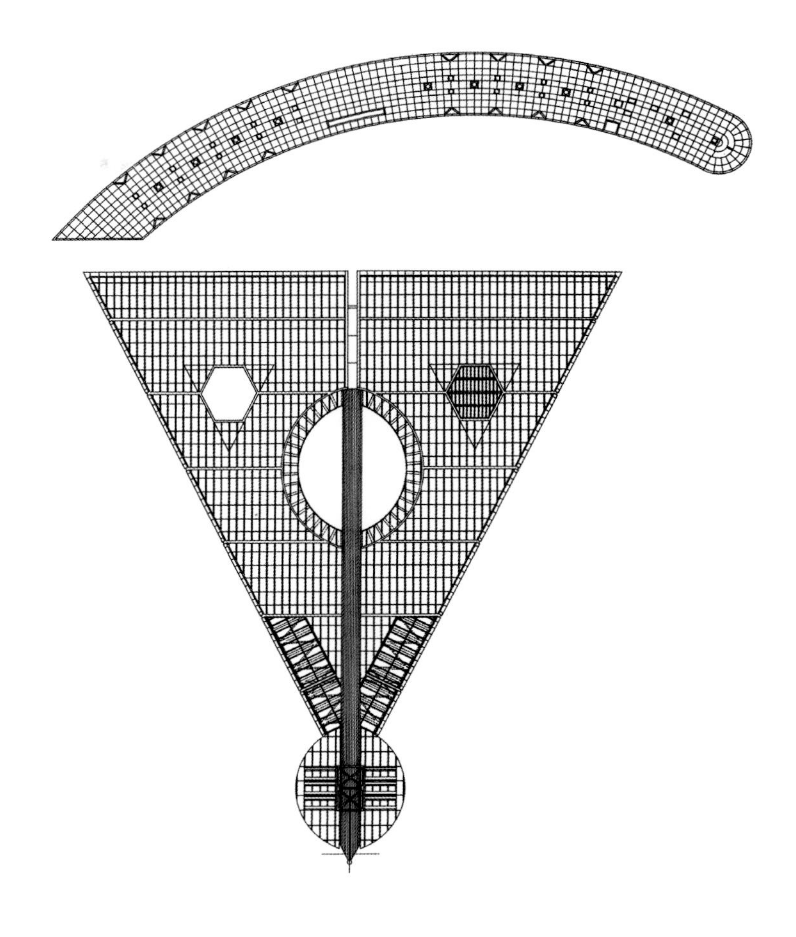

Plans for the roof and the ground floor of the building.

建筑屋顶与一层平面图。

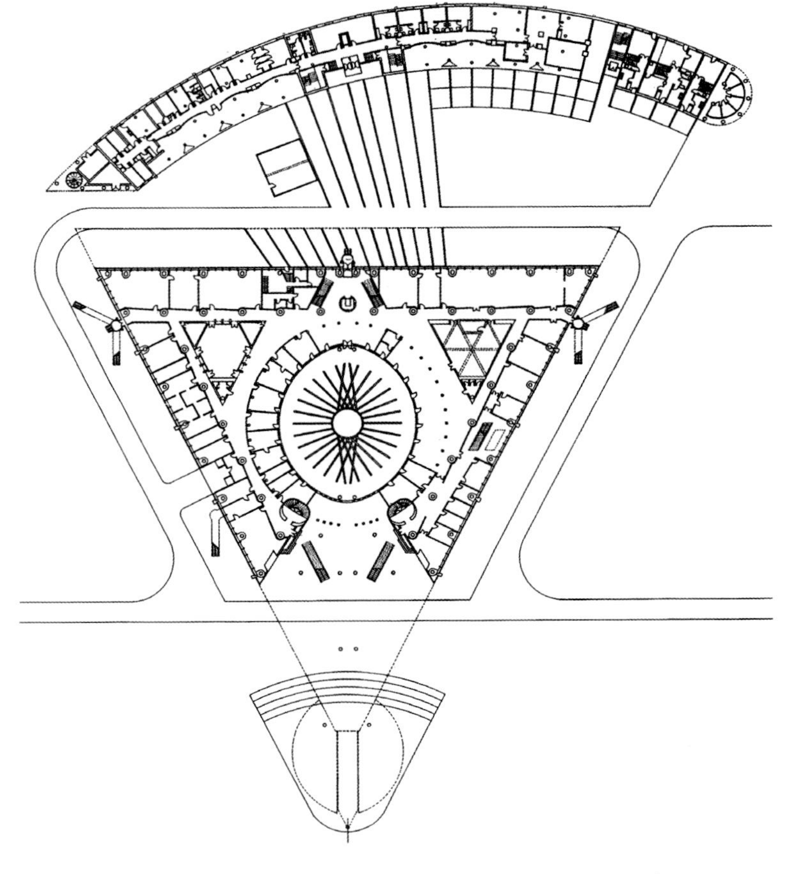

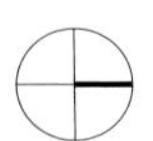

PASTB1

Arab World Institute
With Jean Nouvel, Gilbert Lézènes and Pierre Soria
Paris, France
1981/1987
27 000 m²

阿拉伯世界研究中心
与让·努维尔，吉列博·雷泽能和皮尔埃·索利亚共同设计
法国，巴黎
1981/1987
27 000 m²

Who remembers the bus depot by the side of the river Seine, right on the spot where you can now admire the Arab World Institute (IMA)? The IMA seduces because, a building all of light, it complicates our cultural enlightenment. Reflecting the diversity of origins that constitute Paris as a metropolis and the French metropolis's identity as a city, the IMA, built in the mid-1980s, was ahead of its time, being a wake-up call on the subject. The expression given to the place of Arabic culture in the world and in France had to go through a degree of mystery with a visual play between transparency that blurs into opaqueness and opacity that opens up to light. The north façade of the edifice, all in glass, is the museum. The south façade, protected from the sun by a mechanical screen, reminiscent of the traditional mashrabiya, is the library. A deep rift divides and unites the two parts. This elegant hybrid playfully mixes ideas: a cross between the Moorish lattice screen and camera shutters to control the natural light; a spacing of transparent layers which obscure by addition; a cross between an escalator and a camera; aircraft on the inside, flying carpet on the outside. Finishes worthy of the most sophisticated airliners make up the profiles of the ceilings: they render the wiring invisible and allow the gradual spread of artificial light automatically. Night or day, one can only cherish the IMA, stubborn as a mule in its political optimism, beautiful in its dual, mulatto, identity. H.L.

谁还记得位于塞纳河畔的巴士站？那里是观赏令人赞美的阿拉伯世界研究中心的最佳地点。阿拉伯世界研究中心吸引着我们，因为它是光的建筑，它使启蒙文化变得复杂。反映巴黎大都市的身份离不开多民族的融合，建于20世纪80年代中期的阿拉伯世界研究中心走在了时代的最前沿，带来了相关的反思。阿拉伯文化在法国乃至全球的发展都带有其独特的神秘感，建筑设计则让透光度和不透光度成为一场视觉游戏。建筑物的北向立面采用全玻璃设计，包裹着内部的博物馆。而南向立面则采用类似阿拉伯传统镂窗的控光装置，为内部的图书馆遮挡阳光。深层裂缝将建筑分为两个部分又将其有机结合起来。这个优雅的结合混合了下列想法：阿拉伯传统镂窗与控制自然光的相机光电传感器相融合；透光空间的明亮与遮蔽带来的模糊和黑暗相结合；电梯与影院的配合使用；代表现代文明的飞机与代表传统文化的飞毯的对比。天花板的完美品质如飞机的内壁般精致，同时将控制智能式灯光的众多电缆隐藏。斗转星移，人们依旧赞叹阿拉伯世界研究中心具有双重文化的美，并对其促进文化交融的坚持感到钦佩。H.L.

Floor plan.

楼层平面图。

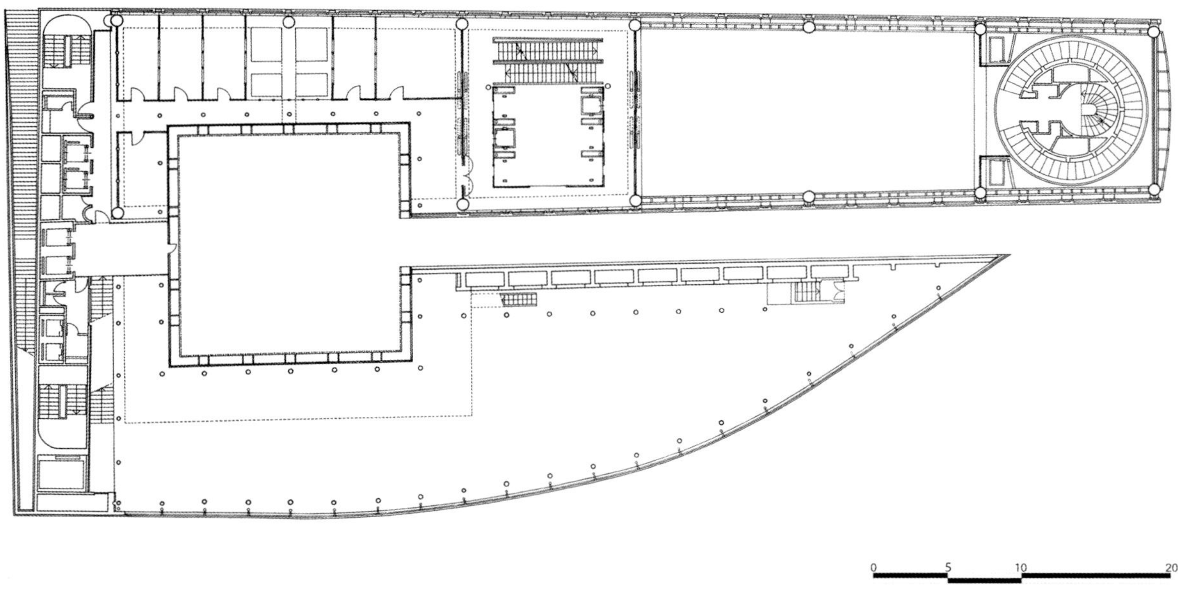

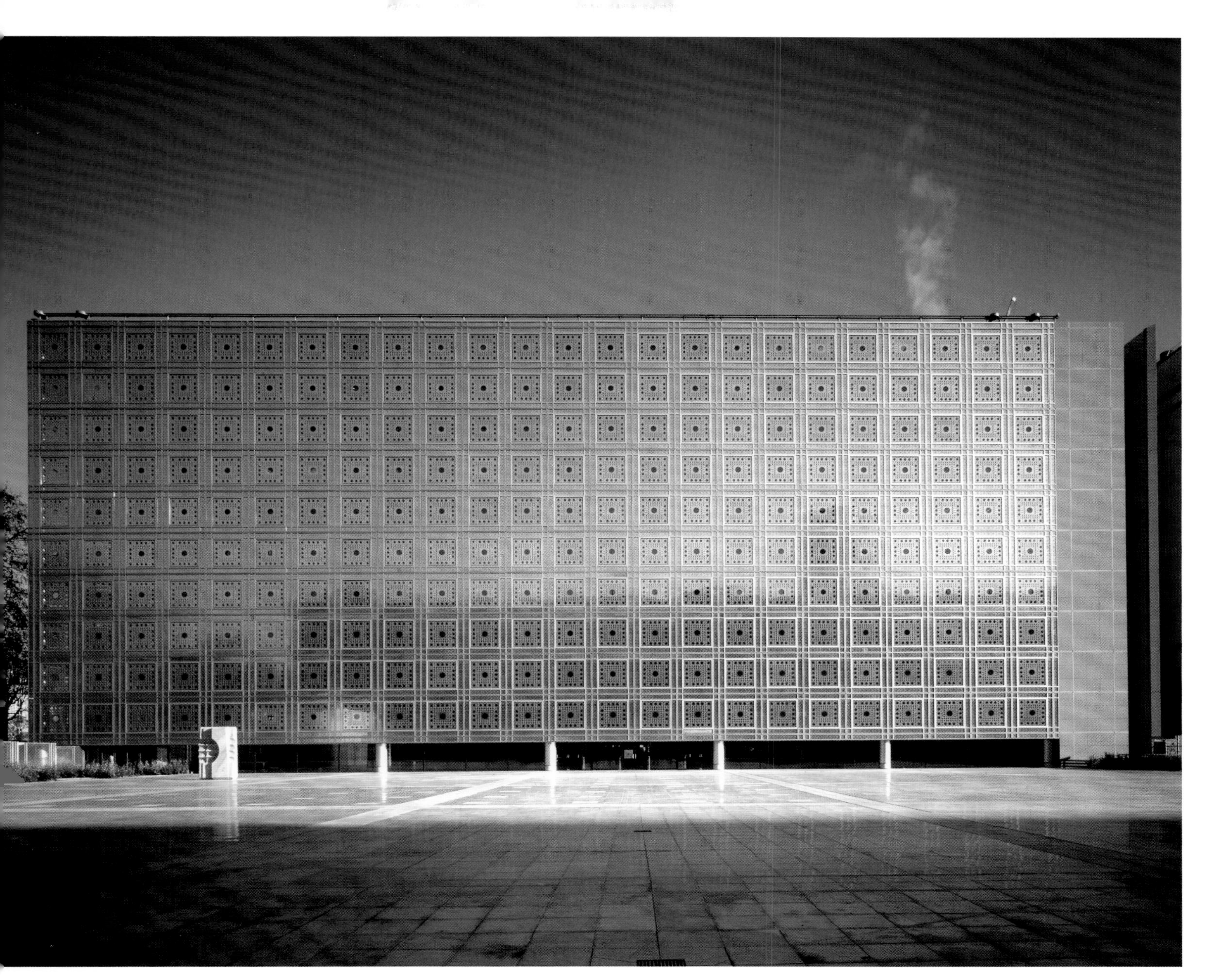

The southern façade covers the library and utilizes traditional geometrical patterns.

使用传统几何图案的南向立面为图书馆遮阳。

PACR1

University Residence
rue Francis-de-Croisset
Paris, France
1989/1996
11 000 m²

Francis-de-Croisset街学生公寓
法国，巴黎
1989/1996
11 000 m²

To build comfortable and innovative studios on an 11 000 m² plot in the heart of an asphalt environment, under heavy traffic pressure – suffocating exhaust pipe emissions and deafening decibels – leaves no choice but to fight: conquer or perish. So far as residents are concerned, it is just a matter of how happy you are to be living there. Architecture faces a big challenge when designing for students, who by dint of their position are often more critical and take a forward-looking attitude. A shield – 30 meters high and 100 meters wide – cuts off the noise. Under its vigil this hall of residence – with its three curvy rooflines, its metal outer skin as polished as a black pearl, its bright red-framed windows – provides amazingly lush peacefulness. From the driver's point of view, nothing looks less boring than this great anti-noise screen beside the motorway. From the student's, this is not so much an oppressive wall as a protective barrier bordering a promenade offering dizzy perspectives on the traffic, not unlike Piranesi's etchings of *Imaginary Prisons*. Sharing a common space is both beneficial to people in terms of human experience and to buildings in terms of thermal resources and ventilation as required in a buffer zone. After socializing you go back to a neat 20 m² room with ingenious storage space. Twenty square meters of guaranteed comfort, which no student that has had the opportunity of staying there will ever forget. H.L.

在沥青大马路旁建造11 000平方米的公寓楼，面对交通重压带来的汽车排放污染和噪声污染，不去征服这样的环境就只能被其所灭，除了战斗，我们别无选择。这也将决定建筑是否能为入住的学生带来舒适愉悦的住宿环境。设计学生公寓时建筑师面临着一大挑战，即大学生那更为严苛和具有前瞻性的态度。一面高30米，宽100米的屏障能够隔绝噪音，在隔音屏障的保护下，学生公寓具有三个弯曲的屋顶轮廓线、黑珍珠般光滑的金属外壳，以及明亮生动的红框窗户，为公寓内部提供了可贵的安静环境。从驾驶员角度来看，没有什么比高速公路旁边抗噪声的巨大屏幕更能引起注意的了。从居住学生角度来看，巨大的隔音墙体更像是为室内的通道和回廊做的防护，这样的视野让人不禁联想起皮拉内西的蚀刻版画作品《想象的监狱》中的天马行空。私人空间之外的共享活动空间一方面能够保证更优质的供暖和通风效果，另一方面为学生提供社交的场所，在聚会完毕后，学生则回到各自房间，体验20平方米且配备充足收纳空间的房间带来的住宿舒适度。对于每个在此居住的学生，这些都是非常难得的住宿体验。H.L.

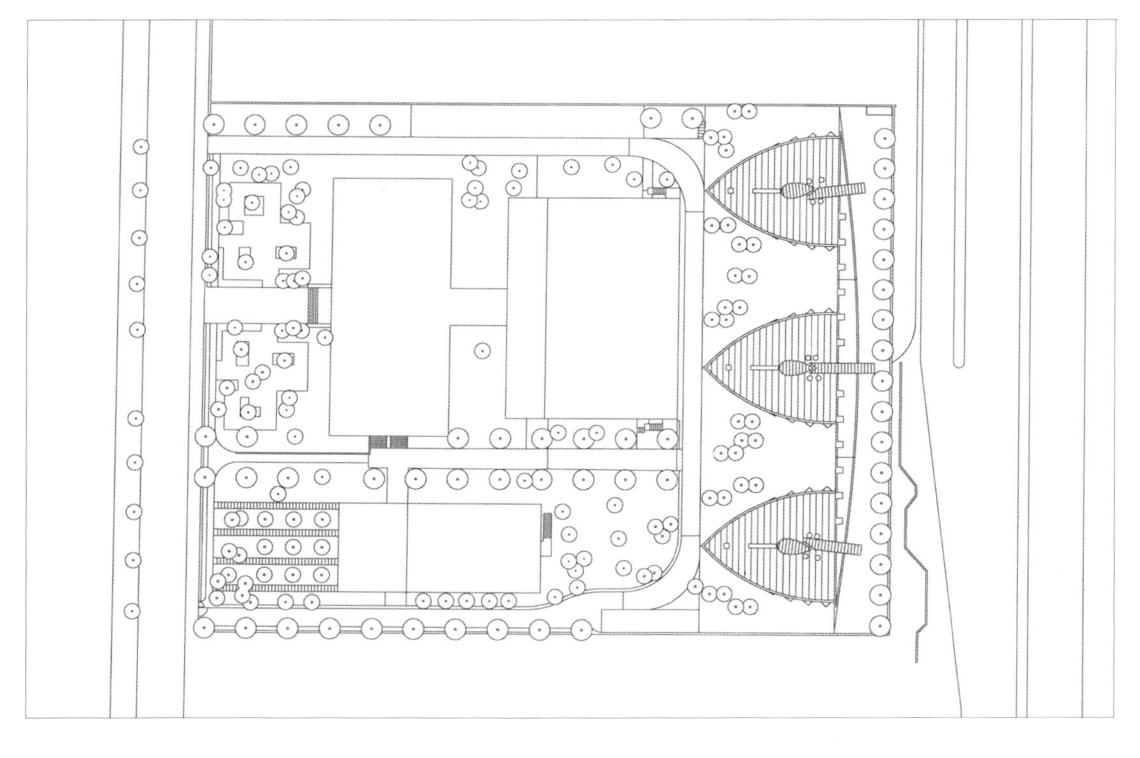

The ground plan of the university residence and the adjoining garden.

大学宿舍楼与附属花园的平面图。

On the north side, the building is a curved wall: a signal built into the kinetic landscape of the city as seen from the ring road.

建筑北侧呈现弧形墙体：从环形公路上看这个墙体与城市动态景观相得益彰。

On the Paris side, three curved eleven-floor buildings emphasize its walls and location in a garden.

建筑面向巴黎市区的一侧，三幢11层的弧形建筑突出了建筑的墙面及在花园中的位置。

DQ1

Citadelle University
Dunkirk, France
1987/1990
15 000 m²

城堡大学
法国，敦刻尔克
1987/1990
15 000 m²

With a view of the beautiful bricks and steel landscape of Dunkirk harbor, this University, designed by Architecture-Studio, seems to be advancing towards the waterfront in the guise of aluminum surf. The building houses scientific activity related to the maritime economy: biochemistry and energy studies. The issue at stake here is the re-actualization of the already existing buildings, bringing the traditional into the contemporary and vice versa. Bricks turn into fiber-reinforced concrete which, in turn, don the color of local heritage. Gables recovered from this former depot are like an old hair-comb stuck in a new hairstyle – a design so lavish it can dress both the original building once dedicated to the tobacco industry and the new parts oriented towards the economy of the sea. A ghost and a newborn baby have merged into one edifice, thus creating a fruitful confrontation between past and present and resulting in a narrative with Gothic suspense – secret passageways and an atmosphere full of mystery. English culture is close and Flemish influence is seeping through from across the border. One suspects that the architects have borrowed from the exuberance and transformative energy of the city's traditional mid-Lent festival to add flavor to this building. H.L.

红砖与钢的完美结合是敦克尔克港的特色景观，法国AS建筑工作室设计的城堡大学坐落在这里，朝向大海，迎接银白的海浪。建筑的功能与其滨海环境息息相关，是一所进行海洋生物化学和海洋能源等海洋经济科研的校园。这个项目是对已存在建筑的更新，并将传统的元素融入当代建筑中。传统红砖变成了纤维性混凝土，并涂上了当地文化遗产的代表颜色。旧址仓库保留的墙面与新建筑的结合好比插在时尚发型上的古典发饰——墙面既是曾经繁华的烟草业的见证，又为新生的海洋经济作出了贡献。就像是一个古老的灵魂与一个新生儿同时融入一个建筑，在过去与现在之间构建了显著的反差。尤如通过建筑讲述一个带有哥特式悬念的故事——秘密通道带来充满神秘感的氛围。敦克尔克因其地理位置与英国文化紧密相连，又有边境线上弗拉芒语的影响。人们都说，建筑师借鉴了这座城市在严肃的大斋期的中期举办热闹狂欢仪式的作乐心态，为原本沉闷的建筑增添了乐趣。H.L.

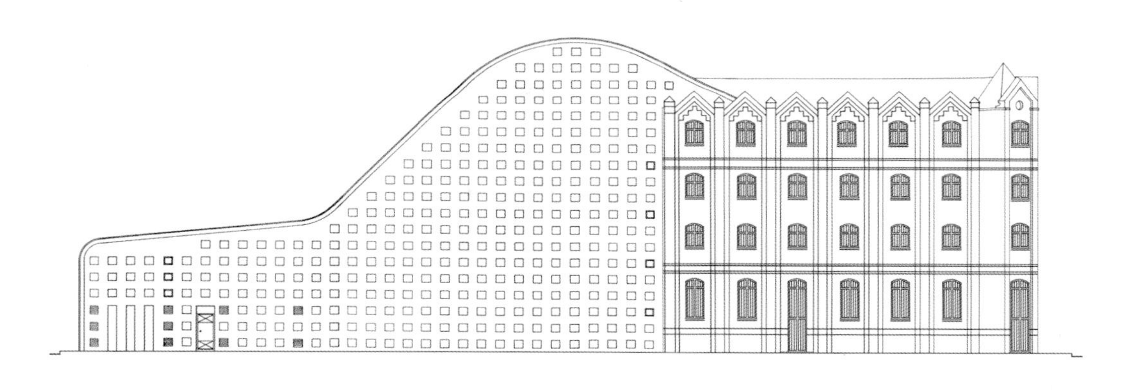

Elevation showing the relationship between the new building and the tobacco warehouse.

立面图显示了新建筑与旧烟草仓库之间的关系。

02

ARCHITECTURE – STUDIO
AS.ARCHITECTURE-STUDIO
生态城市
la ville écologique
LE FORT D'ISSY
UN PATRIMOINE EN DEVENIR
Archibooks
AS.
ARCHITECTURE-STUDIO
PARIS-SHANGHAI-PEKIN-VENISE
INAUGURATION
www.architecture-studio.fr
DANONE
design focus
recycle
CA'ASI
LA MAISON
Son nom est MARTIN ROBAIN,
il est le co-fondateur d'Architecture Studio

TROPHÉES FRANCE CHINE
AS.ARCHITECTURE-STUDIO
Fountain · Surge
La puissance créatrice du collectif: la ligne singulière d'Architecture-Studio
Valérie Jouve
Zhenfei Wang
Amaury Wenger
Gordon Matta-Clark
Didier Faustino
30.05–31.08.2013
CENTRE CULTUREL REGIONAL DE JINAN
中国山东
省会文化艺术中心
三馆
VILLES AFRICAINES
MOUVEMENT
CA'ASI
as. ARCHITECTURE-STUDIO
LA RECUPERACIÓN DE BARRIOS A LO LARGO DE CHILE
Pôle culturel de Saint-Malo
Le Chantier
Pôle culturel de Saint-Malo
La Grande Passerelle
INSTITUT DU MONDE ARABE
YOUNG ARAB ARCHITECTS
LA NOUVELLE ARCHITECTURE ARABE
العمارة العربية الجديدة
NEL PALAZZO DI CRISTALLO
Very interesting!
Grazie per tutto. Era molto affascinante!

02 WINDOWS ON THE WORLD

02 OUVERTURES SUR LE MONDE

世界之窗

The façade of the CA'ASI on Campiello Santa Maria Nova, Venice.

坐落在圣玛利亚诺瓦小广场的CA'ASI艺术展览馆外立面。

The CA'ASI, founded by Architecture-Studio is "aimed at promoting dialogue around contemporary architecture by means of a specific cultural venue", on the Campiello Santa Maria Nova in Venice, Italy. The goals of the CA'ASI, "Architecture-Studio's open house in Venice", include organizing and welcoming exhibitions and seminars, publishing and disseminating ideas, and providing accommodation for authors and architects. This 400 square meter space is located in the Cannaregio area, not far from the Rialto Bridge. The refurbished venue on three floors of a medieval palace includes an exhibition gallery at ground level that opens directly onto the small square. A lounge and a reception space are on the first floor. Also on the first level there several guest rooms as well as a two-room flat in the attic of the building.

Events are organized in conjunction with the Architecture Biennale, often focusing on "young emerging architects or the architecture of developing countries... offering an off-beat style as compared to the main themes presented during the Biennale". At the time of the Art Biennale (which alternates every two years with the Architecture Biennale), the CA'ASI welcomes artists whose work is related to the area and the city. Since 2010, the CA'ASI has presented one or two exhibitions a year, with subjects ranging from Young Chinese Architecture (2010), to the aerial photographer Alex MacLean (also 2010), to subjects on the emerging architecture of the Arab world (2012) and Africa (2014, 2015).

Given the thought that goes into the work of Architecture-Studio, it is not surprising that the partners consider the CA'ASI to

CA'ASI艺术展览馆是法国AS建筑工作室在意大利威尼斯创立的"为促进当代建筑对话与发展的文化场所"，坐落于圣玛利亚诺瓦小广场。法国AS建筑工作室在威尼斯的CA'ASI艺术展览馆向公众开放，通过组织承办展览、研讨会、出版等活动进行思想的碰撞与理念的传播，邀请作家和建筑师一同探讨，并为他们提供住宿。这座占地400平方米的建筑位于距里亚尔托桥不远的卡纳雷吉欧区，在一座三层的中世纪宫殿之中。场馆经翻修之后，一层变为直接对小广场开放的展览廊；二层含有会客大厅以及几间用于接待艺术家、建筑师等特邀嘉宾的客房和套房。

CA'ASI艺术展览馆组织的展览和活动与威尼斯建筑双年展主题配合，已数次聚焦"发展中国家涌现出的年轻杰出建筑设计师"，并在威尼斯建筑双年展期间为他们的作品提供展示机会。在威尼斯艺术双年展期间（与威尼斯建筑双年展轮替举办），CA'ASI艺术展览馆为艺术家提供场地，组织与城市空间相关的艺术作品的展览活动。从2010年建立以来，CA'ASI艺术展览馆每年承办一到两次展览，主题从中国新锐建筑创作展（2010）到航空摄影师艾利克斯·麦克莱恩展（2010），再到阿拉伯世界（2012）和非洲青年杰出建筑师展（2014），以及变化中的非洲城市（2015）等。

合伙人无不将CA'ASI艺术展览馆视为法国AS建筑工作室工作理念在艺术和哲学方面的延续。原因是显而易见的，但用将该理念付诸现实的合伙人的回答或许是最好的解释。第一个发言的是罗多·蒂斯纳多，他明确指出："这是我们工作的基础要素。我们一直致力于创造出最

be very much a product of their philosophy. The reasons for this are relatively clear, but are best explained in the words of the partners, who took the time to define the concept. The first to speak is Rodo Tisnado, who states unequivocally that "this is a founding element in our work. Our idea was always to create the best architecture possible and then to participate in making it known. The CA'ASI also has to do with the idea of presenting young architects, and perhaps to preparing future partners. Venice is the meeting point of East and West. We participated in numerous competitions in Italy, and won some of them, but we never built anything there. Finally we bought this place, which is part of a Gothic palace and, in 2009, we renovated it. It serves to think, to show architecture and art, and to be in contact with the activity that surrounds the Biennale." Another of the very early partners, Jean-François Bonne, also present on this occasion, expands on the thoughts of his colleague: "The idea is to place ourselves in a professional context not only amongst ourselves of course, but to open ourselves to the world. It is a matter of sharing architecture. From the outset, there has been an attention to and an opening to the exterior. We are seeking to intervene even beyond the pure exercise of the profession. The CA'ASI is very much in the spirit of this openness, which is for us essential."

Just as Architecture-Studio has reached out more and more to the international scene, the CA'ASI has served not only to enrich the thinking of members of the team, but also to bring others, for example architects from emerging countries, toward this Venice location. Roueïda Ayache explains this outward looking philosophy of the office: "We do wish to share about what we do," she says, "but there is an element of reciprocity that we seek as well; we wish to receive, but we also want to give others the occasion to transmit their own thoughts and work. This reciprocity is part and parcel of the ambitions of the CA'ASI. We work in certain countries and we also want the young architects of those countries to make their work known close to us, in Venice, which is at once near to Paris and open to the world. We are not in Venice just to make our methods known, but also to learn about how things are changing and how new ideas are coming forward. The presence in Venice gives us a certain distance from the day-to-day issues of our own work, which allows the give and take that we seek. It is a real matter of exchange and reciprocity."

Despite the continual emphasis of Architecture-Studio on the idea of returning architects to the central role that they used to play, there is also a thirst for new ideas and approaches that makes itself evident in Venice. Amar Sabeh el Leil says "The CA'ASI allows us to think about architecture, but also about contemporary art for example. The intersection between contemporary art and architecture is of great interest to us."

Beyond the question of pure architecture, Architecture-Studio is interested in new issues that arise elsewhere in different ways than in Europe. They ask young architects from given regions to submit their work for the exhibitions they organize in Venice without imposing any pre-conceived ideas on them. They receive about 200 submissions for each show and an international jury selects approximately 15 for each exhibition, which is to say three winners and 12 "honorable mentions". Roueïda Ayache explains: "We want to see emerging architecture, but also the issues that inspire it. The city and new architecture are obviously seen differently outside of France and that is what we are interested in. This fits in with the philosophy of Architecture-Studio – it is an opening towards what is possible." As a result of these exhibitions, Architecture-Studio has considered associating itself with a number of the young architects who have emerged in this context, in particular those from China.

The Venice exhibitions are an echo of the so-called *Mercredis d'AS* (AS Wednesdays) encounters where Architecture-Studio asks artists, architects or designers such as Erwan Bouroullec, Didier Faustino, Manuelle Gautrand or Alain Moatti to give a

好的建筑，并希望人们去了解这些建筑。CA'ASI艺术展览馆致力于发掘和推荐世界各地年轻有才华的建筑师，在未来我们也可能跟他们有更多的合作。威尼斯是东西方文化交汇的代表城市，我们在意大利参与了许多项目的竞赛，也获得了一些奖项，但是我们在那没有建设过任何建筑。后来我们买下这个场地，这属于哥特式宫殿的一部分，并在2009年对其进行了翻新。这个艺术展览馆为反思创造条件，在威尼斯双年展的大背景下，通过展览和各种活动展示建筑和艺术。”另一位早期的合伙人让-弗朗索瓦·博内也出席了此次采访，他进一步补充：“这么做是为了将我们置身于一个专业的开放环境中，不局限在自己的小世界中，而是向全世界开放，与之共享建筑理念。从一开始，我们就关注外部世界，现在我们更是寻求跨专业的体验。CA'ASI艺术展览馆拥有的是开放的精神，这对我们至关重要。”

随着法国AS建筑工作室越来越多地出现在国际舞台上，CA'ASI艺术展览馆不仅丰富了自身团队成员的思想，也将来自其他新兴国家的建筑师的新想法带到威尼斯的展览场地。罗伊达·阿亚斯阐明了法国AS建筑工作室对待外界的哲学理念：“我们确实希望与外界分享我们的作品，但是我们也在寻求相互交流；我们希望得到关注，也乐于为他人提供可以相互交流思想与展示作品的机会。这种相互促进机制是CA'ASI艺术展览馆远景的一部分。当我们在某些国家开展设计工作时，也希望这些国家的杰出的青年建筑师们的作品可以通过我们让大家认识，从威尼斯可以迅速传播到巴黎并走向世界。在威尼斯，我们不仅将自身理念推向世界，而更乐于看到世事变迁，从如雨后春笋般涌出的新想法和新事物中学习。威尼斯CA'ASI艺术展览馆的活动使我们从日常工作中释放出来，让我们有机会与更多的同行或艺术家对话，交流和相互促进，这是设计创作中很重要的。”

法国AS建筑工作室除了一再强调建筑的重要性，也同样期待新的创意和方式在威尼斯得以发扬。艾马·萨布埃雷说：“CA'ASI艺术展览馆给了我们对于建筑学和当代艺术进行思考的机会，我们同样对两者的关系充满兴趣。”

在单纯建筑学之外，法国AS建筑工作室还热衷于了解欧洲之外的各处涌现的新潮事物。他们鼓励来自指定地区的年轻建筑设计师们向他们在威尼斯组织的展览竞赛上提交作品，竞赛对创意没有设置任何束缚。每次竞赛都会收到约200件作品，国际评委团从中选取15个作品参与展览，其中包括3个获奖作品和12个入围提名作品。罗伊达·阿亚斯表示：“我们希望看到新颖的建筑风格，当然还有孕育和激发这些新建筑的现实背景。那些与法国建筑风格和城市规划有显然不同的外部世界正是我们的兴趣所在。这也符合法国AS建筑工作室的哲学——对一切可能性都秉持开放的态度。”通过一系列的展览，法国AS建筑工作室自认与众多年轻优秀的建筑师建立了良好的关系，特别是来自中国的青年建筑师。

conference in their offices in Paris. Originally programmed in direct connection to the projects or current issues being faced by the office, these weekly meetings have changed over time. A partner says, "In fact, we changed the AS Wednesdays after we began this work in Venice. We found it was very interesting to place more emphasis on the work of outsiders as opposed to the subjects previously dealt with that were more directly related to our own work. In our architecture, we always attempt to integrate what we do into its context. It is this motivation that you can see at work with the CA'ASI as well. We think of context across a very large range, including ideas and other forms of creativity such as art." Clearly, the AS Wednesday program was born of the same sense of openness and readiness to enrich architecture and urban design that animates the CA'ASI's exhibitions and events. Above all, this is an international and innovative approach to the development of architecture and as such it is very much in the image of Architecture-Studio, and in keeping with what is called here "the DNA of the office".

Jean-François Bonne continues in this vein. "Venice has permitted us to build another impression of how architecture can be made, through the contact that it provides with the outside world in a context that is not directly related to our project work. Openness is, in fact, something that many architectural offices preach, but the CA'ASI allowed us to act on our convictions. Our meeting with African architects, for example, allowed us to reconstruct our own impressions of the relationship of architecture to the world that is emerging. They seemed to be at the polar opposite of our concerns for what we frequently call globalization." The partners found in the case of the exhibition that they organized on emerging African architecture that the priorities of the participants were quite radically different from those currently seen in Europe. Roueïda Ayache explains: "The Africans architects clearly have other priorities – rather than seeking global connections they are addressing themselves to the minimal conditions that often exist where they are building. They also alerted us to the rarity of many resources. Recycling and reuse are their rule, expressing as much as possible with as little as possible." Clearly, in areas such as Africa, there is a great urgency for architecture to play a constructive role in development, a trend that Architecture-Studio has clearly understood.

As the presence of Alejandro Aravena from Chile, the 2016 Pritzker Prize winner and director of the 2016 Venice Architecture Biennale, testifies, new trends and new parts of the world are emerging in a significant way in contemporary architecture. In conclusion, Roueïda Ayache states: "To work on an international scale as we do, it is necessary to be aware of what is emerging in the rest of the world, far beyond the borders of France and Europe. We are in a polycentric world and this activity is a reflection of that reality." Very much in the spirit that animated the very first work of the office, such as the SAP project in Poitiers – where the partners of the time made the effort to bring together the potential residents of the social housing they were building – a similar if culturally even more expansive has been launched with considerable success in Venice. This is what they call "La maison commune" or the open house of Architecture-Studio. It is a place where urban, architectural and artistic concerns are taken into account and approached with an open mind. It is a place with windows that are open on the world.

威尼斯的展览与在巴黎举行的"周三相约AS"（选择一个月中的任意一个周三）讲座是相互呼应的，法国AS建筑工作室邀请如Erwan Bouroullec、Didier Faustino、Manuelle Gautrand和Alain Moatti等艺术家、建筑师或设计师到位于巴黎的办公室进行讲座和研讨。最初这些每月一次的研讨会旨在分享公司内部的新项目和新进展，但随着时间的推移而有了改变。一位合伙人说："事实上，从我们在威尼斯设立CA'ASI艺术展览馆以来，'周三相约AS'就变为邀请工作室以外的同行或相关行业代表进行讲座的场合。因为我们发现更多地聚焦其他建筑师的设计作品比只了解我们自己的设计或相关的项目要更有趣。我们一直希望将我们的项目融入它的背景环境中，这也是我们开展CA'ASI艺术展览馆工作的动力所在。我们认为环境所涉及的含义很广，包括理念和诸如艺术等其他创作形式。"很显然，"周三相约AS"通过讲座和交流促进城市规划和建筑设计发展的意愿，与CA'ASI艺术展览馆开展展览和活动的目的是一致的。最重要的是，这是一项具有国际性和创新性的促进建筑发展的方式，符合法国AS建筑工作室的理念，也是工作室一直传承的基因。

让-弗朗索瓦·博内补充道："通过在威尼斯与外界其他建筑师、艺术家的交流，让我们看到我们自己设计的建筑项目以外的世界，认识到建筑创作的新方式和新可能。事实上，开放性是许多建筑同行所推崇的，但是CA'ASI使我们可以演绎自己的方式。例如，我们与非洲建筑师们的会谈使我们重新审视新兴地区和建筑的关系，在全球化的进程中，他们似乎与我们有着完全不同的担忧。"在非洲青年杰出建筑师展览中，合伙人发现非洲建筑师所考虑的内容与欧洲建筑师所考虑的内容截然不同。罗伊达·阿亚斯表示："显然非洲建筑师有其他的优先关注点，他们时常需要在资源极度有限的条件下完成建设，而远不是寻求全球化。他们同时告诫我们要注意珍惜资源的使用。他们注重回收再利用，尽可能少地依赖资源。"很明显，像非洲这样的地区，建筑在其发展中承担起建设性作用是非常迫切的，法国AS建筑工作室清楚地意识到这一趋势。

来自智利的Alejandro Aravena获得2016年普利兹克建筑奖并担当2016年第十五届威尼斯建筑双年展的总策展人，见证了新趋势和新兴地区在当代建筑界中脱颖而出。罗伊达·阿亚斯总结说："在世界范围内从事工作，需要特别关注在欧洲以外地区涌现出来的新事物。我们处于一个多极世界中，这样做才符合现状。"这样的精神其实在法国AS建筑工作室最早期的设计作品中已经有所体现，例如普瓦捷的共享活动空间项目，那时合伙人将所建社会住房的未来房客们召集起来。在威尼斯，他们也采用相同的方法，通过文化活动将更多不同领域的人召集起来，因此CA'ASI艺术展览馆也被称为法国AS建筑工作室的"共享之家"。这是一个集城市规划、建筑设计和艺术创作于一体的开放空间，犹如一扇世界之窗。

2010

New Chinese Architecture
CA'ASI, Venice, Italy
Club Baiziwan 21#, Beijing, China
2010

中国新锐建筑创作展
意大利，威尼斯CA'ASI艺术展览馆
中国，北京百子湾路21号俱乐部
2010

"People Meet in Architecture" was the theme of the 2010 Venice Architecture Biennale. Architecture-Studio , grateful to China for the number of projects developed there, wished to give back some of the energy it got from working there by inviting young Chinese architects to showcase their talent at its Venetian palazzo, a magnificent building dating from Marco Polo, author of the first Western Chinese travelogue. The palazzo hosts the CA'ASI, Architecture-Studio's "open house", and functions as a laboratory of ideas and future projects ensuring the primacy of the design process over the final product. A competition, that was launched from Europe for Chinese candidacy, attracted some 150 entries. An independent jury chose 11 projects as worthy of display at the Venetian venue. From the shortlist, a further three were selected for their designs' outstanding visual quality. "The laureates were Dong Gong, a member of Vector Architects, Beijing, for "Monmentary City"; Liu Yichun and Chen Yifeng, of Atelier Dessau, Shanghai, for a kindergarten design; Wang Zhenfei, of HHDFun, Beijing, for YJP administrative building. A definitively different outlook on the world. H.L.

"相逢于建筑"是2010年威尼斯建筑双年展的主题。法国AS建筑工作室历年来在中国开展了不少项目，希望作出一些回馈，因此邀请年轻的中国建筑师来到威尼斯这座马可·波罗时代的宫殿中，向欧洲展示来自东方的才能，就像当年马可·波罗将在中国的新奇见闻带到西方一般。位于宫殿中的CA'ASI艺术展览馆是法国AS建筑工作室在威尼斯的"开放空间"，它就像一个艺术和建筑的灵感碰撞的实验空间。欧洲向中国建筑师发出的竞赛公告最终吸引了超过150个参赛作品，独立评审团从中选择了11个优秀项目在威尼斯CA'ASI艺术展览馆进行展出。并在入围的作品中，选出三部评委一致认为极具美感的获奖作品，分别是北京直向建筑设计事务所董功设计的"瞬间城市"、上海大舍设计事务所柳亦春和陈屹峰设计的嘉定新城幼儿园以及北京华汇设计王振飞设计的于家堡工程指挥中心。这些设计作品无疑为西方带来全新的视野。H.L.

Following its presentation at the CA'ASI during the 12th Architecture Biennale in Venice, and the award ceremony hosted by the French Academy of Architecture in Paris, the exhibition "The New Chinese Architecture" was displayed at the Club Baiziwan 21# in Beijing.

入围作品于第十二届威尼斯建筑双年展期间在CA'ASI艺术展览馆展出，并在巴黎法兰西建筑学院举行了颁奖典礼，之后又在北京百子湾路21号俱乐部举办了"中国新锐建筑创作展"。

2010

Alex MacLean: Paris La Défense Seine Arche
CA'ASI, Venice, Italy
2010

艾力克斯·麦克莱恩：巴黎拉德芳斯—塞纳轴
意大利，威尼斯CA'ASI艺术展览馆
2010

Alex MacLean is renowned for his humanist approach to capturing urban transformation. Throughout his career he has tracked the traces of contemporary design with a very sharp eye. His series on La Défense district, west of Paris, testifies to the French government's will to have a business district as prestigious as London's and New York's. The American photographer recorded every stage of the project's bumpy road from the original building site to the final erection of the famous Grande Arche. The photos shown at the CA'ASI document the amalgamated options that had been born out of an elevated esplanade with a strong structural axis between the Arc de Triomphe and the Grande Arche stretched like a string between two wooden pegs. The Défense Seine Arche district remains a monster of triumphant eclecticism. Alex MacLean rightly underscores the contradictions between this type of urban planning and the concerns of the local authorities in Nanterre, upon which it depends, and which, deliberately shunning the dry functionalism that was the norm in the 20th century, is now engaged in sustainable development projects. It is the account of an artist at the heart of the problem. H.L.

艾力克斯·麦克莱恩因捕捉城市转型的人文主义方式而出名。在他的职业生涯中，他用非常敏锐的眼光关注了当代设计。他的巴黎西部拉德芳斯区系列作品反映了法国政府希望打造与伦敦和纽约一样著名的商业区的意愿。这位美国摄影师记录了该区域从原始的建筑工地到著名的新凯旋门最终落成的崎岖之路上的每个阶段。在CA'ASI艺术展览馆展出的图片记录了尤如两个木桩之间细绳一般的凯旋门直至新凯旋门的轴线，以及轴线上摩天大楼兴建等有关的一系列工程细节。巴黎拉德芳斯—塞纳轴线的整体是折中主义的表现，摄影师突出了拉德芳斯商业区规划与南泰尔区当地机构所关注的不同重点。而南泰尔如今也有意回避20世纪标准的机能主义而更加注重可持续发展。这是艺术家触及了核心问题。H.L.

The exhibition of aerial photographs of the area of La Défense by Alex MacLean was accompanied by a film about the Terrasses de Seine Arche, a set of architectural models, video interviews, and an explanation of the photographic campaign.

展览除了有艾力克斯·麦克莱恩拍摄的拉德芳斯区的航拍摄影，还有关于塞纳轴线区域的电影、一套建筑模型、视频采访及拍摄过程记录。

2011

In the Crystal Palace
CA'ASI, Venice, Italy
2011

水晶宫殿
意大利，威尼斯CA'ASI艺术展览馆
2011

The city is the place where mankind can find both political emancipation and social alienation. The movie industry has never ceased to illustrate America's unstoppable progress toward ever-higher buildings and today's computer-generated images now feed the wildest fantasies of verticality across the globe. This is why the CA'ASI wished to exhibit a new generation of visual artists and photographers who deliberately call into question this urban dimension through iconic representations. The city has always generated mixed emotions. Dostoyevsky reacted to the new Saint Petersburg in his time with anguish. Contemporary German philosopher Peter Sloterdjik, on the contrary, expresses sympathy for the place where intimacy dissolves into the city's crystalline bubble. Following the logic of film-editing, the aforementioned artists reflect on the successive collages which have made up the city. The group exhibition curated by the CNAP, the French National Center for the Visual Arts, questions through filmic diaries the individualistic values of modernity and records the marks of urban life which are impressed upon our bodies. It also shows how the city has transformed our idiosyncrasies into less romantic, less bourgeois and more anonymous identities. H.L.

人类能在城市中经历政治解放与社会疏离感。电影行业从未停止过描述美国在建设更高的建筑方面的发展，如今的数码技术图片展现了世界各地对建筑垂直高度的狂热追求。这是CA'ASI艺术展览馆希望为新一代视觉艺术家与摄影师举办展览的原因，这些人通过其图像作品审视城市。城市总会引起复杂的情绪。作家陀思妥耶夫斯基用文字表达他由于新圣彼德堡而感到苦恼。与他不同的是，德国当代哲学家皮特·斯劳特戴克对让隐私消失在高楼大厦中的城市表示理解。就像电影剪辑一样，上述的艺术家反映了构成城市记忆的不同而又连续的碎片。由法国国家视觉艺术中心发起的联合展览通过电影日记的形式对现代个人主义价值提出了质疑并记录人们城市生活的重要标志。展览还展示了城市是如何将人变得缺少浪漫和小资精神后，进而变得没有特点的。H.L.

Louidgi Beltrame, *Katashima Torpedo Base*, 2010; Super 8 film converted into a video. Courtesy of the artist and the Gallery Jousse Entreprise.

Louidg Beltrame，Katashima岛鱼雷基地，2010；超8毫米胶片转换的视频，由艺术家Louidgi Beltrame和Jousse Entreprise画廊提供。

2012

Young Arab Architects
CA'ASI, Venice, Italy
Arab World Institute, Pari, France
2012

阿拉伯青年杰出建筑师展
意大利，威尼斯CA'ASI艺术展览馆
法国，巴黎阿拉伯世界研究中心
2012

Architecture-Studio has many good reasons to appreciate the Arab and Islamic world. Some are related to its culture and some to the international make-up of its team. Others stem directly from its work, which includes among other achievements the Arab World Institute. Having building sites from Muscat to Mecca, to Bahrain, to Qatar, to Jordan, not to mention projects in Palestinian territory, has energized the studio's architects. So it was only fair to give young Arab architects their chance by inviting them to enter a prestigious design competition and enjoy the limelight in an exhibition showcasing their work. There were more than 140 participants, representing 21 countries, some of which are at war. In the aftermath of the Arab Spring in 2012, questions concerning the organization of these different projects arose – what are the opportunities or the risks underlying a building project "abroad"? Would there be sufficient "common ground", a theme which gave its title to the Architecture Biennale that year? Would they cooperate or would they each go their own way? Architecture has never ceased being instrumental in the expression of Arab culture. However, the modernity of certain forms has also often clashed with their more traditional expressions. To prove it, there are these masterpieces of concrete: the Volubilis Archeological Museum impeccably designed by Moroccan Tarik Oualalou; Lebanese Youssef Tohmé's USJ new campus of Innovation, Economics and Sport; and the powerfully orthogonal Memorial for the victims of the Agadir earthquake conceived by the female architect Chamss Oulkadi. H.L.

法国AS建筑工作室有许多钦佩阿拉伯世界与伊斯兰世界的原因。一方面与其文化有关，一方面与其团队的国际化有关，也有直接与建筑项目挂钩的原因，如其巴黎阿拉伯世界研究中心所取得的成就。其从马斯喀特到麦加、巴林、卡塔尔、约旦等地均有建筑项目，更不用说在巴勒斯坦境内的研究项目，这些都激励着工作室的建筑师。邀请年轻的阿拉伯建筑师参加一场有格调的设计大赛并让他们的作品在展览中成为焦点，是作为对阿拉伯世界的回馈。参与竞赛的有超过140个团队，分别来自21个国家，其中包括一些战乱中国家。在阿拉伯之春刚过去的2012年，建筑项目面临着这样一些问题——在"海外"的建筑项目存在哪些潜在机遇或风险？是否有充分"共同基础"（当年威尼斯建筑双年展的主题）？他们会合作还是各走各的路？建筑一直是展示阿拉伯文化的重要途径之一，然而某些外观现代化的建筑又与他们更传统的表达方式发生冲突。以下具体的获奖作品是最好的说明：摩洛哥的Tarik Oualalou设计的瓦卢比利斯考古遗址博物馆；黎巴嫩Youssef Tohmé设计的创新学院、经济学院及体育学院校区；女性建筑师Chamss Oulkadi设计的阿加迪尔地震受害者纪念馆。H.L.

The second international architectural competition organized by the practice, "Young Arab Architects" was displayed at the CA'ASI as part of the 13th Architecture Biennale in Venice, and at the Arab World Institute for its 25th anniversary.

由法国AS建筑工作室组织的第二届国际建筑竞赛"阿拉伯青年杰出建筑师展"入围作品在威尼斯第十三届建筑双年展期间于CA'ASI艺术展览馆展出，后来又在阿拉伯世界研究中心第25届周年纪念日之际进行展出。

2013

Construction Ahead
CA'ASI, Venice, Italy
2013

先锋建筑
意大利，威尼斯CA'ASI艺术展览馆
2013

"Construction Ahead" is typically the wording on a road sign signaling that you are about to enter a building site. Usually, the public is not allowed in. The architects of Architecture-Studio wanted to let the public have a glimpse of three of their major sites–the tower of the Rotana hotel in Jordan, Jinan's Cultural Center in China, and the French broadcasting house Maison de Radio France in Paris–through the eye of a trio of contemporary photographers. Valérie Jouve in Amman, Wang Zhenfei in Jinan and Amaury Wenger in Paris all give an artist's vision of reality that differs from that of an architect. The Rotana Tower has become something of a beehive where the bees are actually individuals. The Cultural Center, echoing the Chinese market's expansion, has the intricacy of a cobweb woven by a giant spider, or rather a tower crane, a machine so human-like that it looks as if it has a conscience. The Maison de Radio France has been explored and photographed as if it was a cave, humanity's archaic abode. As a counterpoint to these thought-provoking images of the work in progress on construction sites, Gordon Matta-Clark's *Conical Intersect* is shown at the CA'ASI in Venice: the 1975 film documents the artist's spectacular "anarchitectural" gesture consisting of a spiraling "cut" into two derelict seventeenth-century Paris buildings adjacent to the construction site of the Centre Pompidou, thus illustrating a rebellious hand-to-hand combat with architecture's established order. Adding to this precious anarchy, architect and visual artist, Didier Faustino, has installed wooden pallets and scaffolding posts like those found on a building site for the exhibition's scenography. H.L.

"前方施工"通常是路标用语，表示即将进入建筑工地，然而公众是不被允许进入施工现场的。法国AS建筑工作室的建筑师想通过三位当代摄影师的视角让公众看到他们的三个重要建筑项目——约旦安曼的罗塔纳酒店大厦、中国济南的山东省省会文化艺术中心以及法国巴黎的法国广播电台大厦。王振飞在济南的摄影、Valérie Jouve在安曼的摄影、Amaury Wenger在巴黎的摄影，通过各自对艺术的呈现表达与建筑师不同的视角。罗塔纳酒店大厦的工地如同蜂巢，而劳作的蜜蜂就是每个工人。山东省省会文化艺术中心的施工地的复杂程度堪比由巨型蜘蛛编织的蜘蛛网，又或者说像是一台拥有人类意识的塔式起重机，它反映出中国不断扩大的市场。法国巴黎广播电台大厦工地的探索和拍摄呈现出来的是一个洞穴、一座人类的古老居所。与这些引人深思的工地照片做对比，Gordon Matta-Clark的作品《圆锥介质》也同时在威尼斯的CA'ASI艺术展览馆展出，这是1975年的记录影像，记录了艺术家完成的壮观"无建筑"装置，即在两座位于蓬皮杜文化中心施工场地旁的废弃的十七世纪建筑上切割出许多圆锥形大洞，表现出叛逆的建筑设计与既定的建筑风格之间的交锋。建筑师兼视觉艺术家Didier Faustino加入这场难得的无序中，安装了木质展板和脚手架，将展览现场打造得像建筑工地一般。H.L.

Valérie Jouve, *Rotana Hotel, Écarts, 2013* (Assistant Walid Husseini); 4.5 inch film. Courtesy of the artist and Xippas Gallery.

Valérie Jouve，罗塔纳酒店，对比，2013（助理Walid Husseini）；由艺术家和Xippas画廊提供。

2014

Young Architects in Africa
CA'ASI, Venice, Italy
arc en rêve, Bordeaux, France
2014

非洲青年杰出建筑师展
意大利，CA'ASI艺术展览馆
法国，波尔多建筑中心arc en rêve
2014

Who is building the future Africa? Today 40% of its population lives in cities but the number will triple over the next 40 years. So how much weight will the continent's own architects be able to throw in the face of such a construction boom, considering that there are three architects for every 100 000 inhabitants, while Europe's ratio is one per 2 000? To assess the situation, this design contest launched in Paris and Venice was aimed at attracting architects, aged under 45 and with less than 15 years' work experience, from 46 African nations. Theirs is an architecture both vernacular and incredibly imaginative, integrating natural landscapes and responding to the contextual sobriety. Their architecture is poor in its means but conceptually rich, as was the Italian movement, Arte Povera, with its wealth of avant-garde thinking. The jury appointed three joint winners: South-African Christensen and Droomer for their minimalist and socially aware project built in a remote tourist destination, their huts and fence posts function as an interface between villagers and random visitors; Kenyan Urko Sánchez for his palm-trellised cover of a house "designed for the available space in between the trees" and itself fragmented into little cubicles; and, finally, the Architects of Justice – Granicki, Rassmann and Lacovig – who have replaced the typical "book cages" so common in South African schools with more pleasing bookshelves, deployed in two shipping containers ready to become replicable models. H.L.

谁在建设未来的非洲？今天，40%的非洲人口居住在城市，但在未来40年内非洲城市人口数量将翻三倍。而非洲每十万居民拥有三位建筑师，欧洲却是每2 000人中就有一位建筑师，这片大陆的本土建筑师将会为当地高速发展的建筑业作出多少贡献呢？为了对此情况作出评估，法国AS建筑工作室在巴黎和威尼斯发布了建筑设计大赛，旨在吸引46个非洲国家的45岁以下且工作经验少于15年的建筑师。他们设计的建筑作品既本土化又具难以置信的想象力，不但整合了自然景观，又符合当地严峻的环境条件。他们的建筑方式很单调，但概念却很丰富，就像用于许多前卫思想的意大利贫穷艺术运动。评审团评出了三个奖项：南非的Christensen和Droomer为在偏远的旅游目的地设计建造的具有极简主义和社会意识的项目，小屋和围栏的特殊设计成为村民和普通游客之间交流的端口；肯尼亚的Urko Sánchez设计的坐落在树木丛空地的房屋，被棕榈外层覆盖并可分隔为小间；Architects of Justice——Granicki、Rassmann和Lacovig用更悦目的书架取代了南非学校最常见的典型“书笼”，并将书架摆放在两个集装箱中。H.L.

More than 200 candidates from across Africa participated in the "Young Architects in Africa" competition. The award-winning architects were presented and awarded in Venice, as part of the 14th Architecture Biennale, in Namibia, and at arc en rêve in Bordeaux, where this rising generation of architects took part in an international symposium.

超过200个来自非洲的参赛团队参加了“非洲青年杰出建筑师”竞赛。获奖建筑师在第十四届威尼斯建筑双年展期间在威尼斯接受颁奖并参加展览，获奖作品随后在纳米比亚和波尔多arc en rêve建筑中心巡回展出，新生代的非洲建筑师因此参与到国际性的建筑文化活动中。

2015

African Cities in Motion
CA'ASI, Venice, Italy
2015

变化中的非洲城市
意大利，威尼斯CA'ASI艺术展览馆
2015

The two architecture exhibitions staged at the CA'ASI in 2012 and 2014 drew an audience which shared its keen interest in the architectural future of Africa with Architecture-Studio. Covering the northern part of the continent in 2012 and its southern part in 2014, the exhibitions showcased novelties that were evidence of African architecture's professional commitment and fascinating creativity. In 2015, a photo exhibition prolonged the mind's visit on the continent with Stéphane Couturier whose eye, sharpened by his reading of philosophers Baudrillard and Virilio, gives a new take on an iconic modernist construction – Climat de France, the gigantic ensemble of apartments built by Fernand Pouillon in Algiers between 1954 and 1957 where 50 000 people still live. After several years, the blocks of flats had deteriorated and become a zone of exclusion for the poorest, as is often the case with such models of urban planning. Real chaos was born out of fictitious order. Adding to this Algerian picture, videos sponsored by the French National Center for the visual arts (CNAP) chart a cartography of an African continent teeming with variegated images of graphic forms and shapes, people and spaces. H.L.

Confirming its interest in contemporary photography, the CA'ASI presented Stéphane Couturier's "Climat de France" series before his exhibition at the Maison européenne de la photographie in Paris.

出于对当代摄影的兴趣，CA'ASI艺术展览馆为Stéphane Couturier举办《Climat de France》展览，随后该展览在巴黎的欧洲摄影艺术展览馆展出。

2012年和2014年在CA'ASI艺术展览馆举办的两场建筑展览，吸引了众多与法国AS建筑工作室一样对非洲未来的建筑有极大兴趣的观众。2012年的展览涉及非洲北部地区，而2014年关注的则是整个非洲南部地区，两场展览展示了非洲建筑专业程度和令人叹为观止的创意。2015年，一场摄影展加深了人们对非洲大陆的认识。摄影师Stéphane Couturier通过阅读哲学家布西亚和维利里奥的著作，眼光变得更加犀利，对现代主义建筑的代表——Climat de France有全新的认识。这座巨大的公寓城位于阿尔及尔，由建筑师Fernand Pouillon设计，在1954年至1957年间建造，至今仍生活着5万人。多少年过去了，与许多同时期的规划项目一样，这个公寓街区已经严重恶化，成为最贫困的人聚集的地区。与这组来自阿尔及利亚的照片相呼应的，是由法国国家视觉艺术中心赞助制作的视频，绘制了一个由丰富的图形、形状、人物和空间等多种图片组成的非洲大陆的地图。H.L.

2016

Mission Trans-Missions
CA'ASI, Venice, Italy
2016
Qian Xuesen Library, Shanghai, China
2017

建筑——跨越时空的使命
意大利，威尼斯CA'ASI艺术展览馆
2016
中国，上海，钱学森图书馆
2017

The three competitions organized in China, the Middle East and Africa have confirmed that there is not just one way of envisaging architecture and demonstrated Architecture-Studio's strong commitment to helping innovative architects from emerging countries, with growing populations and endemic difficulties, to have better access to public commissions. How can this commitment carry on? And what have these architects achieved since their projects were last shown at the CA'ASI? Has anything adequate or interesting been built since? The 2016 exhibition, Mission Trans-Missions, links this common project to the 15th Architecture Biennale, "Reporting from the Front", focused on global humanitarian needs and sustainable solutions. This show is to be read as Architecture-Studio's own manifesto within the official manifesto, advocating architectural group creativity. Designed by the team Change is Good, the exhibition trail unfolds in a dual feature of the architects' personal careers and their cooperative work. Their latest ideas, their iconic projects, and the architectural itineraries of their architect friends from abroad are showcased in Venice in parallel with a selection of Architecture-Studio's projects. Some projects are situated next to locations where the guest architects work and will hence encounter some of the same problems. Sustainable growth is everybody's concern. Planet Earth has not yet given up the ghost but so much remains to be done. Beauty and necessary development do not have to be antithetical. Take Kabul and its new district, Deh Sabz, where urban planning, with a humanistic and reconstructive approach, has aimed at healing the wounds. H.L.

在中国、中东和非洲举办的三场竞赛已经证实，建筑设计不是只有一种方式，也表明了法国AS建筑工作室致力于助力新兴国家创新建筑师的强大信念，随着人口的不断增加和地方性的困难，需要有更好的方法开展社会工作。如何履行这一承诺？这些获奖的建筑师于CA'ASI艺术展览馆展示项目之后又取得了哪些成就？又建造了更多或更有趣的建筑吗？2016年的展览名为"建筑——跨越时空的使命"，与第十五届威尼斯建筑双年展的主题"前方报道"联系起来，重点关注全球人道主义需求和可持续发展解决方案。此次展览的官方宣言——倡导群体创造力，也是法国AS建筑工作室自己的坚持。布展由"Change is good"团队设计，旨在展现出建筑师的使命和团队合作的成果。包括最新的创意、最具有代表性的项目，来自不同国家的受邀建筑师朋友的建筑规划作品与法国AS建筑工作室的建筑规划作品一道在威尼斯展出。工作室的一些项目位于某些受邀建筑师常驻工作地点的附近，因此遇到的问题也有相同之处，可持续发展是每个人的关注点。地球并非无药可救，但还有许多任务有待完成，而美观和发展需求也不一定是对立的。以喀布尔及其新城Deh Sabz为例，城市规划采用了人性化和修复性的方式，旨在治愈城市的创伤。H.L.

The "Mission Trans-Missions" exhibition creates a dialogue between the 27 winners of CA'ASI projects from previous years with the latest achievements of Architecture-Studio. All of these new projects demonstrate the commitment of the architects to a globalized world.

"建筑——跨越时空的使命"展览为二十七名往年竞赛的入围及获奖者的最新作品与法国AS建筑工作室最新的作品之间建立对话。所有展示的项目表达了建筑师在全球化发展中肩负的使命。

03
AFGHANISTAN INFO
Libération
L'espoir de Kaboul
THE BARBER
AS ARCHITECTURE-STUDIO
la ville écologique
ISLAMIC REPUBLIC OF AFGHANISTAN
CONCEPTUAL DESIGN FOR THE DEVELOPMENT OF
NEW DEHSABZ CITY
拉萨经济技术开发区举行 三证合一
一照一码 营业执照首发仪式
关于实名有奖举报 四风 问题的公告

KABOUL
CONCEPTUAL DESIGN FOR THE DEVELOPMENT OF
NEW DEHSABZ CITY
FINAL REPORT
Libération
OCCUPATION
Il y a un an, le coup de Kaboul
CENTRAL PARK
AGRICULTURAL BELT
AGRICULTURAL SCIENTIFIC BELT
SPORTS CENTER
CENTRAL MOSQUE
INDUSTRIAL AREA
CULTURAL CENTER
KABUL
LOGISTIC POLE
NEW INTERNATIONAL AIRPORT
MATCH
New York Retou
DANS KABOUL LIBERE
Le tournant de la guerre
DEHSAB

03 IN THE SERVICE OF THE COMMON GOOD

03 AU SERVICE DU BIEN COMMUN

公共利益服务

Although it is true that the personalization of architectural creation began long ago, during the Renaissance with such figures as Bramante, Brunelleschi and Michelangelo, the cult of the architectural "genius" has taken on a different dimension in the modern era, culminating in the recent wave of "star" architects. Coupled with the increasing specialization of some aspects of design and construction, as well as the tendency of clients to want to control what they pay for, the net result has been one of a quest for iconic images or a recognizable style. In reality, even well-known architects often have less and less to do with the final product of their concepts. The image, or the envelope, has frequently taken precedence over the real architectural content of contemporary buildings. Anxious to impress, both architects and clients have often sacrificed the real relationship of their buildings to society. The economics of architecture have fallen more and more into the hands of accountants and specialized outside companies. Concern, at least in a superficial sense, for ecology and innovation has led to the addition of technological elements to structural envelopes, without any real relationship to the architecture.

Much as Architecture-Studio has eschewed any personalization in their own creations, favoring a system of 12 equal partners, so too the firm has engaged in a real effort to place emphasis on the quality of architecture in its broadest conception, where society, ecology, economics and innovation are considered an integral part of what a building is all about. Style, especially of the repetitive sort, is not on their agenda either; each building has its own logic and requirements. This is not to say that aesthetics are ignored, but that the creation of a recognizable building passes more through method than it does through the visible envelope, or through superficial additions.

Architecture-Studio partner Laurent-Marc Fischer lays down some of the tenets of the approach of the firm, stating, "A primary responsibility of Architecture-Studio, as we have always seen it, is that the amounts invested by the client correspond to what they find in the architecture. We must also be certain that a building will not be overly costly to operate in the future. We are vigilant vis-à-vis significant elements such as the construction cost. Our objective is to improve the experience of those who will use the building. This is one reason that we are always present during the construction process, on the work sites. We are there to ensure that what was agreed on with the builder is in fact done according to specifications. We are intermediaries between the builders and the

尽管建筑设计的个性化兴起于很久之前，尤其是文艺复兴期间的布拉曼特、布鲁内列斯基和米开朗基罗等人物尤为突出，但建筑“天才”的狂热到了现代从另一个维度显现，其顶峰便是最近的“明星”建筑师大潮。伴随着设计与施工方面要求越来越高的专业性，以及客户越来越倾向于掌控自己付费的项目，最终的结果便是对标志性形象或是高辨识度风格的不懈追求。实际上，即使是有名的建筑师，通常情况下，最终作品与其最初设计理念之间也千差万别。建筑的形象或外观在很多情况下，被视为比当代建筑的实际建筑内容更为优先的因素。建筑师与客户急于留下让人印象深刻的作品，往往会牺牲建筑物与社会之间的真实联系。建筑物的经济价值更多地落入专家和专业外包公司的手中。至少从表面上看，对生态和创新的关注使得结构框架被不断加入科技元素，但这些元素或许与建筑本身毫无关联。

法国AS建筑工作室在其自身的创作中尽最大可能避免个人化的倾向。12个合伙人组成的平等体系，为在最初的概念设计中便强调建筑品质的重要性作出努力，确保社会、生态、经济和创新等要素能够真正成为其建筑作品必不可少的一部分。风格，尤其是重复类型的风格并不是工作室所追求的，因为每个建筑物都有各自的逻辑和需求。但这并不意味着他们忽略了美学，一座标志性建筑的创造更多的是靠方法，而非通过其可视化的外形或表层装饰物。

法国AS建筑工作室合伙人洛朗-马克·费希尔列出该工作室部分工作信条，他表示：“正如我们所见，法国AS建筑工作室的首要职责是确保客户的所有投入与其从建筑中所得的成正比，同时也要确保建筑的后期维护费用不得超过预算标准。像建筑成本这样的重要要素，我们必须时刻保持监控。我们的目标是提高建筑使用者的体验感受，确保

As an answer to the international Réinventer Paris consultation, the new Éole-Évangile District is entirely based on the circular economy concept.

作为对“重塑巴黎”国际咨询会的回答，新的Éole-Évangile区完全基于循环经济理念。

Set in a new district of Bordeaux, the Caisse d'Épargne Headquarters of Aquitaine Poitou-Charentes faces the Garonne River. A rhythm of transparency and opacity is organized by the parametric rhythm of the bay windows.

阿基坦普瓦图—夏朗德的两省级储蓄银行总部位于波尔多新区，面朝加伦河。凸窗营造透明又模糊的感觉。

client." In fact, Architecture-Studio is seeking to re-establish the essential links between architecture and its process and end results, links that have been broken over time by changing influences and the rising complexity of construction, particularly in dense urban environments. Laurent-Marc Fischer continues, "We have a responsibility toward the client, of course, but also to the future users and to the city where a building is inserted. What makes us different from other architects is our will to master the budgetary process. Our role goes far beyond the appearance of the building. A successful project is an entire process, not just the initial design, even if that design takes into account many factors." Architecture-Studio has placed an emphasis on the users of their buildings, but also the immediate urban environment in which their projects are located. This aspect of their work escapes the more traditional economic analysis based only on the interests of the client. They believe that the interests of the client are, in fact, to be brought into line with those of users and of the city in general. Their concept of a responsible form of architecture, of course, heeds the economic constraints imposed by the client, but they very often succeed in adding elements that were not specifically part of the program but which are beneficial to users. Ultimately, this kind of attention to the real use of buildings is of course also beneficial the client, who sees the structure play a real part in its neighborhood.

By definition, this kind of hands-on approach distances Architecture-Studio from what some might call the myth of the architect as artist. "We have always refused the idea that we are artists," says Laurent-Marc Fischer. "It is possible to evoke the art of the engineer or of the architect, but we are not artists." For Mariano Efron, the task undertaken by the firm is nothing less than a reintegration of often disparate elements in design and construction into a coherent whole. "We believe in the revalorization of the architect in the construction process. We believe that a close control of the economic aspects of a project is also related to its ecological profile, and ultimately to the evolution of our profession. We are involved in many projects where our mission is not confined to design but also involves construction; this enables us to better master the entire lifecycle of architecture, and its global cost."

In societal terms Architecture-Studio recognizes its responsibility to clients, of course, but also toward the user. In a hospital, for example, the hospital authority thinks first of those who are

与建筑方协商约定好的内容能够百分之百地按照规范落实，这也是我们坚持在工地监控整个施工过程的原因。我们是建造者与客户之间的中间人。"

实际上，法国AS建筑工作室一直在寻求重塑建筑本身、建筑过程以及最终结果之间的根本联系。由于受到不断变化因素的作用，以及建筑物复杂度不断增加的影响，尤其是那些在市区密集环境中的建筑物，这种联系已经随着时间的流逝断裂。洛朗–马克·费希尔说道："我们需要向我们的客户负责，同时也要对未来的建筑使用者以及建筑物所在的城市负责。我们区别于其他建筑工作室的根本在于我们对整个流程的预算控制的决心。我们在整个流程中的角色远在建筑出现之前就开始了，一个成功的项目是一个完整的过程，而不仅仅是初始的概念设计，尽管设计也包含了众多需要考虑的要素。"法国AS建筑工作室不仅非常重视建筑物的使用者，更看重其建筑工程所处的城市环境，这样能够摆脱仅根据客户的利益做预算分析的传统做法。他们坚信，客户的利益与使用者，以及整个城市的利益实际上是休戚相关的。一个负责任的建筑设计必然要配合客户提出的经济预算上的限制，但很多情况下，工作室都能成功的在建筑中融入一些原工程规划以外，但却能为使用者带来便利的内容。这种对建筑物实际使用的关注，也将最终有益于客户，因为对于客户来说，建筑与周边环境是不可分割的。

这种躬亲的工作方法使法国AS建筑工作室区别于一些人主张的"建筑师即艺术家"的理念。"我们一直都不赞成别人把我们称作艺术家，"洛朗–马克·费希尔表示，"工程师或建筑师的艺术创造热情或许能被唤起，但我们并非艺术家。"在马里亚诺·艾翁看来，工作室承接的项目基本上都是将零散独立的设计与建筑元素重新整合为一体的重塑

The Kama hotel chain introduces a new sustainable offering of solutions for tropical and Sahelian countries. The inert casing of the building – consisting of compressed or adobe clay – offers thermal comfort for its users. The solar-paneled canopy provides shade and guarantees energy independence.

Kama连锁酒店为热带地区与萨赫勒地区国家引入了新型可持续性解决方案。由压密黏土或土坯形成的建筑保护层为用户营造了舒适的环境。太阳能板顶棚在遮阴的同时实现能源自给。

The new commercial center of Vigie Illkirch-Graffenstaden (Bas-Rhin) creates a space where cars disappear for the benefit of the landscape. In this new district, the project boosts the surroundings to become one of the structural elements.

伊尔基希—格拉芬斯塔登市（下莱茵省）的Vigie新商业中心创造了一个看不见车辆的空间，让整体景观得以提升，项目成为周围环境的重要构成。

to be treated and then of the personnel, but, in reality, the majority of users are visitors. These visitors are very often under stress out of concern for their loved-ones. "Our role," says Laurent-Marc Fischer, "is to take the visitor population into account even if these persons are not at the heart of the program. The same logic applies to schools, where we must also think of the parents. This is actually true of almost every building; the user is not the client. This is increasingly difficult because the economy and society press us to listen only to those who pay." Marie-Caroline Piot refers to another specific example of the ways in which the office tries to improve on a program for the benefit of the users. "We have often added intermediary spaces," she says. "In the recent renovation of the Jussieu University campus, we proposed a light covering for certain existing patio areas. This was relatively inexpensive but allowed for the creation of places where people can gather, even if it is raining, for example." Architecture-Studio also enriched the ground-level slab of the university complex by placing lightweight planting in an area where they had only been asked to be involved in sealing the surface. For an underground lecture hall that had insufficient ceiling height to allow for good acoustics, the architects proposed and realized an increase in the volume that generates a change in the topography of the slab above, enriching the experience of those walking above while improving the lecture hall.

The role of architecture in the areas of cost and economics, as well as the place of buildings in society is tied by Architecture-Studio to their interest in environmental issues. Laurent-Marc Fischer states, "Technical aspects of a design also participate in the overall impression of quality, and thus finally in the architecture of a building." Underlining the early and logical interest of the firm in energy consumption, Mariano Efron points out that "buildings such as the Arab World Institute had an ecological consciousness that we have carried forward with changing times. Computers have become an integral part of the life of buildings. The solution to ecological problems, like other ones, is not an added-on element, but a real part of the architecture." As opposed to firms that have recently discovered the benefits of reducing energy consumption, Architecture-Studio has taken a rather militant stance. Laurent-Marc Fischer says, "The response of Architecture-Studio is that the solutions to problems must come intrinsically through the architecture itself. We have most often been defenders of what can be called an appropriate, passive response to

过程。"我们认为建筑过程是对建筑师的重新估价。我们相信，对于工程项目经济预算方面的严密控制是建筑生态性的一种体现，也是我们专业性不断发展的表现。在很多工程项目中，我们的任务并非仅限于设计，还包含建筑施工，这使我们能够更好地掌控整个建筑的建设周期以及对整体成本的控制。"

从社会学角度出发，法国AS建筑工作室承担起对客户以及用户的责任。比如在医院项目中，医院的管理者首先考虑到的是接待的病人，然后是医院的工作人员。但实际上，探视人也是医院的主要使用者，这些探视人往往是出于对所爱的人的关心而忧心忡忡地来到医院。洛朗–马克·费希尔表示："我们的任务是将探视人群也考虑在内，尽管这些人并非医院项目的核心要素。同样的逻辑也可以适用于学校，家长也是我们必须要考虑的元素。这个道理基本对于所有建筑物都是适用的，使用者并非客户本身。但由于经济与社会的压力迫使我们不得不尽量满足于出资方，因此上述逻辑也越来越难实现。"玛丽卡·碧欧在这里提到了工作室竭力提高项目使用者利益的另一特殊案例。"我们经常会在项目中增加媒介空间"，她说道："在最近的巴黎第六大学Jussieu校园改建工程中我们提出对几处原有的露天空地进行透光性遮盖。这个做法相对成本不高，但却可以在下雨天为人们提供遮雨聚集的场地。"法国AS建筑工作室还在建筑的一层空间周边设置轻型植物带，以丰富整个校园的景观，而最初业主仅要求将一层密封起来。又比如建筑师提议并实现增加因屋顶高度不足导致音效不佳的地下演讲厅的容积以改变地面的地形，在丰富行人体验的同时提升演讲厅的音效。

在法国AS建筑工作室看来，环保因素与建筑成本、建筑的整体经济以及建筑在社会中的角色都是密不可分的。洛朗–马克·费希尔说："设

At the gates of the historic city, the plateau to the northeast of Chartres is being regenerated through the urban planning and environmental approach of Architecture-Studio.

位于历史名城的城门口，沙特尔东北高地正在因为法国AS建筑工作室的城市规划与环保措施而变得焕然一新。

The University Hospital is organized into three compact and scalable buildings, which appear to extend the contours of the hill through a large canopy connecting the different medical sections.

大学医院由三个紧凑的建筑组成，将不同医务部门连接起来的大型遮阳篷犹如起伏山丘的延伸。

the environment. The architecture itself – through its own constitution, its orientation, its geometry, or its materials – responds to the stated objectives." The partners make a distinction between elements, be they environmental or purely aesthetic, that are added onto the architecture, as opposed to real responses to programmatic and energy issues that they seek to integrate into the architecture itself.

Marie-Caroline Piot gives a specific example of how the partners work on issues related to context and society. "We do feel that the less technology there is in a building, the better the architecture," she affirms. "In the Mirecourt High School, in an area of Lorraine where it is cold in winter and densely forested, instead of increasing the thickness of the insulation, we created intermediary spaces that serve to create a very efficient form of insulation while also improving the architecture. We go beyond the norms; in fact, that is what we try to do in every aspect of our work. Although we were asked to use wood in the building, we decided to do it almost entirely in solid, untreated wood. This decision was based on ecological concerns, in particular, but also to support the local wood industry. We made use of locally harvested wood from the Vosges area. We consciously chose to use as much solid wood as possible instead of glulam that requires the use of polluting glues."

The partners insist that their approach to added technology in many buildings has nothing to do with their openness to innovation. "We are not 'low-tech' when it comes to conceiving our buildings," states Marie-Caroline Piot. "On the contrary we use the latest tools. Parametric design allows us to take into account such elements as light, noise, wind and the uses that the building will be put to in a much more efficient way than we could have just a few years ago. Our goal remains not to use technology that is likely to not work or that is simply a add-on as opposed to being an integral part of the architecture." Mariano Efron provides a specific example of this kind of approach, referring to the facades of the Headquarters of the Aquitaine Poitou-Charentes Caisse d'épargne in Bordeaux. These were conceived with the assistance of parametric design to express the rapport between the light that comes in and the thickness of the structure, implying its ability to filter incoming light. "The façade," he concludes, "is thus the expression of the energy consumption of the building. We worked with engineers specialized in energy to locate sources of consumption such as artificial lighting. We imagined forms that were compatible

计的科技要素也会影响建筑整体品质印象，并最终影响建筑的造型。"马里亚诺·艾翁特别强调工作室在早期就对节能表现出浓厚的兴趣，他指出："像阿拉伯世界研究中心这样的建筑体现了我们历经时代变迁一直在坚持执行的生态意识。电脑如今已经成为管理建筑物的组成部分。建筑生态问题的解决方法并非通过附加元素实现，而是通过建筑本身的一部分实现。"与一些近期才发现节能重要性的工作室不同，法国AS建筑工作室对于节能的立场更为激进。洛朗–马克·费希尔表示："法国AS建筑工作室的立场是问题的解决方法必须从建筑自身内部寻找。而在很多情况下，我们都是被动式解决方式以及依据环境本身思考的拥护者。"建筑可以通过其自身的构造、朝向、几何原理或是建筑材料等方面达到既定的目标。工作室的合伙人严格区分两种做法，一种是出于环境或美学等各种原因而在建筑上附加内容，一种是有规划地将能源等因素融入建筑本身。

玛丽卡·碧欧就合伙人在不同情境和社会背景下如何处理相关议题，列举了一个具体案例。"我们确实感觉到，建筑物科技元素越少建筑本身越成功。"她论证道，"比如位于洛林的米尔库中学项目，那里植被覆盖率高且冬季寒冷，我们并没有增加建筑物的保温层厚度，而是创造一部分媒介空间以实现良好的保温效果，同时优化了建筑的构造。我们打破常规，这是我们在工作方方面面努力追求的目标。尽管起初我们接到要求是在建筑某些部分使用木材，而最终我们几乎在整个建筑中使用了原生态木材。这项决定主要是出于生态学考虑，同时也是为了支持当地的木材工业，我们使用的是距离项目地点最近的木业森林所生产的木材，并且我们有意识地尽可能选择实木，放弃使用含有污染性黏合剂的胶合板。"

合伙人坚信，避免在许多建筑上附加科技元素的做法与其追求创新的理念并不矛盾。"在构思建筑方面，我们并非'低科技含量'，"玛丽卡·碧欧说道，"相反，我们使用的都是最新的研发工具。参数化设计帮助我们在设计时将光线、噪声、风力以及比以前更有效的方式等要素统统考虑在内。我们的目的不在于使用可能无效的技术或者简

The University of Science and Technology Campus of Hanoi is located in an exceptional natural environment. Surrounded by lakes, a belt of interconnecting buildings creates an architectural line in the landscape of gentle, soft traffic between the high-tech equipment.

河内科技大学校园位于独特优美的自然环境中。相互连接的建筑群四面环水，在高科技设施之间营造了一道柔和的风景线。

The Pergamin buildings mark the entrance of the eco-neighborhood of the Green City of Zurich, on the site of the former Manegg paper mill. Integrated into the valley's natural environment, the triple skin of these buildings offers contrasting materials and mixed usage.

Pergamin办公楼建筑群位于苏黎世"绿色城市"生态街区的入口，旧时的Manegg造纸厂区域。办公楼的三层外立面让建筑具有明显的材质对比，透明的表面让建筑很好地融入河谷的环境中。

with low energy consumption. Using less electricity for artificial lighting implies bringing in more natural light." Quite obviously, although the final result has an aesthetic appeal, its origin lies in a technologically advanced study that leads to a purely architectural solution.

Along with their effort to use appropriate passive methods, the office has been very active in keeping aware of the latest developments in building and environmental technology. They created a small structure that is called Intenciti, a kind of think tank, that brings together businesses and start-ups that are specialized in environmental issues. "This allows us to be in touch with technological developments and in some instances to propose innovative solutions to our clients," says Alain Bretagnolle. "We have the ambition, for example, in Africa to create buildings in certain locations that are entirely autonomous, producing their own electricity and recycling water. We are working on a university in Saudi Arabia where we are integrating a solar canopy, particularly because the country is asking for solutions that do not have to do with the use of hydrocarbons, seen as a finite resource. We will do this with local firms that are capable of local manufacturing for the project. This has to do with a global approach to economic issues, of course." Architecture-Studio were amongst the first to work on the idea of a double roof for energy reasons, as was the case with the Lycée du Futur, or a double facade such as that of the European Parliament facing the river. "We have always had a great interest in innovation," states Bretagnolle, "not for innovation's sake but something that serves the purpose of the building or of the architecture itself."

The firm's Saint-Malo Cultural Center is making use of a photovoltaic strip that is part of the architecture, together with green roofs and geothermal energy. This is not gratuitous either; the region has problems with the production of electricity. The photovoltaic strip was not at all part of the competition program, but it also has an urban and architectural function. It links the two main elements of the complex, an art cinema and a media center, creating an arch that has a symbolic relation to the city. "We didn't put solar panels on the terrace – we did not add something that is not part of the architecture," insists Bretagnolle. Laurent-Marc Fischer sums up this attitude: "When we refer to Kierkegaard and 'leaving open the wounds of the possible', it is that we are searching for positive ways to resolve issues and constraints in every project, even going beyond what is asked."

单添加与建筑整体格格不入的元素。”针对以上方法，马里亚诺·艾翁提到了位于波尔多的阿基坦普瓦图—夏朗德两省级储蓄银行总部的建筑外观，这个项目在参数化设计软件的协助下，勾勒出建筑接受外部光线与结构厚度之间的密切关系，使建筑通过自身过滤入室光线的性能达到最佳。马里亚诺·艾翁总结道：“立面成为建筑节能的表现方式。我们与能源工程专家通力合作，确定如人工照明等能源损耗来源。”

“我们设想出符合低能源损耗的形式，尽可能少使用人工照明，意味着需要更优质的自然采光。”尽管最终的解决方案带来了美感，但很显然，其根本在于通过先进的技术以获取纯建筑解决方案。在竭力使用合理的被动式方法的同时，工作室积极关注建筑与环境技术的最新研发成果。他们创建出一个名为“Intenciti”的小型机构，类似一个智囊团，将专业从事环境问题的企业与新兴公司有机融合在一起。“这使我们能够与技术研发保持密切联系，在某些情况下，协助我们为客户提出创新型的解决方案。”阿兰·布勒塔尼奥勒说，“比如，我们计划在非洲某些区域建造能够完全自动化发电并实现水循环的建筑。又比如我们在沙特阿拉伯的一个大学校园项目中运用太阳能顶棚技术，其中的主要原因在于沙特逐渐寻求减少对石油资源依赖的解决方案。我们将与当地有能力实现工程生产的公司合作，当然这又涉及经济全球化问题的解决方式。”法国AS建筑工作室是运用双层屋顶设计达到节能目的的先行者之一，比如未来中学项目，又或者是欧洲议会中心项目（建筑面向河的一面设置双层立面）。“我们一直对创新抱有浓厚的兴趣。”阿兰·布勒塔尼奥勒声称，“但这并非仅仅为了创新，而是为了能够真正服务于建筑本身。”

工作室所承接的圣马洛文化中心项目使用光伏板作为建筑材料，与绿化屋顶以及地热能源一样，都是建筑自身的组成部分。当然这也是有原因的，该项目所在地区的电力生产存在问题。光伏板原本并不是该工程的组成部分，但因其具有促进民生的价值和建筑功能，被采纳为建筑的一部分。它将整个建筑物的两大部分——艺术电影院和媒体中心巧妙联系到一起，创造出城市标志性建筑精品。“而我们没有在露

The architecture of the Saint-Denis prison in La Réunion aims to create conditions for humanized detention, and are adapted to the cultural, topographical and climatic characteristics of the island.

留尼旺岛圣·旦尼监狱的建筑设计旨在创建人性化的拘留环境，并与岛屿的文化特点、气候特征相适应。

Architecture-Studio has made use of some of its precepts on an urban scale such as their 3-D Simulator for the Sustainable City of Santiago, Chile. But here too, as Laurent-Marc Fischer points, out, "the sense of our work on a building in terms of energy use is the same as in our urban design. We naturally look to societal issues in the broad sense, and of course economic and environmental constraints. The city, unlike a given work of architecture, is in constant movement. For urban design we seek keys or mechanisms rather than designing fixed architecture, as may have been the case in Brasilia or Chandigarh. Architecture and urban planning are both related to a certain idea of ethics, of the city in the Greek sense of the word polis."

Marc Lehmann insists on the importance that the office places on urban development schemes such as the large Parc Marianne–Port Marianne project in Montpellier. "Yes," he says, "we are involved in urban planning. We are conscious of the much greater limits placed on urban design than on architecture. We are interested, too, in the differences between what can be planned and what is spontaneous. When we think on an urban scale, we try to take into account the variable aspects of the design. When we conceive a building, on the other hand, we insist that the final result must correspond to what we designed. The intellectual approach is the same. First we have to understand all of the constraints so that we can make them into opportunities. At the outset, it was thought in the Modernist movement that the city could be strictly definaed; liberal economics saw it as being self-regulated. We think of the city as being able to be planned, but it first has to be thought out. The city is about politics in the most noble and original sense of the term. What is the relation between the idea of the city and its structure, that is what interests us. The city has a time element in it that cannot be dealt with the same way in architecture. The city is in perpetual movement."

It was Plato who said that best form of city is that which leads to the common good. In setting aside the issues of personality and ego that have come to dominate contemporary architecture, and in willfully delving into the connections between their work and economic, societal and environmental issues, Architecture-Studio proposes a kind of return to basics, a restoration of the common sense whose ultimate goal is the common good. While the office has often succeeded in innovating in their design and construction methods, they have never allowed themselves to make such

天平台安装太阳能面板，因为我们不会在建筑物上添加原本不属于它的东西。”阿兰·布勒塔尼奥勒再次强调。洛朗–马克·费希尔对以上的工作态度进行了总结：“当我们提到克尔凯郭尔和‘使充满可能性的开口始终保持开放’这句话时，这表示我们积极寻求应对每个项目的限制条件的最佳方法，有时甚至会超越最初的要求。”

法国AS建筑工作室在城市规划中运用模拟的概念，比如智利圣地亚哥的可持续发展城市的3D模拟器。正如洛朗–马克·费希尔在此提出的理念，“我们在建筑设计上的节能意识，也充分体现在城市规划项目中。我们会自发从广义角度分析社会化问题，当然也包括经济和环境限制要素。城市规划不同于既定的建筑工程，它是一个动态的过程。进行城市规划时，我们寻求与一成不变的建筑物设计不同的解决途径和机制，比如巴西利亚或昌迪加尔的案例。建筑与城市规划均与道德标准，或者说与古希腊城邦准则有着千丝万缕的联系。”

马克·莱曼指出工作室对城市规划发展项目的重视，比如在蒙彼利埃的玛丽安公园—港口区规划中所扮演的重要角色。他说道：“是的，我们投身参与城市规划建设。我们意识到，相比建筑设计，城市规划所受的限制更大、更多。但我们也正视可规划方面与自发现象之间的差异。当站在城市的宏观角度进行思索时，我们尽力将设计的方方面面考虑在内。而在构思某一建筑时，我们会坚持尽量使最终结果与原有设计保持一致。规划采用的方法是一致的，首先，我们必须了解存在的所有限制性条件，这样才能将其转化为机遇。现代主义认为城市的功能分区能被严格设定；而自由经济主义则认为城市具有自我调节的功能。我们认为城市能够按照设定规划进行建设，但前提是规划时必须有大量的参数研究。城市的最初的定义是政治政策的直接表现，然而它让我们真正感兴趣的是城市理念与其构架之间的关系。城市发展包含时间要素，始终处于动态发展之中，因此无法照搬建筑设计的解决方法。”

柏拉图曾经说过，最好的城市是能维持公共利益的城市。法国AS建筑工作室避开当代建筑领域主流的个性与自我理念，致力于研究项目与

The women's and children's hematology center at the Hospital of Caen renews the hospital's image. A succession of inner courtyards and the choice of colors form agreeable surroundings, bathed in natural light.

卡昂大学医疗中心的妇幼及血液病中心改变了医院的固有形象，一系列的内庭院与颜色的搭配构成了惬意的氛围，使整座医院沐浴在阳光中。

The new University Department of Psychiatry at the Hospital Sainte-Marguerite in Marseille creates a balance between protected treatment spaces and open spaces. The harmony between the architectural style and its gardens allows the use of timeless and living materials.

马赛市圣玛格丽特医院的精神科的设计保持了受保护的治疗空间与开放式环境之间的平衡，建筑与花园和谐、融洽。

originality into a stylistic foible. Rather, their conception is that of architecture which takes into account its environment in the broadest sense. Saying that it is up to the architecture itself to resolve environmental issues to the greatest extent possible is an affirmation of the deep connection between a building and the place and people it is designed to serve. This is not architecture conceived in the service of a reputation, but rather in the service of the community, the common good.

经济、社会以及环境要素之间的关系，并在此基础上提出一种回归基本、恢复重视公共利益的理念。尽管该工作室多次实现设计与建筑方法的创新，但他们从未让这种原创性成为风格上的缺陷。相反，他们始终坚持，建筑和规划必须从最广义的角度将环境考虑在内，让建筑本身成为解决环境问题的驱动，在最大程度上维护建筑物与使用者的深层关系。这种建筑并非为了寻求名誉，而在于服务社区，创造公共利益。

The "TIKOPIA" housing concept was born from reflection on the "mini housing tower", which forms the basis of sustainable architecture through its intrinsic nature (its density and compactness). It proposes an urban habitat of high architectural and sustainable quality. This is a prototype designed to integrate into various urban contexts.

"TIKOPIA"住宅楼的建筑设计理念来源于迷你住宅塔楼，最终，通过其内在本质（密度与密实性）构成了可持续建筑的基础，这是具有优质设计和高效可持续发展性的城市住宅楼，是可以与各种不同的城市环境相融合的原型。

PLS1

Danone Vitapole R&D Center
Palaiseau, France
2000/2002
30,000 m² building/9.8 ha ground

达能研究发展和质量中心
法国，帕莱索
2000／2002
30 000 m²建筑面积／9.8 公顷园区占地

The contracting owner of Danone's Research and Development Center deemed the project manager's offer of a High Quality sustainable plan as befitting the company's ethics, i.e. healthy food. The building borrows its engaging versatility from nature, combining chance and necessity through unpredictable flowery forms contrasting with no less spontaneous symmetrical forms. Wood – posts, cladding, roof – was used right from the start. This material, all too often neglected, reduces the carbon footprint of human settlement, and connects our modern souls much closer to the countryside and wild life than the Modernist, Le Corbusier, could ever have imagined. Rain water is recovered and recycled into a natural body of water that irrigates the vast grass lands that link Vitapole to Palaiseau. As start-up companies in the Silicon Valley well know, pleasure is beneficial to good work. In the summer, lush vines growing on the buildings' façades give a sensation of coolness to researchers and employees working inside. In other seasons, the large glass surface allotted to windowpanes allows the sun to heat the rooms, thus saving considerable energy. Visual comfort in the studios, good acoustics in the offices, olfactory isolation from the lab activities and an organic vibrancy that runs through the whole edifice are keys to this project's success. H.L.

达能研发中心的业主认为项目负责人的建筑规划提议符合高质量可持续发展计划要求和公司道德规范，即生产健康食品。建筑借鉴了自然界的多样性，通过变幻的浪花造型与鲜明的对称形式的结合，预示偶然性与必要性的共存。从建筑设计的最初便决定使用木柱、木质隔层、木质屋顶等这些经常被忽视，但能减少人类定居点的碳排放的材料。木质材料的运用使我们现代的灵魂更接近农村和野外生活，这甚至超出了现代主义者勒·柯布西耶的想象。雨水被回收再循环到一个自然水体中，用于灌溉从达能研发基地到帕莱索的广阔草地。就如硅谷的初创公司都知道的——身心愉悦有益于提高工作效率。夏天，生长在建筑物外墙上郁郁葱葱的葡萄藤给在里面工作的研究人员和雇员们带来了一丝凉爽。在其他季节，大面积玻璃窗立面可达到阳光加热房间的目的，从而节省大量的能源。工作环境的视觉舒适性、办公室的良好音效、实验室的气味隔离以及整个建筑的有机活力是该项目成功的关键。H.L.

At the research, development and quality center of Danone Vitapole in Palaiseau, located in the south of Paris, the architecture is closely linked to the landscape. The scalability of the interior spaces of the research center is ensured through a loose internal structure.

位于巴黎南部帕莱索的达能研发和质量中心的建筑与景观密切相连。研究中心的室内空间是能够确保室内扩展性的宽松结构。

The natural environment seeps into the building through two landscaped areas, providing organic vibration in the interior spaces.

自然环境通过两个景观区域渗入建筑中，为室内空间提供有机共鸣。

The wooden façade
on the entrance canopy.

支撑入口顶棚的木质立面
结构。

Guy Dolmaire Secondary School
Mirecourt, France
1999/2004
10 000 m²

居・多尔迈勒中学
法国，米尔库
1999/2004
10 000 m²

Preparatory work on Collège Guy-Dolmaire started in 1999 and the building, housing 800 students, was completed in 2004. Located in the heart of the mountain woodlands of the Vosges, it was built with imported Black Forest wood because the French wood sector, which has much improved since then, was unable to provide the material in time. The closeness of a forest impacts upon its style; taking into account the leafy environment, this building has been designed as a shelter amid the trees. The school's roof spreads out like a parasol making it look like an Italian stone pine (*Pinus pinea*) and it stands out against the regional hegemony of the silver fir (*Abies alba*). Does it mean that ecological conformity need not pander to a regressive "cult of the roots"? No doubt. Ironically, local mountain architecture has not used the resources of the forest as much as Architecture-Studio did for this project. Not all colleges in the region have an astronomical observatory either. Minor unorthodoxies in the building are just a coy counterpoint to the sturdy orthogonal base of a wood structure worthy of carpenter Joseph's humbleness and the solidity of Noah's Ark. Sports-mad students will be able to sweat away in the well-ventilated premises equipped with 2 000 mobile valve elements. Good thermal isolation, based on the principle of passive solar building design, cuts energy consumption by 50%. There is high environmental quality here. Equally, there is also high architectural quality. H.L.

规划能容纳800名学生的居・多尔迈勒中学的筹备工作始于1999年，于2004年竣工。建筑位于孚日山脉林地的中心地带，由于法国木业部门当时处于改革初期，无法及时提供材料，因此建筑用进口的黑森林木材建成。建筑毗邻森林，建筑风格也因此受到影响，考虑到枝繁叶茂的环境，这座建筑被设计成在树林中的居所。学校的屋顶如遮阳伞一样散开，看起来像是一棵意大利的五针松，挑战着银杉在当地的区域性统治权。这是不是意味着生态一致性无需迎合过时的"传统崇拜"？毫无疑问。具有讽刺意味的是，当地的山中建筑并没有像法国AS建筑工作室的作品那样充分使用森林资源。而且在该地区，也并非所有的中学都设有天文观察室。在建筑中接受教育的具有活力的未成年学生，与体现木匠约瑟之谦卑和诺亚方舟之坚实的建筑方正木质结构形成有趣的对比。热爱体育的学生能在因装有2000个通风阀元件的通风良好的场所中大汗淋漓地运动。基于被动式太阳能建筑设计原理，建筑具有良好的隔热性，能将能源消耗降低50%。周围环境质量优良，同样，建筑质量也很高。H.L.

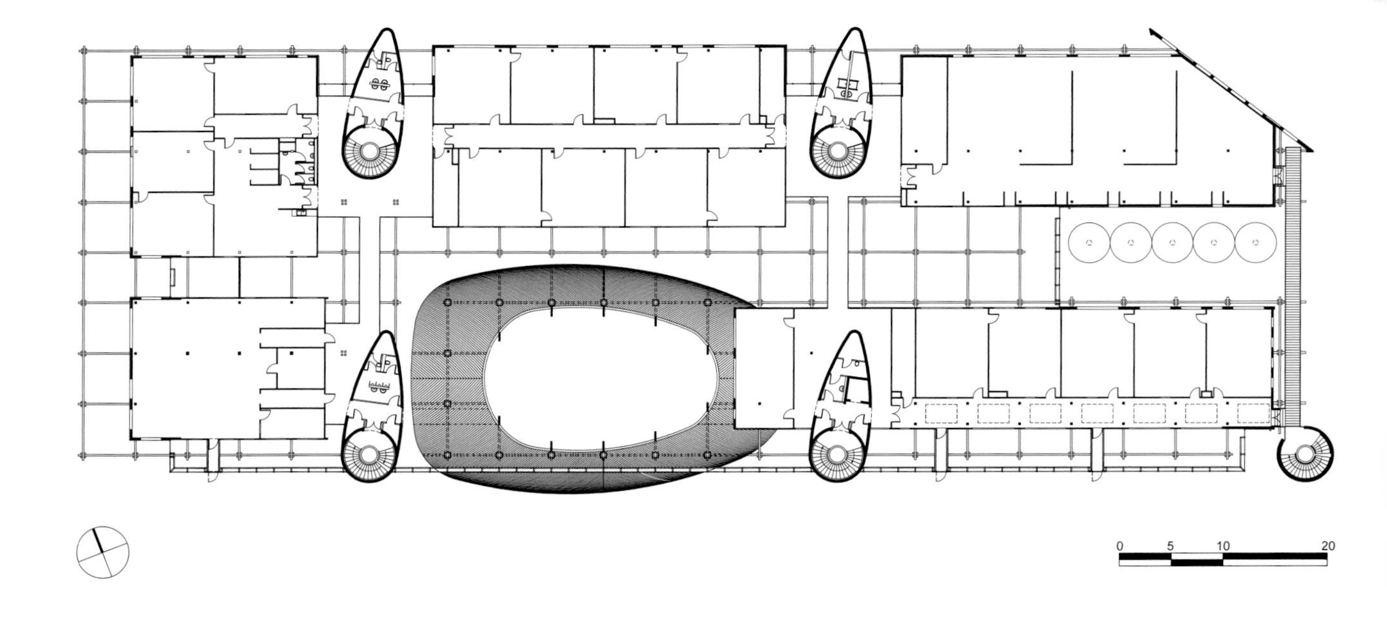

The first floor plan showing the concrete cores, the wooden structure and the volume of the host courtyard.

二层平面图展示了混凝土核心筒、木质结构与主庭院的大小。

South façade.

南向立面。

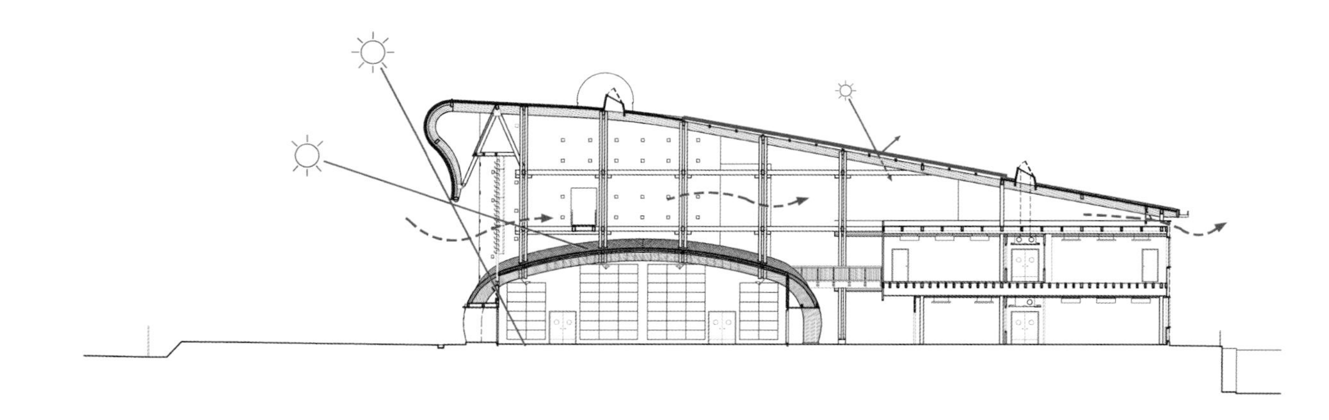

Transverse section: the building in the wind and the sunshine. In winter, the interior air volume is heated by the solar rays crossing the glass façade. In summer, the glass panels can be opened, allowing effective natural ventilation, which, together with the "parasol" effect of the roofing, shelters it from the heat.

建筑剖面图：建筑充分利用风能与光能。在冬季，透过玻璃立面的阳光能够对室内空气进行加热。在夏季，玻璃面板开启后能够促进通风，与巨大屋顶所形成的“遮阳伞”相配合，防止建筑过热。

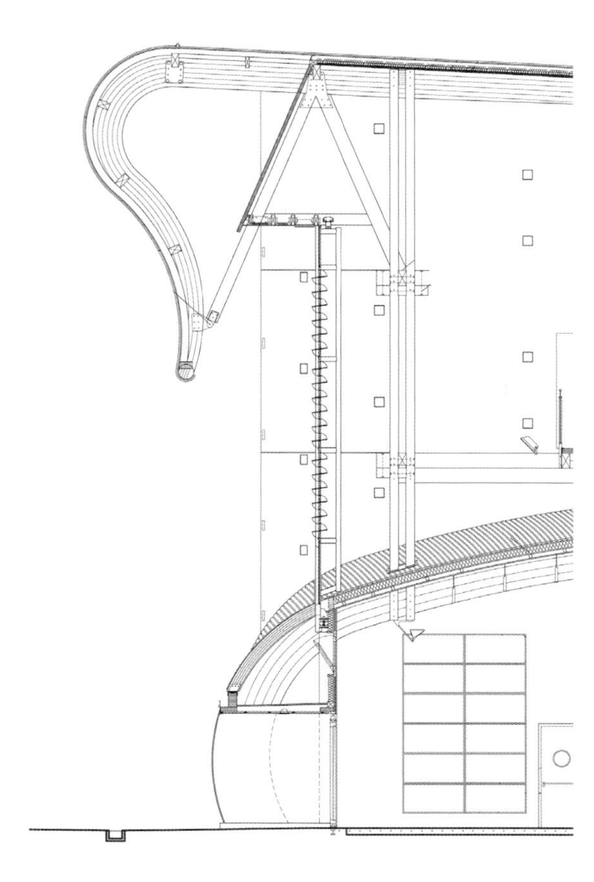

Under the wing of the large roof, the high school is organized as an educational village with an environment open to nature.

在巨大屋顶的保护下，中学被规划为教育新村，与自然融为一体。

Detail of the south façade.

南向立面的细节图。

Conceptual Design for the New City of Deh Sabz
Kabul, Afghanistan
2007/2008
40 000 ha

Deh Sabz新城概念规划设计
阿富汗，喀布尔
2007/2008
40 000 公顷

Kabul's infrastructures, intended for its 700 000 inhabitants in the 1978 urban plan of the Soviet era, were partly destroyed. Hence the total inadequacy of the public services which are now unable to provide for the population that has risen to 4 million. In 2007, Architecture-Studio and its partners won the international competition, organized by the Afghan government, to design a new city on the 40 000 ha site of Deh Sabz, 10 km north-west of the historic city center (which will also be renovated). The Taliban's destruction of the land registries did not help. In a region where people suffering from endemic power cuts are used to burning plastic as fuel, sustainability becomes a challenge as well as the spine of the project. Deh Sabz offers eco-friendly solutions for the economy, as well as for schools and universities and cultural projects. Nothing came to fruition at once, of course. The different stages involved financing, surveying the site and training thousands of jobless people in the construction industry. In this urban plan, social diversity functions as the lungs of the city, stimulating an inter-ethnic circulation between communities that were traditionally segregated in different quarters. The urban dimension of equality is illustrated by equal access to public services and equipment. To consider this new breathable district as part of the whole city entails a holistic urban planning that envisages the city as a communal house. H.L.

喀布尔的基础设施是在1978年苏维埃时期为70万居民规划设计的，部分已经遭到破坏。因此，现在的公共服务设施整体不足，已无法满足400万城市人口使用。2007年，法国AS建筑工作室及其合作伙伴赢得了阿富汗政府组织的国际竞标，项目目标是在位于历史悠久的喀布尔市中心西北10公里处的一片4万公顷的土地上设计一座Deh Sabz新城，同时喀布尔市中心也将被重新规划。虽然塔利班破坏了土地登记册让项目变得更复杂但却不能阻止项目的发展。在这个深受断电折磨的地区，人们习惯以塑料作为燃料，因此，可持续发展成为项目的一个大挑战。Deh Sabz新城被设计为一座生态友好型城市，为经济发展提供可持续的环保方案，并兼顾学校和文化项目。当然，城市的建设不可能一次性完成，阶段性任务涉及融资、调查现场、将数千名失业人员培训成建筑工人。在这个城市规划中，社会多元化就像是城市之肺，刺激了传统分居在不同地区社区之间的民族间的交流。整座城市层面的平等体现在居民能够平等享受公共服务和设施。将这个活跃的新区设计作为整个城市的一部分，需要将这座城市设想成一座公共房屋去进行规划。H.L.

Maps of the existing city of Kabul and of the new city of Deh Sabz.

喀布尔市老城区Deh Sabz新城区的规划地图。

North of Kabul, Deh Sabz will eventually accommodate more than three million inhabitants. It will ensure the population has access to health and social facilities, which are currently lacking in the current capital.

喀布尔北部Deh Sabz新城最终将容纳超过300万居民。它将保证居民能够享受到目前缺乏的卫生及社会福利设施。

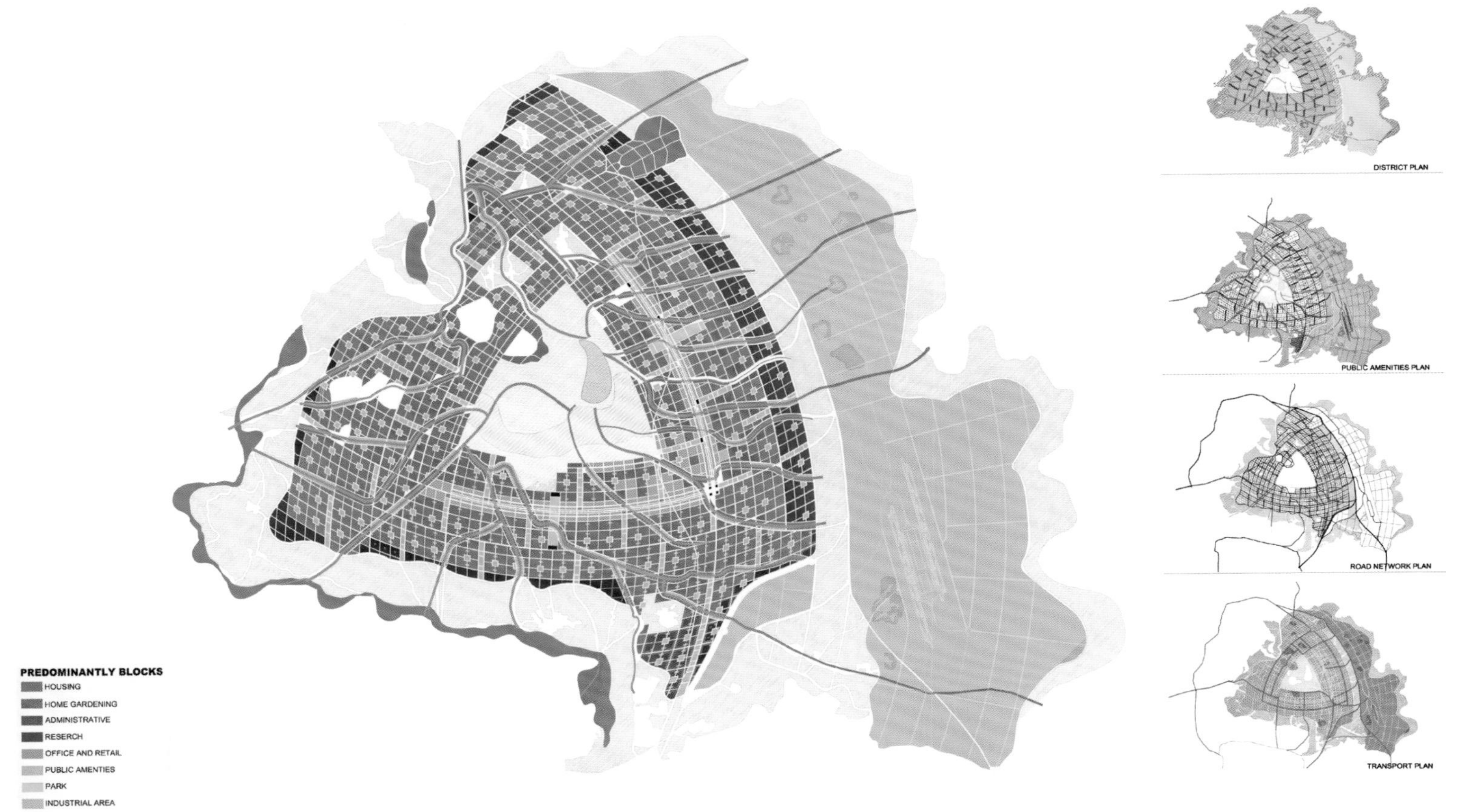

A view of the ravines in Deh Sabz and the transformation of these wetland areas into agricultural and planted corridors within the new city.

Deh Sabz新城的峡谷景观，以及将这些湿地区域转变为农业区与种植区效果。

The new city adopts the form of a Reuleau triangle, the result of constraints generated by the uneven relief of the setting. It is surrounded by a green belt to be used as market gardens and at its center is a large park.

新城规划采用了勒洛三角形的形状，这是地势不均匀所造成的结果。整个新城被由种植园组成的环城绿化带包围，市中心有一个大型公园。

Parc Marianne – Port Marianne Urban Development Area
Montpellier, France
2003/2014
20 ha

玛丽安公园—港口区域城市规划
法国，蒙彼利埃
2003/2014
20 公顷

Montpellier is a big city – France's fifth largest in terms of population. A forward-thinking local council, its Mairie celebrated the country's first gay marriage. In the context of demographic expansion, but always keeping in mind sustainability, Architecture-Studio was commissioned to design the Parc Marianne – Port Marianne district, just 7 km away from the Mediterranean. The nearby river Lez, which is the biotope of a small, threatened, species of fish called the "chabot", runs through the future district. Old, precious plane trees are protected and accounted for in the urban plan. The commissions given to Architecture-Studio by the Municipality consist of the design of two avenues, a street, a square and a body of water on the 80-hectare building plot. Seven building contracts comprising housing, offices and shops were awarded to private contractors acquainted with the architecture. Satisfying the universal criteria of sustainable growth and integrating into the plan a tri-generation plant – simultaneously producing heating and refrigeration and renewable electricity from wood – the architectural vocabulary of this project still speaks with a typical local accent: a promenade under the tramway, terraces, passageways, cloisters and classical attics on the façades. The colors used – white (Spain is to the west) and ochre (Italy is to the east) – do not hamper other sources of inspiration, e.g. different tones of red and bright pink, a true chromatic variety that renews the "visual tools" once honed by Daniel Buren and Niele Toroni. Artists and architects likewise stress the importance of opening up to a site, such as a street, and deploy what they call "*in situ* work". H.L.

蒙彼利埃是法国人口第五多的大城市。该市的地方议会具有前沿性，使得该市成为法国第一个允许同性婚姻的城市。在人口不断扩张的背景下，该市仍然坚持可持续发展道路，并委托法国AS建筑工作室规划设计这个距离地中海仅7公里的玛丽安公园—港口区域。附近的莱斯河生活着杜父鱼这种濒危的小型鱼类，而这条河沿着整个规划区域的边缘流过。古老珍贵的树木也需要在城市规划中得到关注和保护。蒙彼利埃市政府对法国AS建筑工作室的委托包含：在80公顷的建设土地上规划设计两条大道、一条街道、一个广场和一片水域。并与私有地产商共同设计开发七座建筑包含公寓、办公室、商店等。整个区域规划满足可持续发展的普遍标准，并设置了热电冷联产基地——通过对木材的运用达到同时产生供暖和制冷以及可再生电力的效果。该项目中的建筑规划设计具有典型的法国南部特色：步行道穿越有轨电车轨道、露天广场、走道、石廊以及古典的阁楼。而使用的颜色——白色（代表西边的西班牙）和赭色（代表东边的意大利）——并不妨碍其他灵感来源，例如，不同色调的红色和亮粉色实现了真正的色彩多样化，使得丹尼尔·布伦和尼尔·托罗尼打磨出的“视觉工具”重现光彩。艺术家和建筑师一样，均强调空间安置的重要性，就如在一条街道上展示“原位作品”一样。H.L.

An aerial view of Montpellier showing, in red, the perimeter of Parc Marianne – Port Marianne.

蒙彼利埃鸟瞰图，红色圈内为玛丽安公园—港口区域。

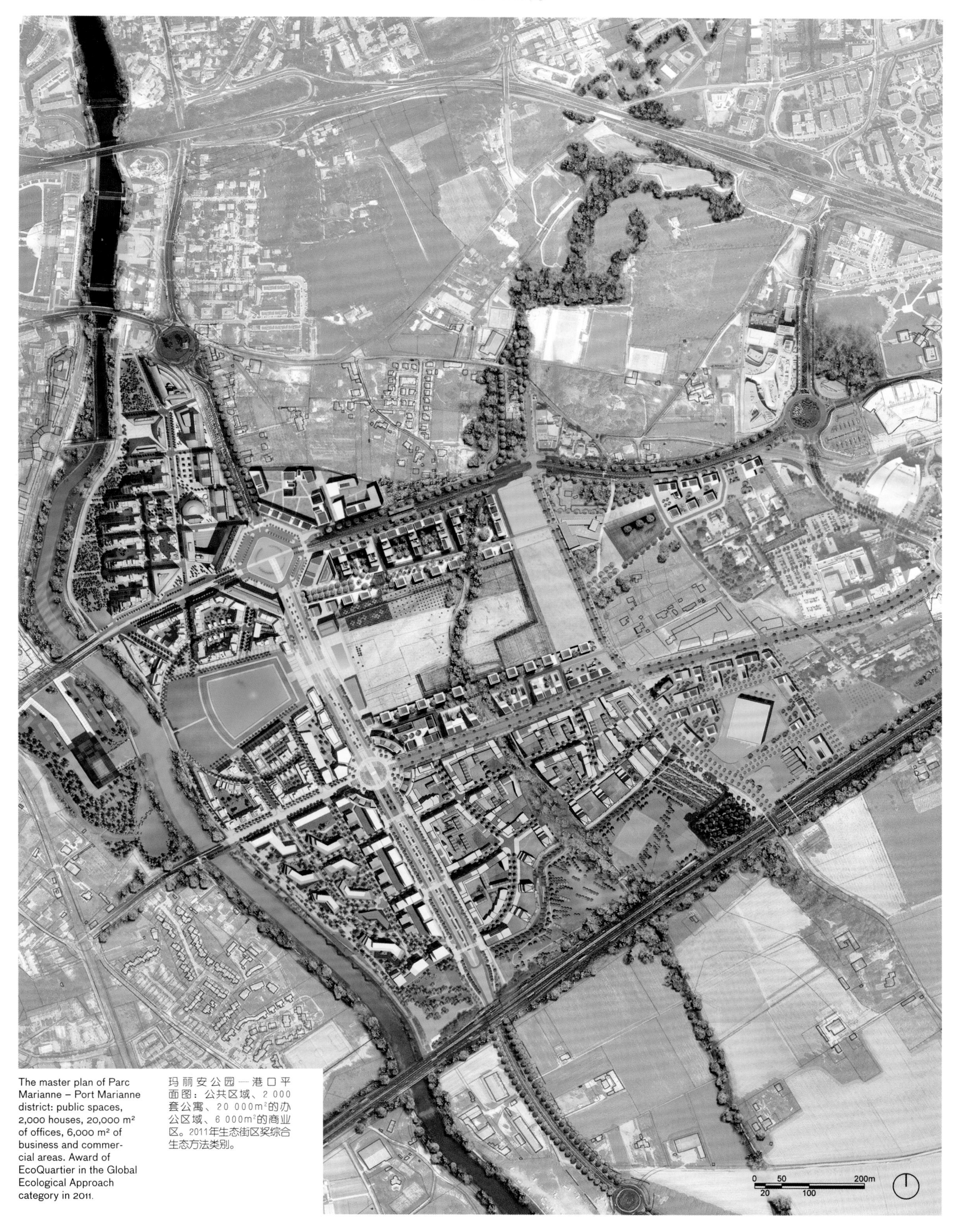

The master plan of Parc Marianne – Port Marianne district: public spaces, 2,000 houses, 20,000 m² of offices, 6,000 m² of business and commercial areas. Award of EcoQuartier in the Global Ecological Approach category in 2011.

玛丽安公园一港口平面图：公共区域、2 000套公寓、20 000m²的办公区域、6 000m²的商业区。2011年生态街区奖综合生态方法类别。

Ernest Granier Square with tramway lines 1 and 3. In the background is the O'zone office building.

有轨电车1号线和3号线经过Ernest Granier广场，广场后方是O'zone 办公大楼。

Raymond Dugrand Avenue, showing the water feature on Stéphane Hessel Esplanade and Georges Charpak Park.

Raymond Dugrand大道，图片展示了Stéphane Hessel休闲广场的水景与旁边的Georges Charpak公园。

The housing complex on Park Avenue and the pedestrian facilities on the Lironde.

Park大道旁的住宅楼群与横跨Lironde小河的人行道设施。

Public spaces. 公共空间。

The Lironde landscape layout. Lironde小河景观布局。

Millenium Avenue housing building.

Millenium大街房屋建筑。

Dor[a]Mar Building:
housing, offices, and shops.

Dor[a]Mar大楼：包括住宅、
办公室与商店。

PHARMACIE de
MAIRIE
Avenue
Raymond
Dugrand

University Hospital Centre of Pointe-à-Pitre / Les Abymes
Guadeloupe, France
2012 / 2021
77 600 m²

皮特尔角城/萨莱比梅大学医疗中心
法国，瓜德罗普岛
2012 / 2021
77 600 m²

Spread out on the ground, the new University Hospital buildings will replace those perched like a tower on a hill. The location of the new site in Abymes, one of Guadeloupe's most populous towns, will place the hospital just 15 kilometers from the airport, a necessity when the medical skills and hospital facilities are required by those from other islands in the Antilles archipelago. "Look at my wounds and the scars of my wounds. See my storms, my flow." The poetry of Edouard Glissant corresponds with the design of a hospital system which aims to be appropriate for both contemporary and future use, at the same time as being able to respond to natural disasters such as earthquakes, cyclones, and the tropical climate. The project's structure will include passive management of earthquakes, distributing the shock waves and suppressing aftershocks. The "Caribbean Street" is the backbone of the building and provides a reception area, circulation zones, administrative spaces and a large, well-ventilated space. The efficient air conditioning of the hospital counteracts the tropical humidity with intelligent, natural ventilation and the building is shaded from the perpetual sunshine by the self-regulating heat shield of the façade. Architecture-Studio has not lost the desire to do more with less and has used digital simulations to optimize the design. The hospital is a place of extremes and these have defined its architecture – advanced technology has been placed in an extreme environment. The friendly hospitality of the Creole culture, *lakou*, is referenced in the architecture. On a human scale, this centre for medical emergencies is identified with the cultural symbols of the community it serves. H.L.

这座全新的大学医疗中心建筑"铺展"在大地上，将取代那些像塔一样坐落在山丘上的建筑。医疗中心所在的莱萨比梅市是瓜德罗普群岛最多居民的市镇之一，医院选址距离机场仅15千米，当安的列斯群岛的其他岛屿需要使用医疗技术和医院设施时，这将成为首选。"看我的伤口和伤口上的疤痕。看我的风暴，我的潮涌。"爱德华 · 格里桑的诗歌与该医院的设计相得益彰。医院系统的建设旨在供当下和未来使用，同时为应对自然灾害作出贡献，如地震、飓风和其他热带气候带来的影响。该项目的结构将包含被动式应对地震的管理方式，分散冲击波和减少地震带来的影响。"加勒比大街"就像是该建筑的脊椎骨，这个大且具有良好通风条件的空间设有接待处、流通空间、管理办公空间。医院运用巧妙的设计营造高效的降温效果来对抗热带的湿热天气，一方面促进自然通风，另一方面利用建筑立面可自动调节的隔热屏阻挡四季常在的阳光照射。法国AS建筑工作室并没有失去使用简单却高效的工具的欲望，运用数字化模拟来优化设计。大学医疗中心所在的极端环境决定了建筑设计师在这种极端的条件下应用极其先进的技术。而在克里奥尔文化中用于款待客人的场所lakou庭院也被运用到医院的建筑设计中。从人性化角度来看，该医疗急诊中心具有其所服务的社区的文化特征。H.L.

Caribbean Street, the beating heart of the hospital.

加勒比大街，医院的核心区域。

293
231
301
291

Solar insulation of the bedrooms and meeting rooms, made from lacquered aluminum strips.

漆面铝制的条状遮阳板为病房和会议室遮阳。

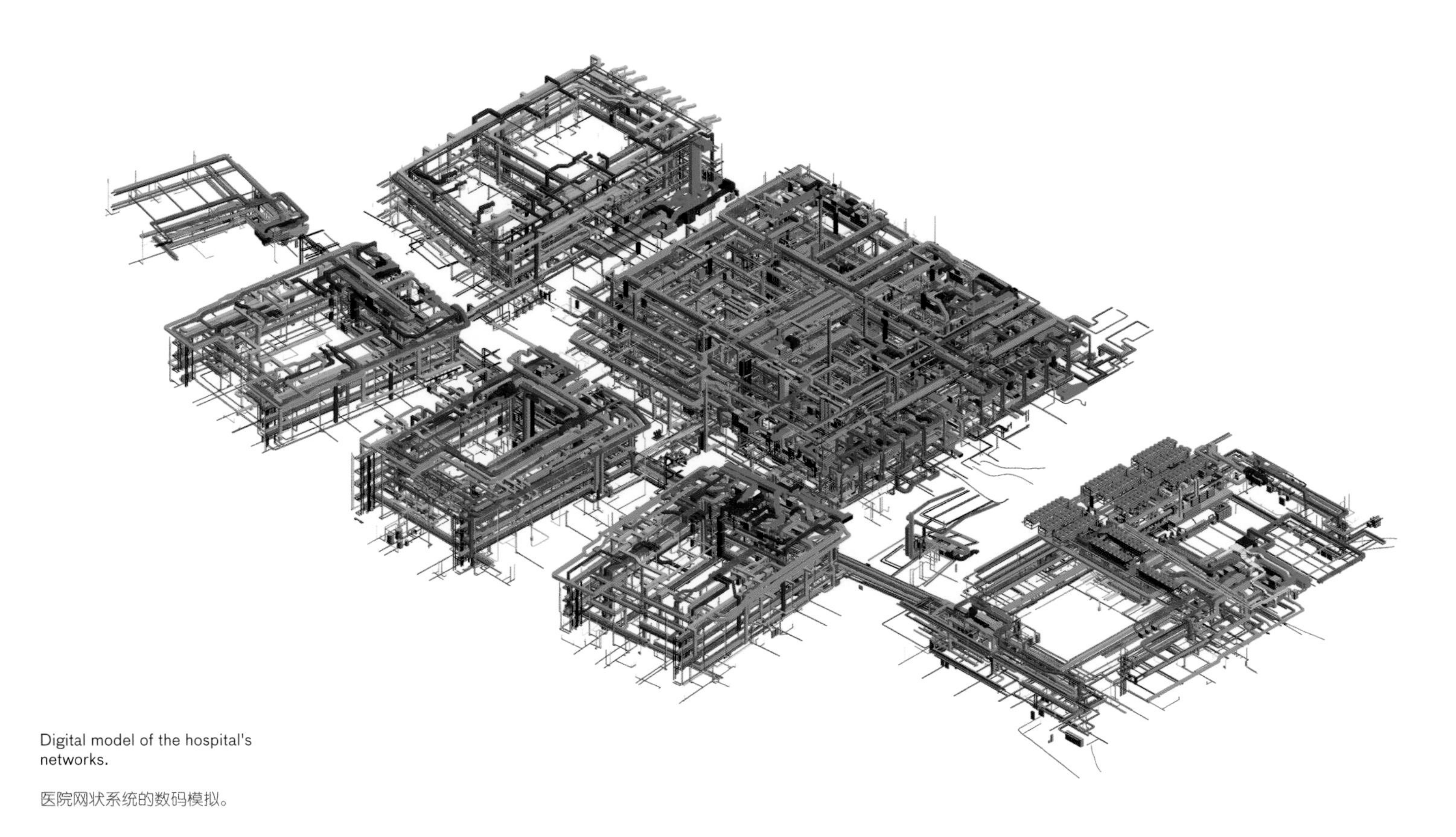

Digital model of the hospital's networks.

医院网状系统的数码模拟。

Issy-les-Moulineaux Fort
Issy-les-Moulineaux, France
2001/2011
120 000 m²

伊西莱穆林诺数码要塞
法国，伊西莱穆林诺
2001/2011
120 000 m²

The mayor of Issy-les-Moulineaux, André Santini, was as determined to turn the town's old military fort into inhabitable housing as was Jack Lang when deciding to erect the Arab World Institute on one of Paris rare constructible plots still belonging to the State. No success is guaranteed without the strong will of the commissioner. Architecture-Studio can boast of an eco-district as ebullient as the town itself and as peaceful as a garden city. This mix is a eulogy on life as it happens. The strength of this project is drawn from its architectural diversity, both a metonym of the variegated urban environment and a metaphor of Eden. Apples and pears are ready to be picked from the orchards which surround 18 buildings. Eighteen cozy cocoons are equipped with the best amenities for their inhabitants – such as a pneumatic system of garbage collection and geothermal heating provided by a natural underground source. The 1 200 apartments, some of which follow the original linear design of this former military base, have generally espoused a curvilinear aesthetic, be it parabolic or oval-shaped. This design, with its seven levels, breaks away from the selfish and rather churlish autonomy of orthogonal alignment. Everything here breathes of flexibility and a graceful lifestyle. Collaborative creativity, which is Architecture-Studio's trademark, has allowed seven teams from other architecture firms to embark upon this great project with them. Within the precincts of the old fort – a pentagon – curves and rectangles suffuse a warm atmosphere of wood and metal with an Asian touch that favors a matte aesthetic over a shiny one. The mirroring pond in the Place du Belvédère illustrates this perfectly. H.L.

伊西莱穆林诺的市长André Santini决定将该市的旧时军事要塞变为居民小区，就像贾克·朗当年决定在巴黎极其珍贵的可建造的国有地块上建造阿拉伯世界研究中心一样。若没有委员的坚持，项目就不能成功的开展。法国AS建筑工作室将原有遗址勾勒为生态小区，使其与该城热情洋溢的气息相匹配，同时又像一座花园城市般安静。这种融合的实现便是对生活的一种颂扬。该项目的魅力来自其建筑的多样，不仅有对多变的都市环境的转喻，也有对伊甸园的隐喻，果园中的苹果与梨都已成熟，等待收割。18座舒适的建筑为栖居者提供最便捷的设施——如气动垃圾回收系统和天然地热能供热系统。包含1 200套公寓的建筑群，其中一些建筑遵循军事基地原有的线条设计，呈现抛物线形或椭圆形的曲线美，进而使这些七层的建筑物摒弃单调、粗犷的直角直线。这里随处都能感受到灵动和优雅的生活方式。协作创造，是法国AS建筑工作室的主张，他们与其他七家建筑工作室的团队一同完成了这个伟大的项目。在旧时军事要塞的五角结构内，曲线和矩形的建筑沉浸在木材和金属的温馨氛围里，为亮面赋予具有亚洲格调的磨砂质感，位于Belvédère广场的镜面水池也得以凸显。H.L.

Ground plan and aerial view of the eco district of Issy Fort. The former military fort has been transformed into a new district containing 1,200 apartments, 1,520 parking spaces, commercial areas, a cultural center, two schools, a secondary school, a gym, a child care center, gardens and orchards. It was given the Ecoquartier Award in 2011, in the category of Urban Requalification and Digital Innovation.

伊西莱穆林诺数码要塞生态小区的平面图与鸟瞰图。旧时的军事要塞已转变为包含1 200套公寓、1 520个停车位、商业区、文化中心、两所学校、一所中学、一座体育馆、一个托儿中心、花园与果园的新小区。获2011年度生态街区奖城市转型与数字创新类奖。

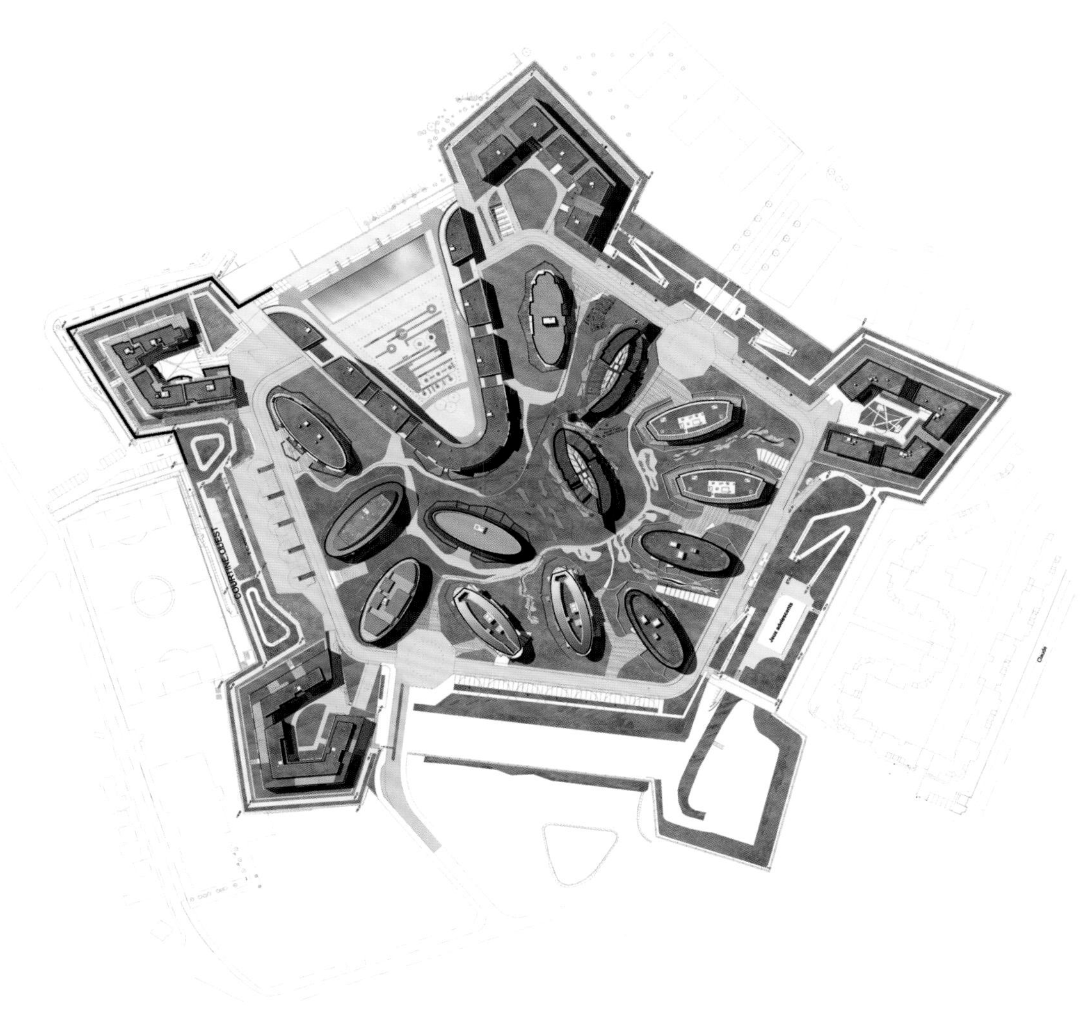

The Belvedere plaza at the center of the Fort is lined by a large building whose parabola-shaped layout opens onto Paris and the Seine valley. The pedestrian public space surrounded by shops, cafés and restaurants is the living heart of the neighborhood.

Belvédère广场位于小区中心，被一个巨型的建筑物环绕，呈抛物线型，开口朝向巴黎和塞纳河谷。被商店、咖啡厅与餐厅环绕的公共步行区域是邻里的生活中心。

21

The façades of Villa no. 5 contribute a kinetic aspect to the building. A planted roof enhances the sustainability of the building. The private gardens on the ground floor are defined by different bamboo hedges.

5号住宅楼的立面，彰显出该建筑动态艺术的一面。绿化屋顶增强了建筑的可持续发展性。一层的私人花园被竹篱笆包围且分隔开。

Covered by a green façade, Villa no. 1 includes 55 social houses. As with all the Fort buildings, the heating and the production of domestic hot water are provided by a heat pump at the water table.

被植物立面环绕的1号住宅楼内有55套社会住房。与数码要塞内的其他建筑一样，暖气和生活热水由位于地下水层的热泵供应。

Brasserie
PARIS'SY
Bar
7

coiffure
CENTRE DE
massot
INOA
11

STGO

3D Simulator of the Sustainable City
Santiago, Chile
2014

可持续发展城市模拟器
智利，圣地亚哥
2014

Architects have put their efforts into the common task of creating a scientific tool, “Santiago Des3aDo”, an innovative product helping decision-making via 3D images and simulations concerning the various aspects of sustainable design and urban planning: mobility, water and garbage recycling, energy, telecoms, etc. The simulator favors and promotes French sustainable solutions for urban planning such as architecture, services, products and technology related to the eco-city. Born out of an agreement signed in March 2014 between the French Government and the Municipality of Santiago, this powerful instrument is a project financed by the French State under the FASEP–Innovation verte provision. The grouping of five French specialist firms (Artelia, Veolia, Architecture-Studio, Arte Charpentier and Siradel) has engaged in a genuine collaboration with local firms concerned with the sustainable rehabilitation of the Chilean capital’s city-center in order to reduce the dividing line caused by the Pan-American Highway. Based on the 11 key points of the sustainable city, the simulator studies 78 strategic indicators and offers solutions to build up a green network/biodiversity in an urban environment. This defines a new centrality, favoring eco-friendly traffic and offering an alternative model for growth. The urban renovation program will give back to downtown dwellers the capital’s lost attractiveness and a long-forgotten quality of life. “Santiago Des3aDo” combines both interactivity and visual utility. H.L.

建筑师们的共同任务是将精力放在打造科学用具“Santiago Des3aDo”上，这是一种通过三维影像和模拟来帮助决策的软件，涉及可持续设计和城市规划的各个方面：流动性、水、垃圾回收、能源、电信等。该模拟器支持并宣传了法国在城市规划方面可提供的与生态城市相关的建筑、服务、产品和技术等可持续性解决方案。法国政府与圣地亚哥市政府于2014年3月签署协议，由法国政府通过FASEP绿能创新基金资助这个强大设备的研发。五家法国专业公司（Artelia、威立雅、法国AS建筑工作室、夏邦杰建筑设计事务所和Siradel）的团队与智利当地企业开启真正合作，为促进智利首都市中心的可持续恢复发展，并减弱泛美高速公路所造成的对城市的分割。基于发展可持续城市的11个关键要素，该模拟器研究了78项战略指标并提供在城市环境中建立绿色网络/生物多样性的解决方案。模拟器重新定义圣地亚哥，支持环保交通并为发展提供了可选模式。城市改造项目将为市区居民重新带来首都曾失去的吸引力和长期被遗忘的生活质量。“Santiago Des3aDo”是结合跨专业和可视化功用的工具。H.L.

3D digital representation of different indicators for a sustainable city, analyzed by the simulator.

模拟器对可持续发展的城市进行了分析，图中是该市3D数字影像。

transit spaces
中转空间

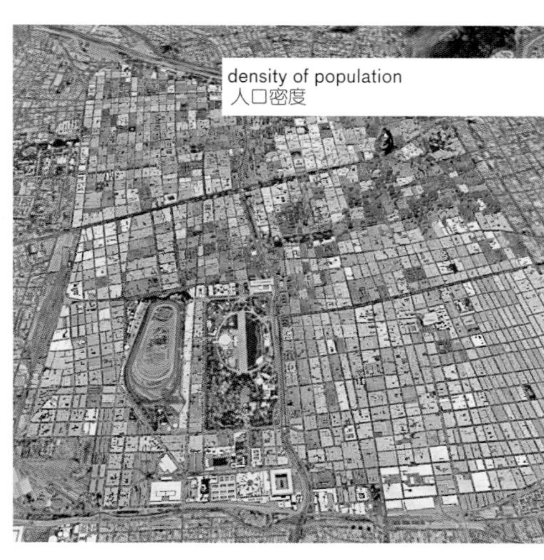
density of population
人口密度

accessibility to public green spaces
公共绿化开放度

economic activities
经济活动

bus network coverage
公共巴士交通网络

revegetation rate
交通拥堵度

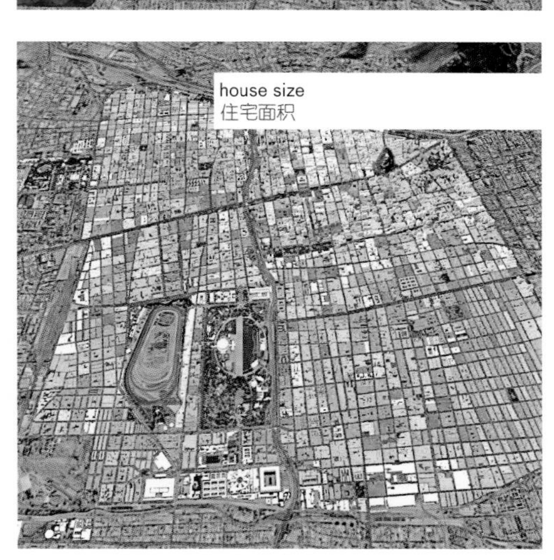
house size
住宅面积

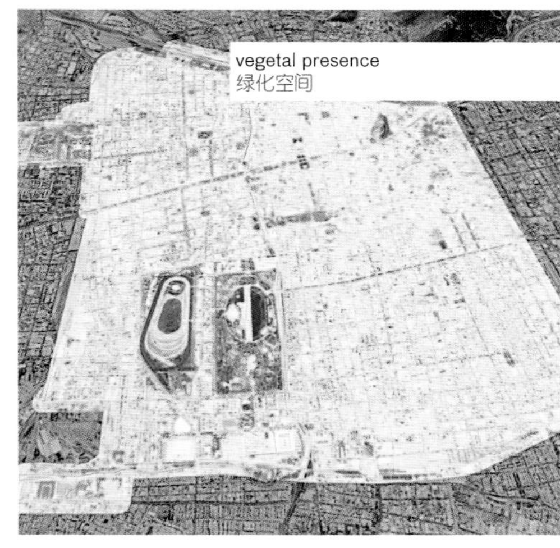
vegetal presence
绿化空间

digital divide
网络覆盖

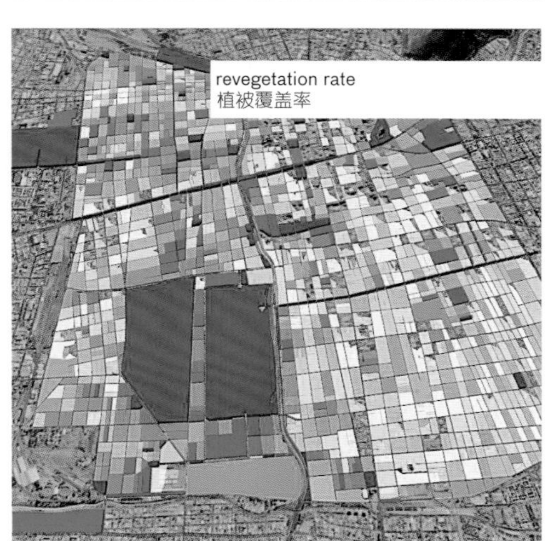
revegetation rate
植被覆盖率

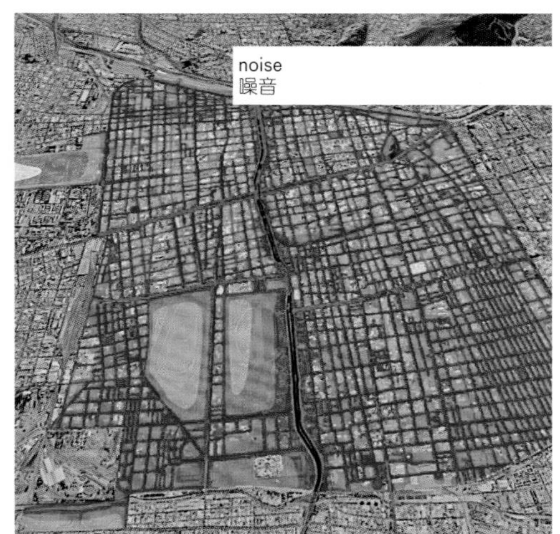
noise
噪音

3D simulation of potential construction (top) and sunlight (below) in the study zone.

研究区域的未来建筑群（上图）以及日间光线（下图）的3D模拟效果。

The simulation allows the presentation of some development solutions for the Autopista Norte site. A continuous boulevard (top) and a succession of planted pedestrian plazas (below) are two possible solutions.

通过模拟可勾勒出高速公路北部区域的一些发展方案，如：延伸的大道（上图）与绿化市民广场（下图）是这两个可行的方案。

04

ARCHITECTURE
ARCHITECTURE ET TECHNOLOGIE
374
MARIO PISANI
ARCHITECTURE · STUDIO
RITES DE PASSAGE
GRAND PRIX
D'ARCHITECTURE
DU MONITEUR
EQUERRE D'ARGENT
1987
INSTITUT DU MONDE ARABE A PARIS
EXPOSITIO
PARIS :
ARCHITECTU
ET
UTOPI
PAVILLON
DE L'ARSENAL
DU 14 DÉCEMBRE 1989
AU 21 FÉVRIER 1990
UN CLOCHER
DE PLUS A PARIS
Architecture Studio constructs cryptic buildings that possess an unimpeachable logic. Whatever the programme, be it a major public amenity, service industry building or factory, housing, school or university; or be it studies relating to a neighbourhood or section of a town; they are "contextual", to quote the architects that make up the Architecture Studio group. The buildings are rooted in our time. Built with precision, using the latest techniques, they are embedded in the town's economic, social and cultural textures, in order to stimulate its development. In addition, they display the pleasure of inventing shapes and spaces which have a sense. They impose material and colour, venture contrasts, signify the ruptures of the modern world. This publication, edited by Techniques et Architecture, is split into three parts. Architecture Studio illustrates its recent realisations and projects that are in progress, with commentary by the architects. Critics analyse the group's creative and theoretical itinerary (texts by Geneviève Brisac, Mario Pisani, Wojiec Lesnikowski, Chantal Beret). Cities Studio presents its urban projects (texts by Bruno-Henri Vayssière). Finally, D. Studio, offers the intimacy of a closer look. Three scales which meet and reinforce one another

REVUE CATHOLIQUE INTERNATIONALE
COMMUNIO
L'église dans la ville
Faut-il construire des églises ?
Que valent les projets récents ?
ARCHITECTURE STUDIO
RODO TISNADO
MARIO PISANI
Perspective
PARLEMENT EUROPÉEN
STRASBOURG
AS ARCHITECTURE-STUDIO
1999_2009
PARLEMENT
EUROPÉEN
STRASBOURG
140
WORLD ARCHITECTURE
ARCHITECT

04 GIVING AND LEARNING

04 TRANSMETTRE ET APPRENDRE

给予与学习

The partners of Architecture-Studio often point out that an international outlook was part of the firm almost since its inception. Indeed, Rodo Tisnado, of Peruvian origin, was one of the very early members of the group. This outward-looking element was confirmed when Roueïda Ayache, of Lebanese origin, became a partner in 2001, followed by Mariano Efron, from Argentina, and Amar Sabeh el Leil, from Syria, in 2009. Both Roueïda Ayache and Amar Sabeh el Leil speak Arabic, a definite advantage in the recent projects that the office has won in the Middle East. As Martin Robain explains, reaching out to participate in international competitions and to design buildings in places like China and Saudi Arabia was not the fruit of a concerted effort to go to these countries, nor was it in any way a result of the variety of nationalities represented amongst the partners. Rather, it was a question of circumstances and the growing reputation of Architecture-Studio, marked first, especially vis-à-vis the Middle East, by their participation in the Arab World Institute project and again, very strongly, when they completed the European Parliament in Strasbourg in 1998. It was the scale of the European Parliament that gave the office the references required in order to be able to work in China, for example.

The Parliament was completed at the time that other significant changes were, in a more general way, allowing information and participation in competitions to become more international, in France and elsewhere. The progressive opening of China strongly stimulated the movement toward globalization, with the nearly simultaneous explosion of the Internet. "The Arab World Institute, of course, opened horizons for us in Arab countries," says Roueïda Ayache, who worked at Architecture-Studio for ten years before becoming a partner. "Paris is considered cosmopolitan outside of France, and that was especially the case for architecture during the Mitterrand years (*Grands Travaux*). There was the construction of Berlin as well." It can be noted in passing that Architecture-Studio co-founded an association called AFEX (*French Architects Overseas*) in 1996. This was the time when Paris was home to a large number of foreign architects, drawn in part by the ambitious construction program of the French President. The creation of AFEX was a useful element in encouraging French architects to participate in international competitions. AFEX comprises 200 members including 120 architectural offices and public and private partners as well. Thus the partners of Architecture-Studio are quite unanimous in insisting that while

法国AS建筑工作室的合伙人经常说，国际化视野从一开始就是工作室的一部分。事实上，团队最早期的成员罗多·蒂斯纳多来自秘鲁，而自从来自黎巴嫩的罗伊达·阿亚斯在2001年成为合伙人后，这种开放理念得以正式确立，后来在2009年又有来自阿根廷的马里亚诺·艾翁，以及来自叙利亚的艾马·萨布埃雷成为合伙人。在合伙人团队中，罗伊达·阿亚斯和艾马·萨布埃雷都会讲阿拉伯语，是执行阿拉伯项目时不可忽视的优点。正如马丁·罗班所说，工作室参与国际竞赛，并在中国和沙特阿拉伯等地设计建筑物，并非公司通力开发这些国家市场的结果，也不是多国籍的合伙人团队带来的结果，而是得益于法国AS建筑工作室日渐提高的知名度。比如工作室在中东的知名度，很大程度上与他们参与设计阿拉伯世界研究中心项目相关；而1998年由他们设计的欧洲议会中心竣工使他们更为出名。正是因为有过欧洲议会中心这样的大型建筑设计的工作经验，使工作室取得了中国业主的信任并开始在中国工作。

欧洲议会中心建筑竣工的时候，正是法国乃至全球的信息传播方式和国际竞赛参与方式产生巨变的时候。中国的逐步开放有力刺激了全球化的进展，同时，互联网的兴起让沟通变得更加便捷。罗伊达·阿亚斯说：“阿拉伯世界研究中心项目无疑将我们介绍给了许多阿拉伯国家。”在成为合伙人的时候，她已经在法国AS建筑工作室工作了10年。“巴黎总是被视为法国之外的世界性大都会，尤其是在密特朗时期他所主持的‘大工程’计划吸引了许多海外建筑师参与设计。”1996年，法国AS建筑工作室联合建立了法国海外建筑师协会（AFEX），成为鼓励法国建筑师积极参与国际建筑竞赛的有利途径。

Situated on the northeastern side of the Yuzhong peninsula, and bordered by the Yangze and Jialing rivers, the Chongqing Science and Technology Museum is a new cultural district landmark, facing the city's historical site.

重庆科技博物馆位于渝中半岛的东北方向，被长江与嘉陵江环绕，面朝城市的历史遗迹，是该文化新区的地标。

The Head Offices of Shanghai Wison Chemical Engineering in Shanghai. A monumental entrance porch leads to a building punctuated by atriums and a garden.

位于上海市的上海惠生化工工程有限公司总部。仪式性的入口通往的是与中庭和花园巧妙结合的建筑内部。

their international outlook was in some sense inscribed in the "DNA" of the office, it was also the result of a very particular set of circumstances and meetings.

The victory of Architecture-Studio in the limited competition to design a building intended to be the very representation of the European Community, as described by the partners, is indicative of their approach and revealing as to the ways that they have progressed. René-Henri Arnaud confirms this: "The fact that the office went toward China had to do with circumstances rather than any grand scheme. Even the Parliament was also the story of a meeting with Catherine Trautmann." Catherine Trautmann was Mayor of Strasbourg and Chairwoman of the Strasbourg Urban Community Council (1989–1997 and 2000–2001), and French Minister of Culture (1997–2000). Arnaud continues, "We had participated in a competition for a public building in Strasbourg and she was a member of the jury. Although we didn't win, it seems that we made a good impression and Catherine Trautmann selected our project among the few architects that were shortlisted to participate in the competition for the Parliament. It was Mrs Trautmann who organized the competition and thus imposed Strasbourg as the location for the Parliament. Brussels was ahead in this race because they had built a convention center that was destined to become the European Parliament. We won the competition that was organized by the City of Strasbourg. At first the European Parliament rented the building and then, only after a few years, bought it. We continued to work on the building for 15 years and assisted in the transitions as the Community became larger."

Winning this competition immediately placed Architecture-Studio in a different category to other Paris offices which had never had occasion to build such significant structures. A similar case, in the same time period, was that of Dominique Perrault, who won the 1989 government-organized competition to design a new building for the French National Library on the banks of the Seine. Shortly after completing the European Parliament building, Architecture-Studio won an even more surprising commission in 2001 against 14 international teams to develop the master plan for Expo 2010 Shanghai. The site of the event was a 5.28 square kilometer area set on either side of the Haungpu River near the center of the city. The actual realization of the scheme was taken on by the Tongji Institute from Shanghai but along the lines set out by Architecture-Studio.

法国海外建筑师协会有200名成员，其中包括120家建筑设计公司和公共建筑机构。法国AS建筑工作室合伙人一致认为，他们的国际视野在某种程度上是工作室的"基因"所决定的，但同时也是一系列巧合与巧识的结果。

法国AS建筑工作室在欧洲议会中心的竞标中获胜，其建筑设计代表的是欧洲联盟，合伙人认为这体现了工作室把握机遇的工作方式，以及所带来的发展路程。勒内—亨利·阿诺对这点进行了确认："工作室进驻中国市场完全是因为机遇而非基于某个宏大的策略。连欧洲议会中心项目也源于我们与凯瑟琳·陶德曼的一次会面。"凯瑟琳·陶德曼曾担任斯特拉斯堡市市长、斯特拉斯堡城市共同体的主席（1989—1997和2000—2001）以及法国文化部部长（1997—2000）。阿诺继续说道："我们之前参加过位于斯特拉斯堡一处公共建筑的竞标，陶德曼是竞赛的评委之一，尽管我们当时没有中标，但似乎我们给她留下了不错的印象。后来她将我们选为参加欧洲议会中心竞标设计的仅有的几家建筑公司之一。最终我们赢得了竞赛，并跟踪这项工程15年，随着欧盟成员不断增加，在建筑竣工后我们仍为建筑的调整转变做了很多辅助性的工作。"

赢得此次竞赛使法国AS建筑工作室立刻跃升到法国各大建筑工作室的顶级行列，而其他大多数的工作室还没有过建造如此具有象征性建筑的机会。相似的例子，还有同时期的同行多米尼克·佩罗，他于1989年赢得了政府组织的在塞纳河畔建造法国国家图书馆新馆的竞标项目。在完成欧洲议会中心的项目不久，2001年，法国AS建筑工作室又在14支国际团队中脱颖而出，一举摘下2010年上海世博会会址总规划竞赛桂冠。规划会址所在地是靠近市中心的一处5.28平方公里的临黄浦江区域，根据法国AS建筑工作室的大规划，由上海同济城市规划设计研究院进行了深化和最终落实。

法国AS建筑工作室合伙人多次强调，在法国以外的地区工作，持续性的学习是必不可少的。如果说2010年上海世博会会址总规划对工作室来说是一个重大项目，那么法国AS建筑工作室在中国完工的第

Located on both sides of the river, these two towers mark the entrance to Ningbo and define a new urban corridor. The architecture consists of many layers, meeting the skyline of the city.

两座大楼位于河的两畔，标志着宁波的入口并定义了新的城市走廊。由多组体量组成，与城市的水平线相呼应。

The partners of Architecture-Studio frequently place emphasis on the learning process that has been an integral part of their work outside of France. Although the master planning for Expo 2010 was a significant project, the first building that Architecture-Studio finished in China was the Wison Chemical Headquarters and Laboratory inaugurated in 2003 in Shanghai. Martin Robain explains: "The President of Wison came to see us. We did the plans here and the client named an engineer in Shanghai to follow the project. There was a white marble in the entrance and they were not able to use the same marble for the entire 500 square meter surface, so the client asked me to come and when I noticed the difference, he ordered that everything should be replaced. We felt that this was an indication of a respect for the architect and for aesthetics that we were not expecting. Ten years later he asked us to design a new headquarters building (Office Park for Wison Headquarter, completed in 2013) that was almost ten times bigger, designed according to the principles of *Feng Shui* and developed in accordance with the working methods of the firm. Now that we know of the importance of *Feng Shui* in architecture and landscape design in China, we have become very attentive to that aspect."

Having already established a reputation in the region through the Arab World Institute was surely a positive element in the invitation extended by Saudi Arabia to Architecture-Studio to participate in a limited competition for the West Entry of Mecca, the King Abdul Aziz Road master plan, in 2002. Architecture-Studio won the competition ahead of three other offices, at the same time as they were working on the Shanghai master plan. Their scheme proposed a kind of "landmark avenue" leading to Mecca. This symbolic entryway ensured ease of public transport for the pilgrims, as well as shaded walkways. This layered approach to mobility took into account the local climate and offered pilgrims different options. As Roueïda Ayache explains, the Arab World Institute won an Aga Khan Award for Architecture in 1989. One of the people familiar with the Award was a member of the Mecca competition jury. As they had in Strasbourg, with Catherine Trautmann, in this instance Architecture-Studio again showed a talent for making a positive impression on the right people at the right time.

Although other offices surely benefit from such coincidences and favorable circumstances, the continuity of the presence of Architecture-Studio in some countries has proven to be a major asset as they have won progressively larger commissions.

一栋建筑，于2003年正式开幕的上海惠生化工工程有限公司总部，则代表了学习中国文化的最初记忆。马丁·罗班说："惠生集团总裁曾经来法国拜访我们，当时所有的建筑图和设计细节都是在巴黎完成的，客户在上海请了一家当地的工程公司落实项目的建设。在我们的设计里，建筑入口规划使用白色大理石，但施工地当时很难找到足够的白色大理石以铺满整个500平方米的面积，因此客户来征询我的意见。当我详细地说明不同大理石的差异后，总裁先生决定将所有不属于原有设计的材料都替换掉。我们能感受到这是对建筑师的一种尊重，也是对建筑整体的尊重，这是出乎我们意料的。十年之后，客户委托我们设计一座新的办公建筑（惠生集团总部园区，2013年竣工），这项工程规模是之前工程的近十倍，并且按照风水学进行设计，将工作室的工作方法与中国传统文化相结合。我们在中国学习并了解到风水学无论在建筑或景观设计中都是非常重要的，因此现在在这方面特别留意。"

得益于阿拉伯世界研究中心为我们带来的在阿拉伯地区的知名度，2002年，我们受邀参与沙特阿拉伯麦加西入口处的阿卜杜勒阿齐兹国王大道总规划方案竞赛。法国AS建筑工作室最终击败其他三家大型建筑工作室赢得了本次竞标，而在那之前工作室刚中标2010年上海世博会会址总规划。该设计方案强调"里程碑大道"特点，大道直通麦加禁寺。这条具有象征性的大道有快捷的运输和遮阴的设计，多层次的交通方案考虑到当地气候，并为朝圣者提供了多样选择。就像罗伊达·阿亚斯所解释的，阿拉伯世界研究中心于1989年荣获了"阿卡汗建筑奖"，而麦加这次竞标的评审团有一位对此奖项比较熟悉的评审。正如他们在斯特拉斯堡与凯瑟琳·陶德曼会面一样，这次法国AS建筑工作室再次展现出在正确的时间给正确的人留下了积极印象的能力。

法国AS建筑工作室在一些国家的持续参与度成为了一个优势，帮助工作室在越来越重要的竞赛中拔得头筹。

工作室在1989年完成了位于马斯喀特的法国驻阿曼苏丹国大使馆项目，在2005年，承接了阿曼苏丹国驻巴黎大使馆的扩建工程。又在

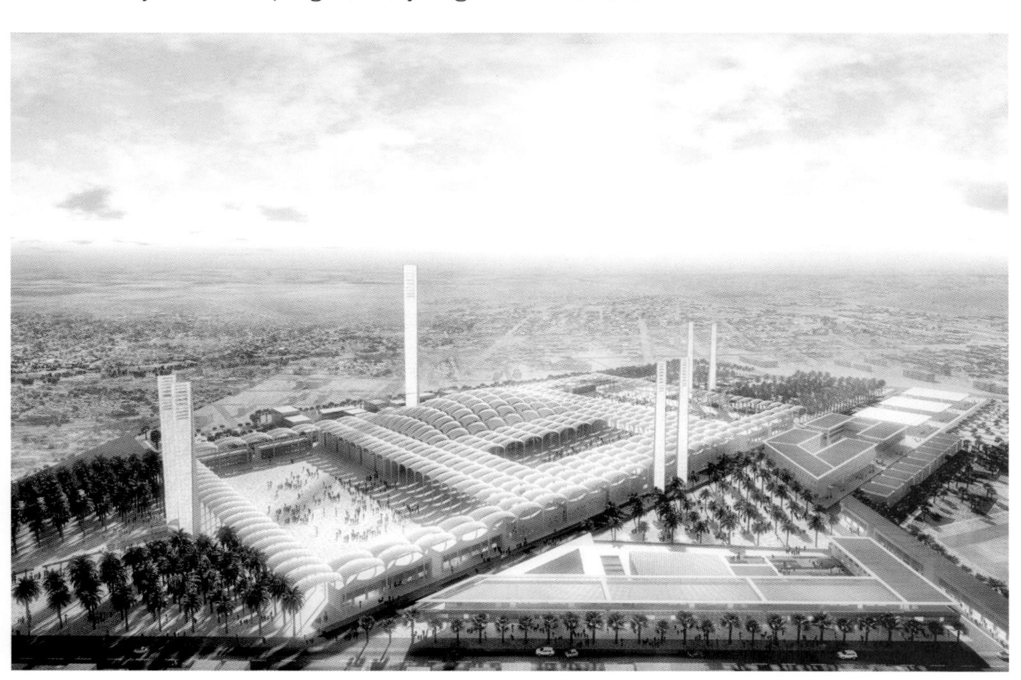

The pilgrim village offers the Muslim community of N'Djamena, and Chad as a whole, a prayer and gathering space of great size. This place of worship will also serve as a leading cultural space, with modern facilities for education, health, and leisure activities.

朝圣者之家为恩贾梅纳，乃至整个乍得的伊斯兰教徒提供祈祷与聚集的场所。这里还将成为文化中心，提供学习、健身和娱乐等相关现代设备。

The office completed the French Embassy in Muscat, Oman in 1989, and then worked on an extension of the Embassy of Oman in Paris in 2005. Their victory in the 2008 competition to design the new 70,000 square meter Cultural Complex Center in Muscat was thus a logical continuation of the good relations that they had established with the earlier work. This project is currently on hold.

Naturally, work in other countries – in particular outside of the European area – has led Architecture-Studio to develop methods to deal with local conditions. René-Henri Arnaud states that "working in different countries we have to accept local methods and even different notions of time. In a country like Oman, the gestation of a project does not follow the time scheme that we are familiar with in the West." Amar Sabeh el Leil confirms the approach of the office: "When we work abroad, we seek first to understand and we never seek to simply impose a specific way of doing things". Nor does some work involve only the concepts and methods of a single country, as Roueïda Ayache points out – on some occasions, Chinese construction companies are involved with the projects in the Middle East.

The group admits and accepts that international work has changed the practice and will continue to do so. Just as they have up until now, Architecture-Studio continues to advance on the basis of meetings and opportunities rather than any master plan for their future. Martin Robain states: "For the moment, Architecture-Studio has functioned by being a bit in advance in our choices. We started doing international work before other French offices; we were in China from 1995. We have made structural choices such as our collective approach and these choices have given good results. But what about tomorrow? We don't know about this."

China has taken a major place in the work of Architecture-Studio, particularly since the end of the 1990s when the office worked on a number of important competitions including that for the Shanghai Expo, which they won in 2001. By 2004, Architecture-Studio had enough work to open an office in Shanghai, and then a second one in Beijing in 2006. Many of the architects working in these offices have come for three month periods to the Paris office, just as team members from France are also assigned to China for several months. Projects were obtained through competition in Chongqing (the Science Museum) and Jinan (the Cultural Center). The work method of the office has evolved along with these projects through the creation of weekly video conferences between Paris and China.

2008年中标面积达70 000平方米的马斯喀特文化中心的建筑设计工作，这都是以往的工作所营造的良好关系的进一步延续。

在其他国家，尤其是欧洲以外的区域工作，使得法国AS建筑工作室探索出以合作的方式并根据当地实际情况解决问题的方法。勒内-亨利·阿诺说道："在文化各异的国家工作，我们必须接受当地的工作方法，甚至是迥异的时间观念。比如在阿曼苏丹国，项目的筹备过程跟我们所熟悉的西方模式有很大出入。"艾马·萨布埃雷就工作方法重申："在国外工作时，我们首先会去学习和了解，而不是单纯将我们习惯的工作方式强加上去。"罗伊达·阿亚斯补充道，"有时候光是学习某一个国家的特色和工作方式还不足够，比如说我们在中东地区的一些建筑项目是与来自中国的施工公司合作的。"

工作室团队承认国际化的工作性质会带来工作方式的持续变化。正如目前为止的经历一样，法国AS建筑工作室着力以相遇和机遇为基础而不断前进，而不是依靠一份对未来的总规划。马丁·罗班表示："目前，法国AS建筑工作室的顺利发展得益于我们过去的选择，我们比起其他法国的建筑事务所更早地开启国际化业务，早在1995年我们便已经踏上中国的土地。我们在公司结构上选择了集体合作模式，也已经带来了很好的效果。但是明天又会如何？我们也不知道。"

中国已经成为法国AS建筑工作室的主要目标市场，尤其是自上个世纪90年代末期以来，工作室接连参与了一系列意义重大的竞标活动，其中便包括在2001年赢得的上海世博会会址总规划的投标。到2004年，法国AS建筑工作室在中国的工作量已足以支撑在上海开设工作室，2006年又在北京开设了第二间工作室。许多在中国工作的建筑师员工会到巴黎总部进行3个月的工作学习，而在巴黎工作的建筑师也会到中国工作一段时间。工作室持续中标重要的项目，比如重庆科技馆项目和山东省省会文化艺术中心项目。工作室的工作方法也随着以上工程的开展而得到进一步发展，除了合伙人在中法两国间的穿梭，还有每周固定的巴黎总部与中国工作室之间的视频会议。

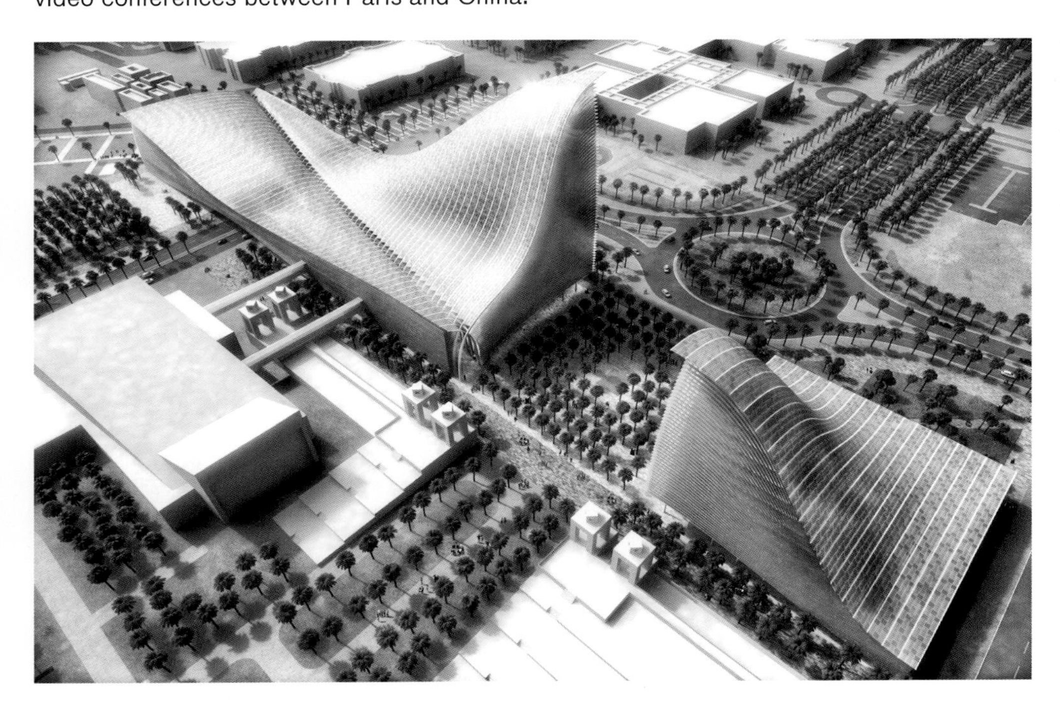

The King Abdulaziz University is an urban landmark in the cityscape of Jeddah and is an opportunity to restructure the spaces of the university.

阿卜杜勒阿齐兹国王大学是吉达都市景观中的标志性建筑，重塑了大学校区的空间。

Despite the origins of the partners and the broad spectrum of work outside of France, the office is pragmatic about the need to balance projects in terms of their geographic spread and the potential dangers that they entail. Roueïda Ayache says it clearly. "There is a higher risk in international work, both economic and intellectual. We are not always in situations that are fully predictable. A firm has to weigh what is the reasonable level of risk. Work in China or the Middle East is now less risky than it might have been, but working abroad in general obviously requires resilience and a capacity to adapt to changing circumstances."

Even the reaction and approach of foreign countries varies according to the region. The partners believe that it is a great advantage to be French and thus associated with a country that has a strong tradition in urban design, for example. "We are still perceived as a French office," states Martin Robain, "and in China we are called on because we are French." Just as they have always put an emphasis on the full environment of their buildings, Architecture-Studio accepts that French architects are perceived as having certain qualities, and they are happy to be called on for that reason, but then each project has its context. Roueïda Ayache explains: "We started working in Qatar because of what might be called the French culture of the city and its public spaces. When we started working in Jordan, one cannot say that it was the great number of towers in France that eased our task! We have worked there on contextual towers, on how the structures can help to develop an urban space, down to the level of the pedestrian. The Rotana Tower is intended as an urban magnet that creates a sense of orientation for public space."

Just as they do in the case of the exhibitions at the CA'ASI, the partners are careful to approach foreign work with a fully open attitude, which means, of course, bringing their own expertise and sense for innovation as well as their unique teamwork to their clients, while also seeking to learn from each experience. Martin Robain states: "What I think is interesting is the change in our way of thinking that work outside of France brings. Working in Africa has brought some fundamental changes in our way of thinking about architecture and the environment. There is a very interesting rapport with what we might call existential issues, or ones of material substance. We don't talk about ecology there the same way we do here – we talk about it because it is related to survival. We will be learning more from Africa."

由于工作室在法国以外的工作区域广大，且合伙人有多国籍背景，因此工作室对于地缘政治和相应潜在的风险会进行非常实际的权衡和考虑。罗伊达·阿亚斯清楚地指出：“在国外工作，经济和脑力输出方面都有更大的风险，因为身处的环境并非完全能够预测，公司必须衡量合理的危险程度。现在在中国或中东地区的工程已经没有以往那么大的不确定性，但整体上讲，承接国外的项目需要更多的韧性和适应环境不断变化的能力。”

根据地域的不同，与各国客户接触的方式以及他们的反馈也会有所差异。合伙人相信，作为一家法国公司，与法国著名的城市规划传统会自然地联系在一起，这为他们带来天然的优势。“人们首先会将我们视为法国的建筑工作室”，马丁·罗班说道，“在中国，的确有过因为我们是法国公司而将项目委托给我们的案例。”就像法国人赞扬他们自己所创造的环境一样，法国AS建筑工作室接受法国建筑师总是被视为法国创造力的代表这一现象。因为这种固有的总体观念而获得了工作项目，他们自然也是高兴的，但其实每个项目的背景又是有所区别的。罗伊达·阿亚斯表示：“我们在卡塔尔接到委托项目，是因为法国城市与城市规划文化的知名度。我们在约旦建设摩天大楼，会研究如何通过建筑构架以达到开拓城市空间的目的，比如人行道等空间的效果。而罗塔纳酒店大楼起初便是作为城市地标所设计，旨在为公共空间创造一种方向感。”

秉持着对海外项目的开放心态，工作室在CA'ASI艺术展览馆所举办的多个青年建筑师竞赛和展览，表明工作室不仅仅与业主分享自身的专业和创造力，也希望从其他优秀建筑师身上学习更多的经验。马丁·罗班说：“在法国以外的国家和地区工作给我们的思维方式方面带来的变化，让我觉得十分有趣。比如在非洲工作为我们对建筑和环境的思考带来了根本性的变化，我们认识到那里的建筑与生存和资源限制问题有着更深沉的关系，他们所说的生态学与我们广泛理解的不一样，对于他们来说生态就是生存的基本。我们希望从非洲学习到更多。”

At Larak Garden, geometry meets the landscape: this ambitious project aims to develop an attractive district at the entrance to Teheran.

在Larak花园，几何结构与靓丽的风景相得益彰：这个项目旨在在德黑兰的入口处开发一个美丽的街区。

STR2

European Parliament
Strasbourg, France
1991/1999
220 000 m²
In association with Gaston Valente

欧洲议会中心
法国，斯特拉斯堡
1991 / 1999
220 000 m²
联合设计: Gaston Valente

As Belgian rock singer Arno once sang: "Sure as hell! Being a European ain't no bad thing after all!" Based in Strasbourg, the most democratic of European institutions is packed full of life – 750 MEPs, 18 meeting rooms with a seating capacities from 50 to 350 places, 1,133 offices for parliamentarians and their staff, a central catering service, a chic popular bar – but despite this you can still enjoy fluid circulation and a real sense of space. TV programs covering the debates of this growing European democracy have popularized the images of the Parliament building. Its architecture is inspired by Europe's stylistic foundations, embodied in the Baroque and Classicism. The classical circular design of the tower integrates the typical baroque curvature formed by its spacious courtyard. It is as if, passing from Galileo's full circle to Kepler's ellipsis, you go symbolically from highly centralized government to a plurality of decision centers – the multi-polarity that defines democracy. Here history gives its best illustration. The curvaceous glass façade echoes the curving arm of the river Ill. Underscoring the full-fledged volume of the hemicycle, the glass is as much an expression of architectural clarity as one of political transparency. Looking up from the street, the building has magic proportions and has already assumed iconic status, much like the Eiffel Tower. H.L.

比利时摇滚歌手Arno曾经唱过："非常肯定！当欧洲人不是一件坏事！"位于斯特拉斯堡的欧洲议会中心，这个最能代表民主的欧洲机构里面有能够容纳750名欧洲议会议员的半圆形议会厅，18间可容纳50至350人的会议室，1133间供议员及其团队使用的办公室，一个餐饮中心以及一个优雅又不失潮流的酒吧。与此同时，在那里面又能感受到植物的绿意以及空气的流通。电视时常报道这里关于欧洲民主不断发展的辩论，也让欧洲议会中心建筑映入人们眼帘。议会中心的建筑设计灵感来源于西方文化的基础：古典主义和巴洛克风格，古典正圆形塔楼和巴洛克式椭圆的大型内庭院相互融合。展现了从伽利略的正圆到开普勒的椭圆，从向心结构到多中心结构，从中央集权到欧洲民主多极性的过渡，历史在这里得到最好的展现。沿伊尔河蜿蜒的外围建筑曲线的大型玻璃幕墙后，显露出半圆形大议会厅，这种透明既传达了一种建筑美学，也象征了政治方面的公开性。从街道向上看，建筑的巨大体量让它拥有和埃菲尔铁塔一样的象征性地位。H.L.

On the Ill river, the glass façade of the European Parliament displays the hemicycle. Its transparency is both architectural and political.

在伊尔河沿岸，欧洲议会中心的透明玻璃外立面显露了半圆形大议会厅，这种透明既传达了一种建筑美学，也象征了政治方面的公开性。

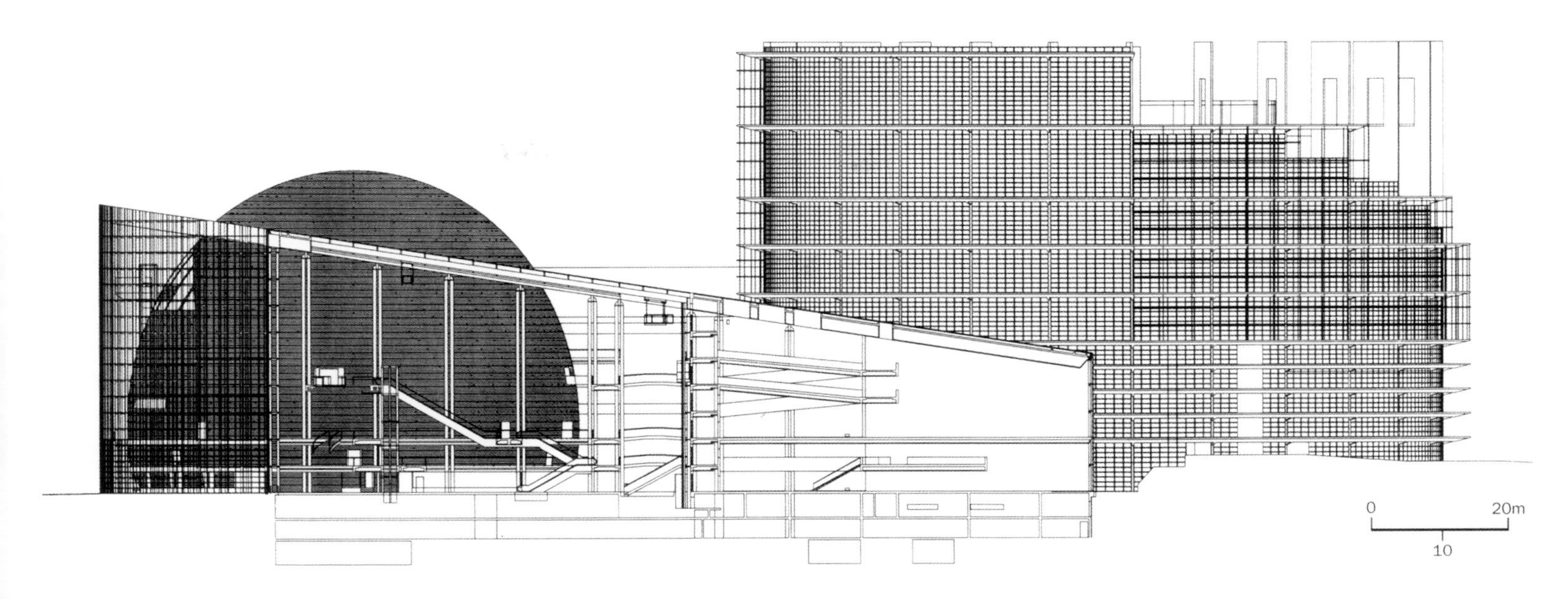

Section and floor plan.

剖面图（上图）与平面图（下图）。

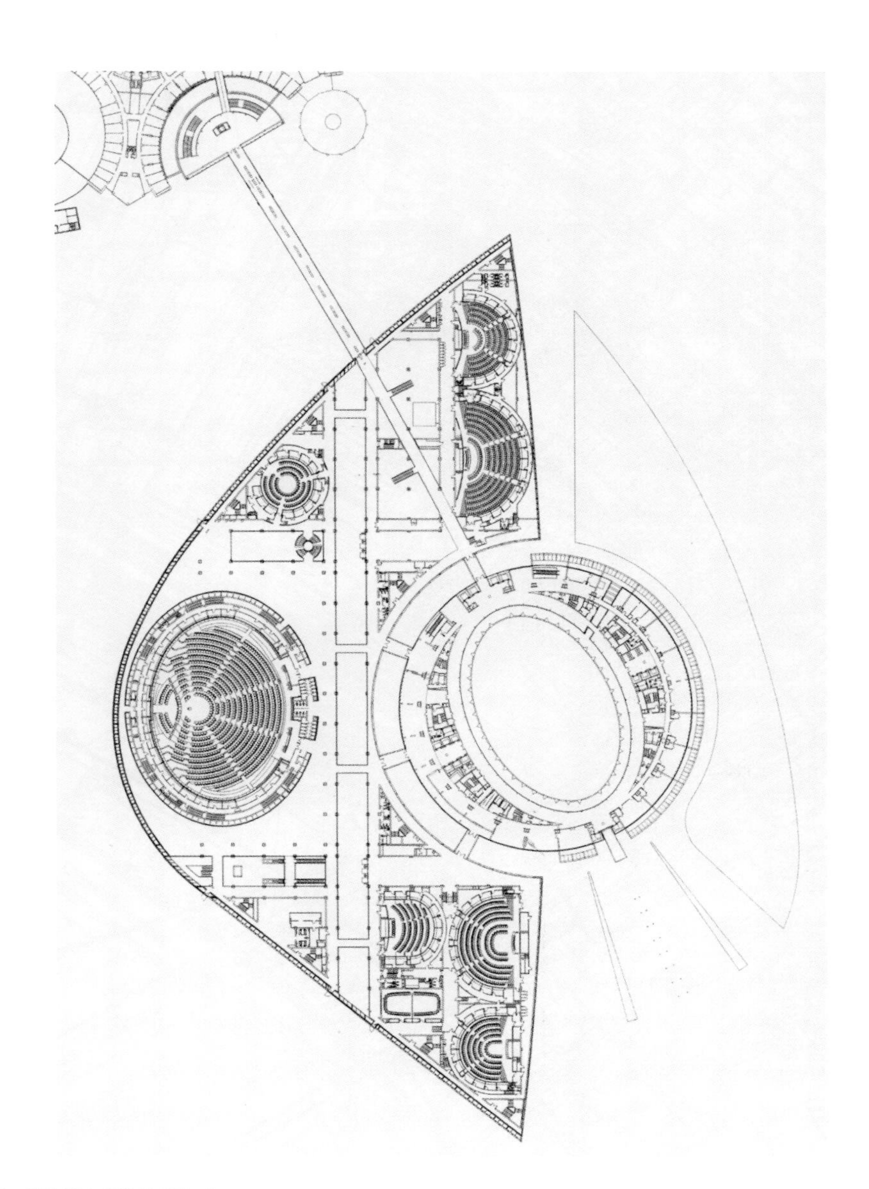

The European Parliament is an iconic, urban landmark for Strasbourg, as well as for Europe and the rest of the world. This contextual building embodies the strength of unity and the openness of democracy.

欧洲议会中心是斯特拉斯堡市的标志性建筑，也是欧洲乃至世界的重要建筑。该建筑体现了团结的力量与民主的开放性。

The official staircase.

礼宾楼梯。

The geometry of the building embodies the foundations of Western civilization, Classicism and the Baroque: the transition from a central geometric structure (Galileo) to the ellipse (Kepler) is also marking the way from centralized power to democracy.

建筑的几何形状体现了西方文化的基础：古典主义和巴洛克风格。从伽利略的正圆到开普勒的椭圆，象征着从中央集权到欧洲民主多极性的过渡。

SHN3

Shanghai World Exhibition 2010
Shanghai, China
2001
9,000,000 m²

上海2010年世博会会址规划
中国，上海
2001
9 000 000 m²

The theme of Shanghai's 2010 World Exhibition was "Better City, Better Life". It was an occasion to reflect upon all that which could be improved in the Middle Kingdom's city life. In 2001, Architecture-Studio won a competition to design its site and create an urban plan for it. Situated in the heart of the southern metropolis, with the Huangpu river running by, the chosen site had been allotted 310 hectares of exhibition space. A further 126 hectares were reserved as access zones for housing employees and storing materials. An architectural concept is long-term – you have to seize the moment and this was a once-in-a-lifetime occasion to define the urban planning of a city with a vision that would bequeath its transformative legacy beyond the life of the exhibition. This construction plan has put sustainable infrastructures to the fore with winding vegetal corridors and a close-knit network of canals. In this way, issues of pollution were partly resolved. The relationship between the city and nature changed completely – both urban concentration and traffic jams diminished. If the 600-meter long Flower Bridge had been built, it would have linked the Puxi district to the Pudong district; at 250 meters high, it would have symbolized both the foundation of the city and celebrated its urban renovation. H.L.

上海2010年世界博览会的主题为“城市，让生活更美好”。这是一个体现中国城市生活质量提高的最佳机会。2001年，法国AS建筑工作室赢得了本次世博会会址总规划的国际竞赛。上海位于江南都市群的中心，所选会址横跨黄浦江两岸，规划了310公顷的会展用地，另外126公顷为通行区、工作人员住宿区和器材储藏空间的用地。规划构思为长远考虑——旨在抓住此次难得的优化城市规划机会，并成为世博会之后仍能够留下城市改造印迹的结构性项目。规划将可持续发展的基础设施建设放在首位，设置植物走廊，紧密的运河网穿插其中，并通过这种方法解决部分污染问题；会址的规划还要解决城市交通、密度方面的问题，让城市与自然之间的关系发生彻底变化。如果规划中的600米长的花桥得以建设，则可以把浦东浦西艺术地联系在一起，250米高的花桥不仅仅可以成为城市的标志，也是城市整改的纪念。H.L.

The flower bridge on the River Huangpu.

黄浦江上的花桥。

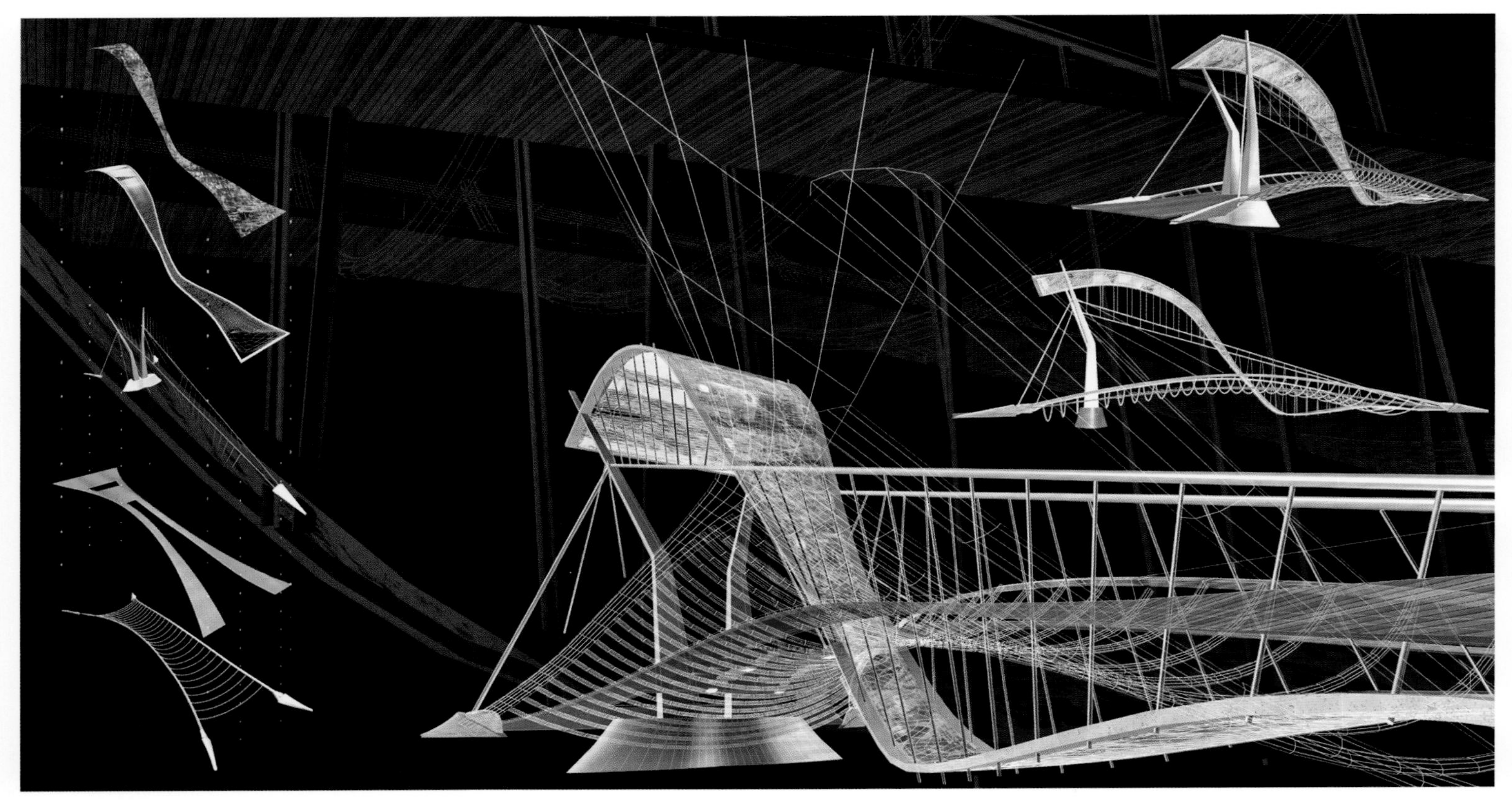

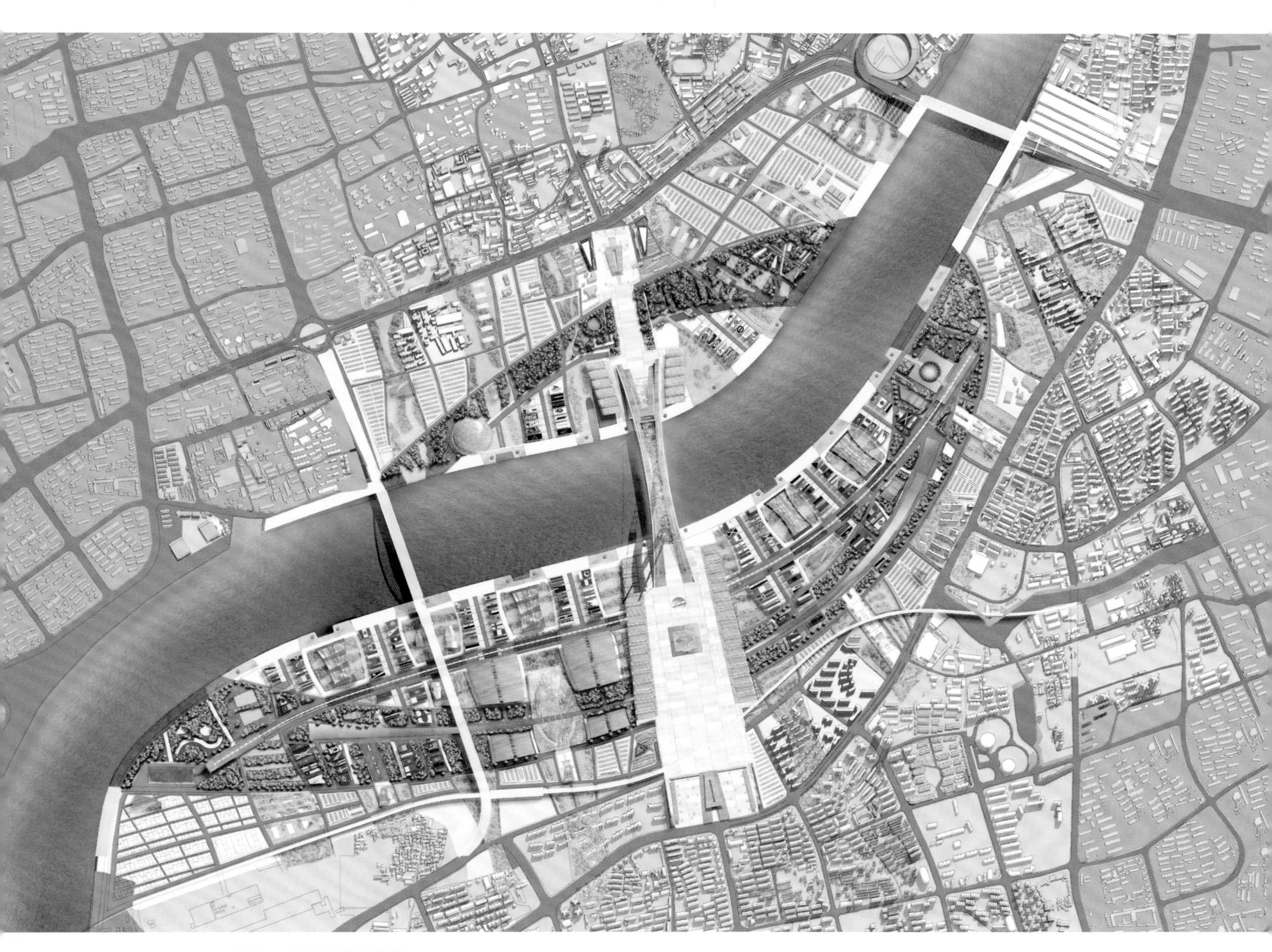

The layout for the Universal Exhibition of 2010 of a brownfield site, straddling the Huangpu River, is part of a sustainable development project for the city of Shanghai.

上海2010年世界博览会规划会址位于黄浦江畔的原工业废地，这是上海可持续发展的重要项目之一。

LMC –

Urban Development
King Abdul Aziz Avenue and King Abdallah Mosque masterplan, Haram extension
Mecca, Saudi Arabia
2004
2 940 000 m²

阿卜杜勒·阿齐兹国王大道、阿卜杜拉国王清真寺和麦加圣寺扩建规划
沙特阿拉伯，麦加
2004
2 940 000 m²

Abdul Aziz Avenue, the rectilinear four-kilometer long axis that runs through the city, combines a high-speed motorway for cars and buses with service roads and pedestrian-scale segments. This work of art was commissioned after a tight selection process through an international competition between the best qualified architects, as befits urban planning in the holiest city of Islam and construction on a most sacred plot of land. Stretching from west to east toward the Masjid al-Haram (the Sanctuary), the avenue embodies the Quran's commandment: *"From whencesoever Thou startest forth, turn Thy face in the direction of the Kaaba"*. This direction is prescribed for the whole of the Muslim community, pointing out the way of pilgrimage to the faithful. This major project would be the Saudi equivalent of the renovation of Paris's Champs-Élysées, New York's Fifth Avenue or Barcelona's Ramblas. Everyday traffic and the ceaseless influx of pilgrims impact the deployment of a space both public and religious. Architecture-Studio has designed the King Abdallah Mosque mid-way, where the former Umm Al-Qura Avenue joins the newly-built road on the one side and the Makkah Central Park on the other. The esplanade of the mosque, a big square with its four 156-meter sides, can welcome up to 20 000 faithful and

阿卜杜勒·阿齐兹大道是一条从市区穿过的中轴大道，长4千米，融合了高速公路、公共交通以及人行通道等元素。这是一个由国际评审团把关的国际竞赛，旨在选出最适合为伊斯兰教最神圣的城市和最神圣地块上进行城市规划和建筑设计的建筑师，最终，法国AS建筑工作室被选为最有设计资格的建筑事务所，并完成了这如艺术作品般的规划设计。规划中的大道贯穿东西，一直延伸到离麦加禁寺仅咫尺之遥的地方，灵感来源于古兰经的戒训：无论处于何处都须保持信仰核心，虔诚地面对属于神圣象征——克尔白天房。这个方向是整个穆斯林群体所遵循的规定朝向，指出了朝圣的方向。阿卜杜勒·阿齐兹大道之于沙特阿拉伯，就像香榭丽舍大道之于巴黎、第五大道之于纽约、布兰拉大道之于巴塞罗那。大道的设计必须考虑城市日常的交通，以及在重大宗教节日期间，朝圣人数激增时，保障城市的正常运转。规划位于阿卜杜勒·阿齐兹国王大道中央的阿卜杜拉国王清真寺，一侧与乌姆埃尔古拉大道相连，一侧与麦加中央公园相连。清真寺的大型广场边长156米，可接纳两万名朝圣者，属于城市干道的中心。

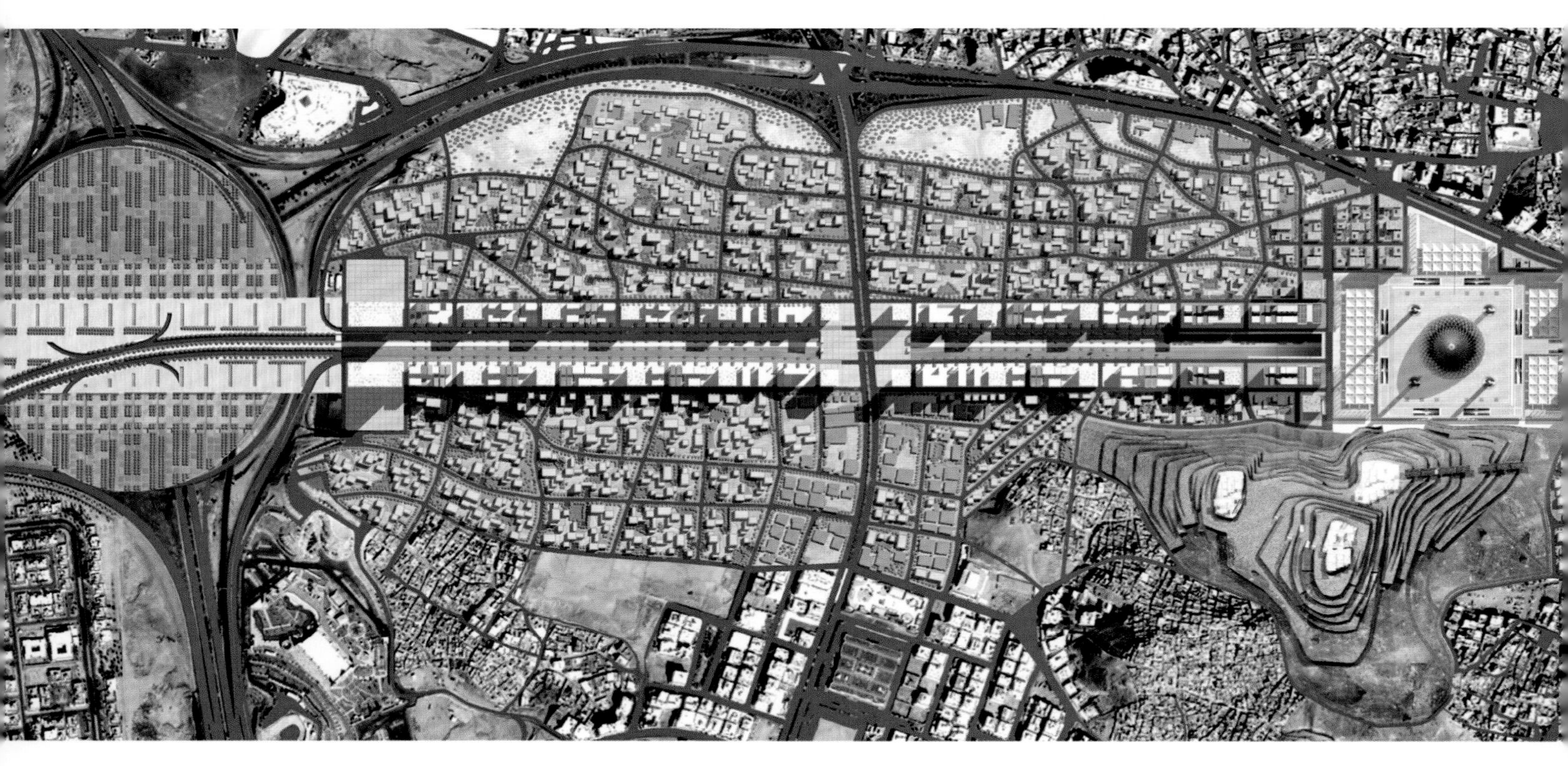

Within the exceptional site of the Mecca, the urban renewal project stretches from west to east towards the Haram, with the objective of structuring flux while creating quality public spaces.

在麦加的这片特殊区域，城市改造项目贯穿东西，一直向麦加圣寺延伸，目标是疏导交通的同时创建优质的公共环境。

is the hub of this urban artery. The round mosque at the center of the square has been conceived as a double shell, a sort of abstract tank, or shield protecting the congregation from high temperatures. It revisits the long history of religious cupolas and domes.
Mecca welcomes a million Muslims during the Hajj period. Due to circumstances, too many have been forced to pray outside the precincts of the al-Haram Mosque. The successive expansions of the city, due to demographic pressures, have intensified construction blocking out the sanctuary from view. To handle this problem, this extension project has devised a circular form inspired by the radiocentric development of the original site and the ritual circulation of the faithful. A first ring, based on the present mosque, puts the edifice back in the center of the city map. Other rings, composed of terraces raised above the urban environment, offer cool zones of shade, sheltering pedestrians from the scorching sun and giving concrete architectural shape to Islamic spirituality. Once the project is completed, the Masjid al-Haram will be able to take in four million pilgrims, thus quadrupling its present capacity. Alongside the religious extension, more constructions are on the way: a new street network, a hospital and a cultural venue.
H.L.

广场中心的圆形清真寺具有双层外壳，是阻挡阳光的结构，可在高温的情况下为集会提供遮阳避暑的保护。

麦加在穆斯林朝圣期间可迎来上百万的朝圣者。由于目前圣寺的空间有限，很多朝圣者只能被迫在麦加圣寺之外进行朝拜。随着麦加人口数量的增加，城市化进程也不断扩张，城市的高密度使得麦加圣寺的可见性不断减弱，它的扩建成为一个刻不容缓的问题。为了解决这个问题，扩建部分建筑呈同心辐射环状散开，第一环是在现有的清真寺基础上改建的，它能够清楚定义圣寺在市中心的位置。其他环则采取了统一的悬浮式的平屋顶结构，提供了一个巨大的遮阳屋顶，保护前来朝圣的朝圣者免受太阳的暴晒；这样的建筑概念也体现了宗教思想。若项目能够按照这样的规划建成，竣工后的麦加圣寺将可以接纳四百万的朝圣者，也就是原本接纳能力的四倍之多。除了宗教场所，还有许多其他建筑也在规划中，如街道网络、一家医院和一个文化场所。H.L.

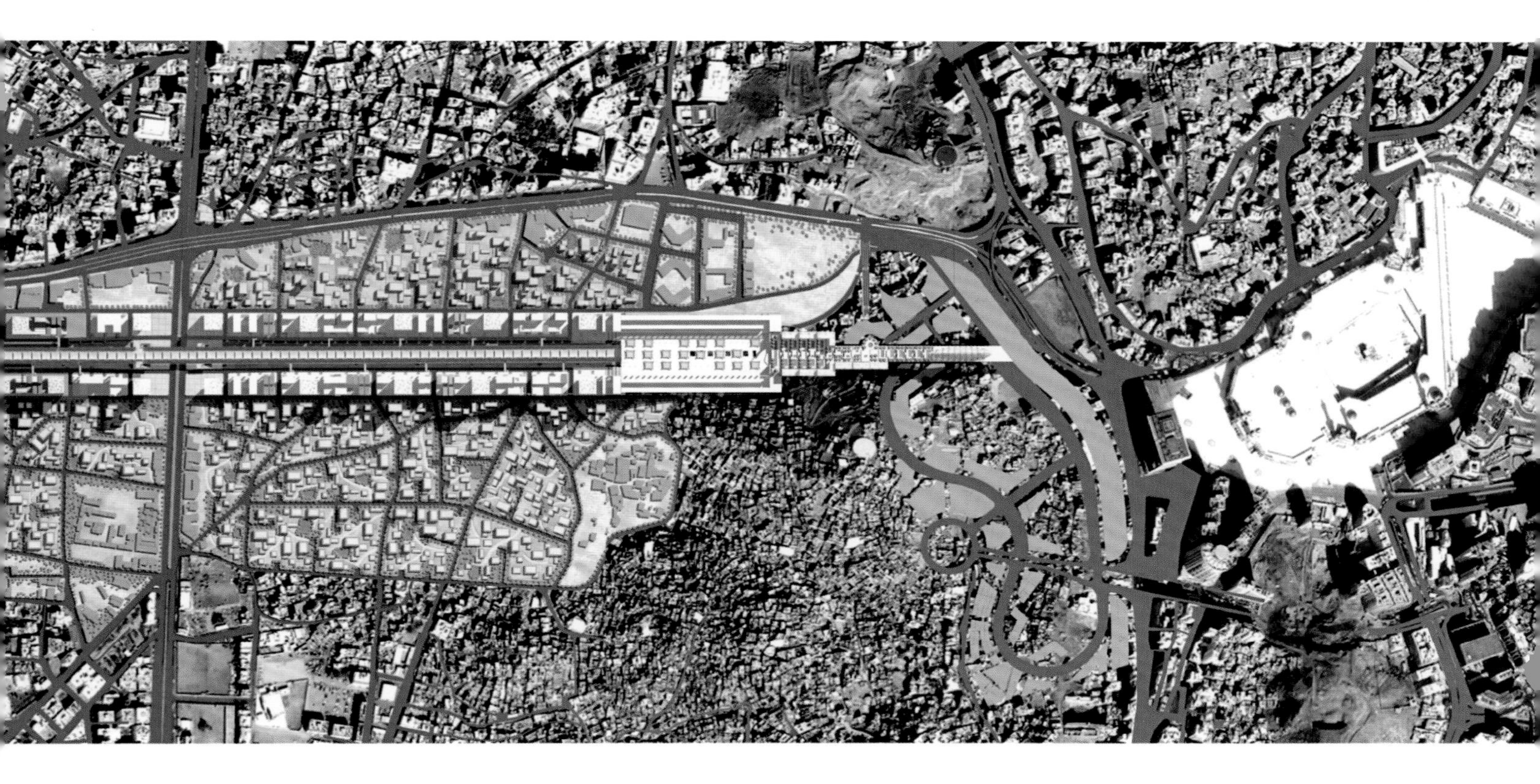

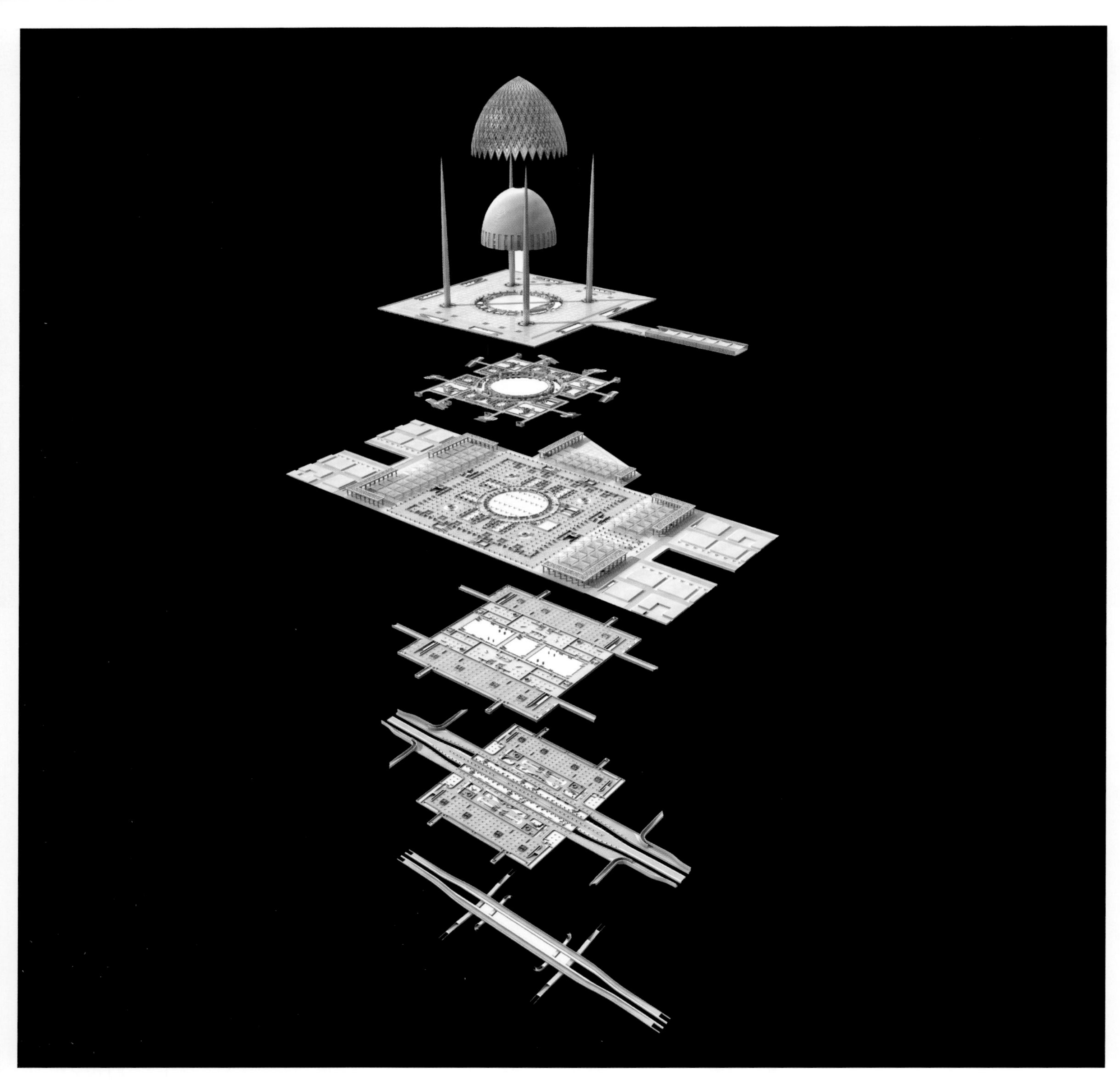

The Mosque's elements are set out on three architectural scales which structure the urban area. Four minarets overlook the esplanade, which serves as a link and as a landmark on the avenue. A convention center is built beneath the Mosque and is dedicated to the diffusion of religious science.

清真寺的四个尖塔立于广场之上，是大道的地标。一个专门服务于神学研究的会议中心建设于广场下方，致力于传播宗教。

The King Abdallah Mosque and its surrounding plazas form a devotional space ready to welcome 20,000 pilgrims.

阿卜杜拉国王清真寺及其广场提供了进行祷告的场所，可迎接两万名朝圣者。

Interior view of the 65 meter-high double-shell cupola.

65米高的双层圆顶内视图。

The major objectives of the Haram extension are to restore the Haram Mosque into a central place in the city, to be able to welcome a growing number of visitors and to improve the pilgrims' experience of the site.

麦加圣寺扩建的主要目标是重振其在麦加市区的中心地位，增加接待朝圣者人数，并提高朝圣者的体验。

Ultimately, the Haram Mosque will be able to welcome four million pilgrims, four times its current capacity.

扩建后，麦加圣寺将能够接纳400万名朝圣者，是当前容纳人数的四倍。

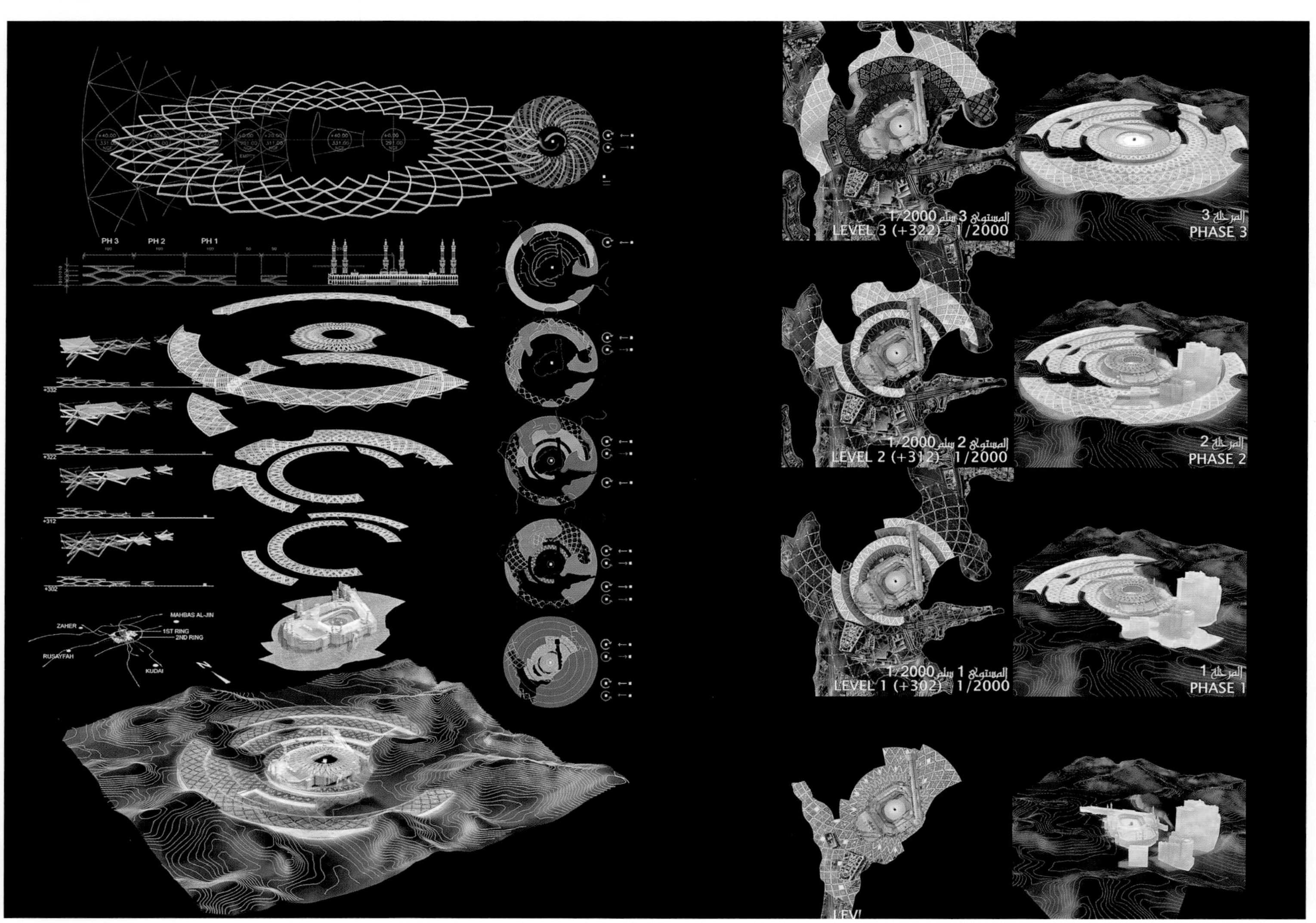

The circular form of the project structures the evolution of the site and the ritual journey of the pilgrims around the Kaaba.

扩建部分建筑呈同心辐射环状散开，象征着朝圣者环绕克尔白天房朝拜之旅程。

MSC2

Muscat Cultural Centre
Muscat, Sultanate of Oman
2008/2016
70 000 m²

马斯喀特文化中心
阿曼，马斯喀特
2008 / 2016
70 000 m²

If you had to name examples of giant pools of light, the Muscat Cultural Center with its undulating mashrabiya canopy comes to mind. But this is no architectural comet fallen from the sky: its aesthetic roots lie deep in the context of an opulent oasis. In the same manner as the indigenous palm trees that form a natural trellis capture light, this building, with its close-knit fishnet composed of small squares espousing the curve of the cover, retains incredible luminosity. Here the traditional mashrabiya is revisited by surrealism. In a cultural venue, the lighting system must be such that it entices the visitor to stay through an alluring interplay of light and shadow – filtering the bright rays of the sun, toying with the cool shade, modeling the dark or letting in a luminous flux setting space ablaze like some exuberant bonfire.
The inside pillars embrace the volume. A central plaza invites the public to linger in this landmark of the Sultanate of Oman, housing two national institutions. One is dedicated to the written word – the library and the archives; the other to theatre and drama. Complementing the three main buildings that integrate these services on different levels, a literary center, a cinema and a flexible exhibition space play their part in the cultural vibrancy of this venue. The light, and the presence of water, reconcile this ultra-contemporary district with the country's own architectural tradition. H.L.

若要说出一个巨大的捕光池例子，我们会想到马斯喀特文化中心起伏的雕刻花窗式顶棚。但这样的结构并不是从天上掉下来的建筑彗星：其美学根源来源于周边华丽的绿洲。与能够捕捉和透光的本土棕榈树一样，建筑的华盖由无数的方形透光细网构成，给室内带来难以置信的光感。传统的雕刻花窗被改造为超现实主义的设计。一个文化场所必须拥有舒适的光线和迷人的光影效果才能让人在此停留——镂空顶棚可以过滤太阳明亮的光线给室内带来阴凉，同时为室内空间带来光影游戏，营造出篝火在夜里闪烁的效果。建筑内部的柱子围绕起来形成空间，在此的中央广场将成为阿曼苏丹国人民的重要场所，因为这一广场将国家图书馆、国家档案馆以及国家剧院等重要机构连接起来。除了这三大主要建筑外，文化中心还设有文学中心、电影院和灵活的展览空间，以发挥促进文化传播的作用。自然光和水池相呼应，马斯喀特文化中心这一当代特色建筑是对当地传统建筑文化的致礼。H.L.

Located near the capital, the Cultural Centre emerges from a unique landscape, between the sea and the mountains.

马斯喀特文化中心毗邻首都，在山峦和海洋间，从景观中脱颖而出。

With reference to Omani architecture, a mushara-bieh canopy covers all the buildings. While filtering the sunlight, this canopy creates plays of light and shadow that echo the reflections and sparkles of the water features.

根据阿曼苏丹国的传统建筑特色，雕刻花窗式顶棚覆盖所有的建筑。镂空顶棚在遮挡阳光的同时，为室内带来光影效果，营造波光粼粼的水面。

SHPDG19

Wison Headquarters
Shanghai, China
2010/2013
139 830 m²

惠生集团张江总部园区
中国，上海
2010 / 2013
139 830 m²

Compact and sturdy-looking on the outside, vertical and multi-storied on the inside, the panoptic design of Wison headquarters in Shanghai is a composition of five different buildings, joined in the middle by a unique central body, symbolizing the company's rich diversity of activities. The façades look as if they have been lightly scratched by some lazy cat, leaving faded horizontal marks. In fact, this climatic sheath has been designed according to the region's weather conditions and takes into account fundamental parameters such as light, topography and energy consumption. The roof and the cover offer a big open-air space, stimulating social exchange. The southern part of the building leads to a piazza on Zhongke Street. The western part opens onto the riverbank and the surrounding vista. The ground level is transparent, offering a beautiful panorama and liberating space for fluid circulation. The façade of the entrance building – a radical polygon whose rhythm is shaken up by a crescent of light – encapsulates the spirit of the whole edifice and leads into ZhangJiang Engineering Park where it is located. This project was named the most beautiful office building in China in 2014. H.L.

上海惠生集团总部的建筑设计表现为外观坚实紧凑，内部垂直多层，由五座具有不同功能的建筑组成，通过一个独特的中心体相连，象征着公司丰富而又多样的业务。建筑的立面看起来像是被慵懒的猫轻触后留下的横向抓痕，事实上，这是根据当地气候所做出的精心设计，集中考虑了光、朝向和能量消耗等基本参数，屋顶和立面的特殊设计为使用者提供更开放的观感。建筑南面的广场连接中科路，而西侧，建筑向河岸及沿河景观带风景最大程度地敞开。建筑底层全部通透，为建筑留出出入口，也给使用者带来视线上通透的感觉。大胆张扬的建筑形体保证了舒适的采光，又体现一种对张江中区现有建筑和景观规划上的连续性和互补性。该项目于2014年被评为中国最美办公建筑。H.L.

The design of the new Wison corporate headquarters and its surrounding gardens is based on *Feng Shui* principles. The bold shape of the building reflects the innovative dynamism of the company and is an expression of the company, with different entities gathered around a central garden.

惠生集团总部建筑及周围花园的设计遵循了风水学原理。大胆张扬的建筑外形体现了公司的创新精神，也象征着不同的业务分支机构环绕着中心花园而建。

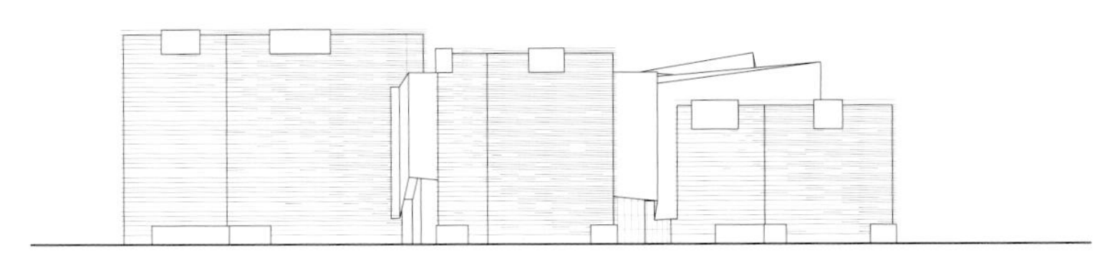

West elevation.
西立面图。

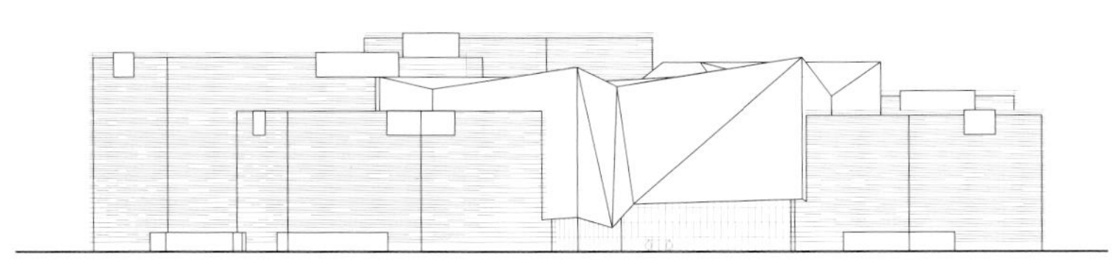

South elevation.
南立面图。

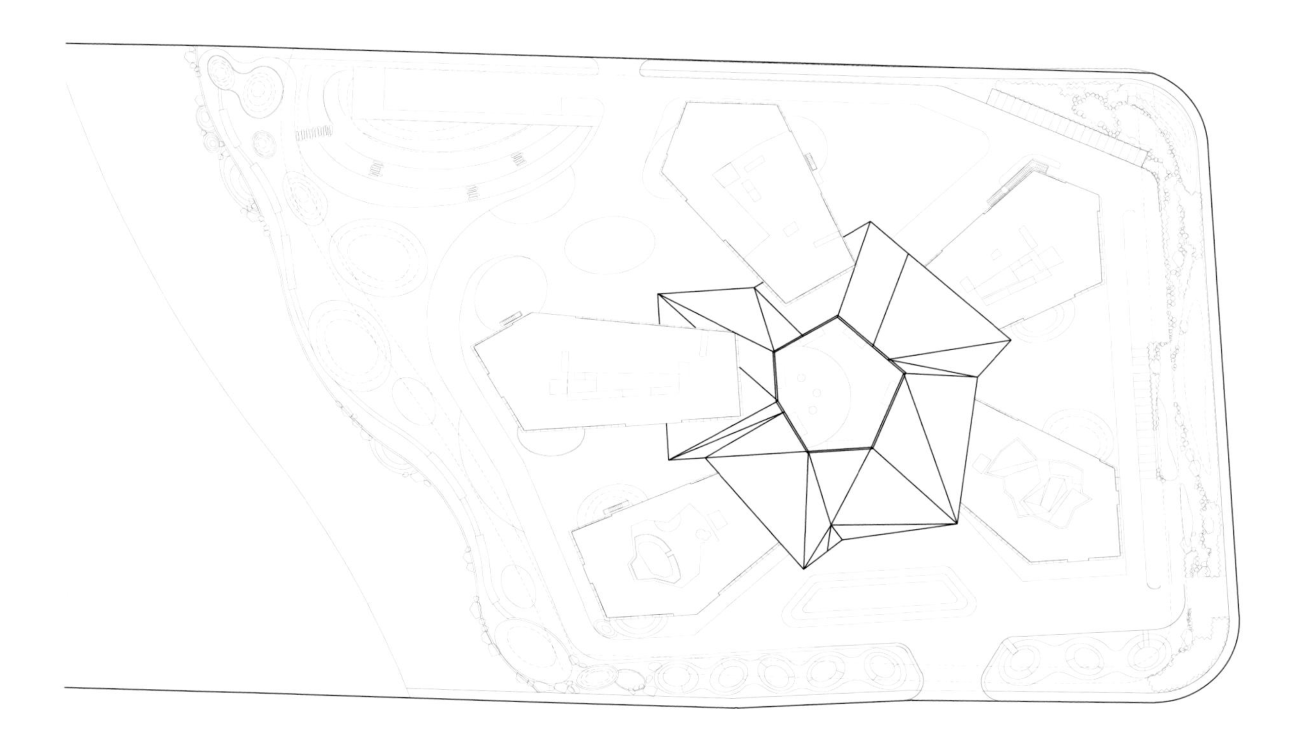

Ground plan.

总平面图。

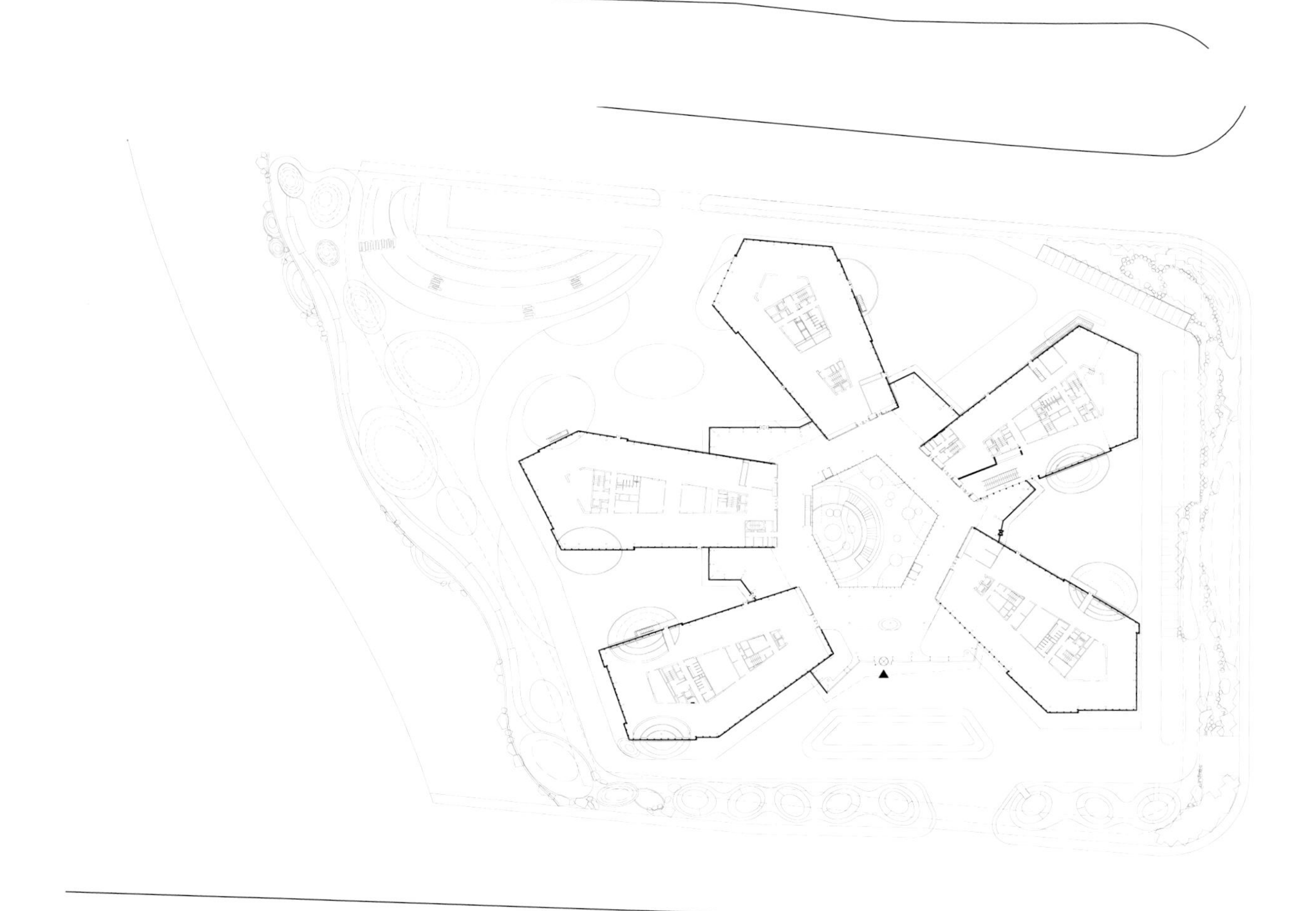

Ground floor plan.

底层平面图。

Exterior and interior views of the atrium.

中庭的外视图与内视图。

LHS1

Tibet Natural Science Museum
Lhasa, China
2009/2016
32 000 m²

西藏自治区自然科学博物馆
中国，拉萨
2009/2016
32 000 m²

From the outset, this architectural project was always tied in with that of the museum: its spaces are flexible and it is equipped with the latest technology and viewing devices. The museum's large central hall functions as a buffer zone, allowing energy saving. This sustainable architecture is a masterpiece of lyricism expressing the beauty of the surrounding nature. The Natural Science Museum of Tibet, which was built at an altitude of 3 650 meters, is a landmark on the roof of the world, integrating the rugged skyline of the mountain crests. The metal mesh enveloping the building, like some good-luck Buddhist scarf, is in dialog with the snowy summits. The different layers of the façade's wrought motifs and the roof filter the sun and modulate the light in order to heat the rooms. This external skin gives lightness to an otherwise imposing structure. Rebelling against the Modernist all-white trend, the moldings are redolent of the nature-inspired design of Tibetan mandalas. The building is furnished with photovoltaic panels and a water-recycling system. With a canal running nearby, the Museum is also close to water, an element that is synonymous with life in Tibetan symbolism. This architectural feat gives global visibility to a venue located in one of the most isolated places on the planet. H.L.

从最初的设计开始，这个建筑项目就与博物馆联系在一起：它的空间灵活，并配备了最新的技术和陈展设备。博物馆的中央通高大厅，沐浴在阳光下，是内外的缓冲空间，并实现节能。西藏自治区自然科学博物馆位于海拔3650米的拉萨市，群山环绕，坐拥世界之极，从周围的山脊汲取灵感，使这座可持续性建筑与自然环境完美交融。建筑的外部形态被设计成两臂伸展的形状，做出热情迎宾的姿态，仿如一条美丽的白色哈达，与雪山景色相呼应。建筑顶部和立面通过图案及材料厚度的变化来控制室内采光和温度，镂空立面让原本庞大的建筑显得无比轻盈。与现代主义的全白建筑不同，建筑立面纹案的基本结构源于西藏文化与藏传佛教的象征元素——曼荼罗和吉祥结的结合。项目配置的太阳能电池板和水循环处理系统，充分体现了其节能特征。依傍拉萨河，靠近代表生生不息的流水，这座建于地球上最“与世隔绝”的高原的建筑向世界展示了它的美。H.L.

West façade of the Museum, facing the water feature.

博物馆西立面，与其水面倒影。

Main entrance of the Museum.

博物馆的正门。

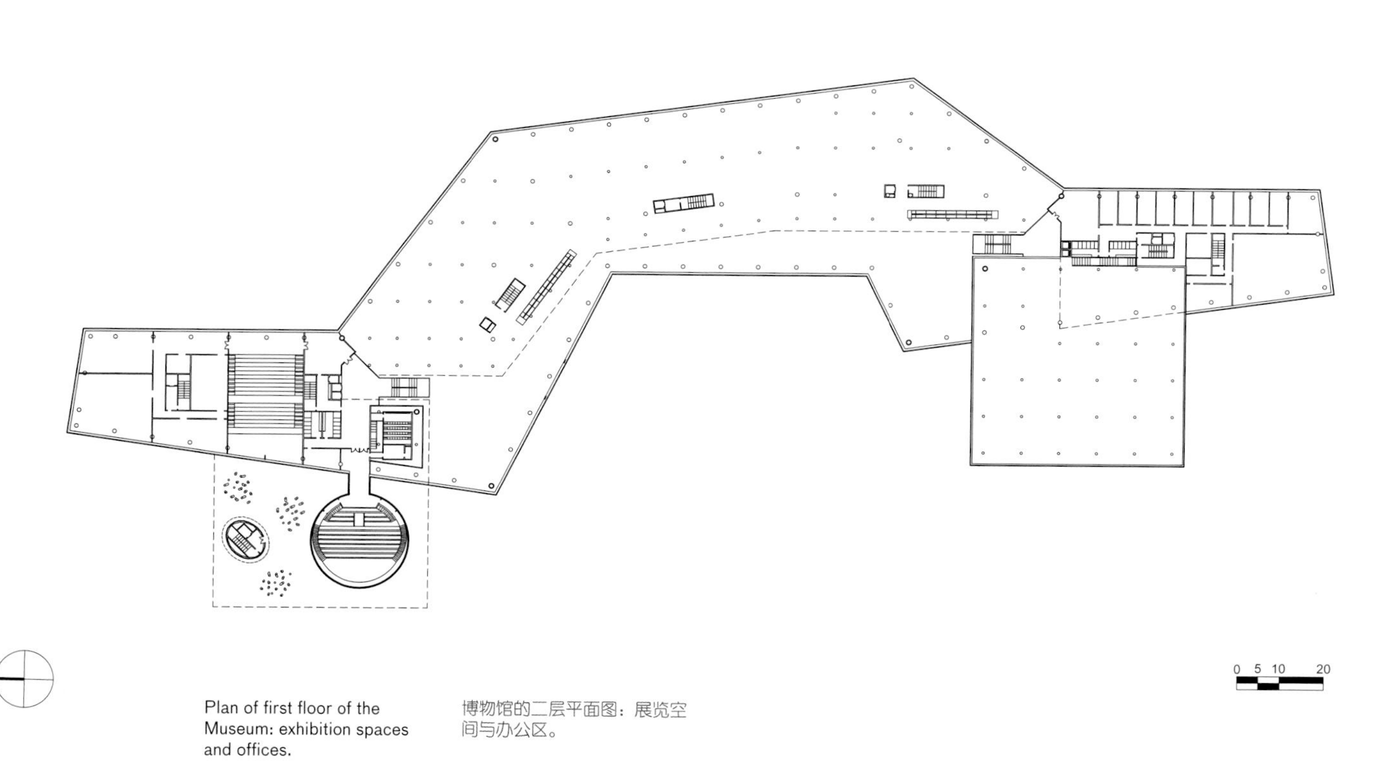

Plan of first floor of the Museum: exhibition spaces and offices.

博物馆的二层平面图：展览空间与办公区。

View of the main lobby.

中央大厅视图。

ATN2

Onassis Cultural Centre
Athens, Greece
2002/2011
20 000 m²

奥纳西思文化中心
希腊，雅典
2002/2011
20 000 m²

Maria Callas would surely have enjoyed performing in the 900-seat theater, giving a press conference in the 200-seat auditorium/cinema, trying out the open-air Greek theater with the same sitting capacity, reading at the library or eating at the restaurant.... This multifunctional building is, above all, beautiful. Its simple volume and abstract design confer majesty on an otherwise tiny plot of land which seemed, at first sight, very difficult to build upon. Raised above ground and erected on a glass base, this aerial body is dressed in Thassos marble and offers façades that can look alternately opaque or transparent, depending on the distance from which they are viewed – the edifice's appearance very much depends on the visitor's standpoint and movements. The venue's setting could have been designed by some kinetic artist, such as Soto, and is based on the principle that a façade is an organic membrane. Adaptive to change, it symbolizes the foundation's activity and its openness to the city and the world at large. But the very core of the project is to be found inside this multi-layered building that looks both rock-solid and delicate. This precious round structure embraces three auditoriums which occupy the interior of the edifice. The empty space around it, not unlike the silence after a diva's rendering of a masterpiece, magnifies the work of art. A backdrop to acting on stage and the play that unfolds, it emphasizes the variety of the foundation's activities at the levels of both the private individual and the wider city. H.L.

玛丽亚·卡拉丝肯定也希望有机会在这个能容纳900位观众的剧院中表演，在能容纳200位观众的礼堂/电影院举行新闻发布会，在拥有相同容量的希腊式露天剧院表演，在这里的图书馆阅读又或在这里的餐厅用餐……最重要的是，这个多功能建筑十分美观。由于地块大小的限制，奥纳西思文化中心所实现的简约大方的外形和抽象的建筑表达如同不可能完成的任务。这座半透明状的建筑建于玻璃底座之上，外立面由达索大理石构成，根据观赏的远近呈现出透明与半透明的透视效果，整个建筑充满动感。立面尤如有机膜层，建筑所体现的城市特色与Soto这样的电影艺术家的成就是离不开的。建筑对外界变化的适应性，象征着奥纳西思基金会对整个城市乃至整个世界的开放。但是这个项目的核心体现在让建筑看起来既坚固又细腻的多层次建筑物内部，贯穿整个建筑的巨大圆形主体容纳了三个音乐厅空间，而四周空间衬托出主体的华丽壮观，像是一场大师级表演后的沉默，更能表现出作品的艺术性。舞台演出和戏剧演绎的背景幕设计，强调了基金会对于个人和公众等不同层面人群所开展的多样性的活动。H.L.

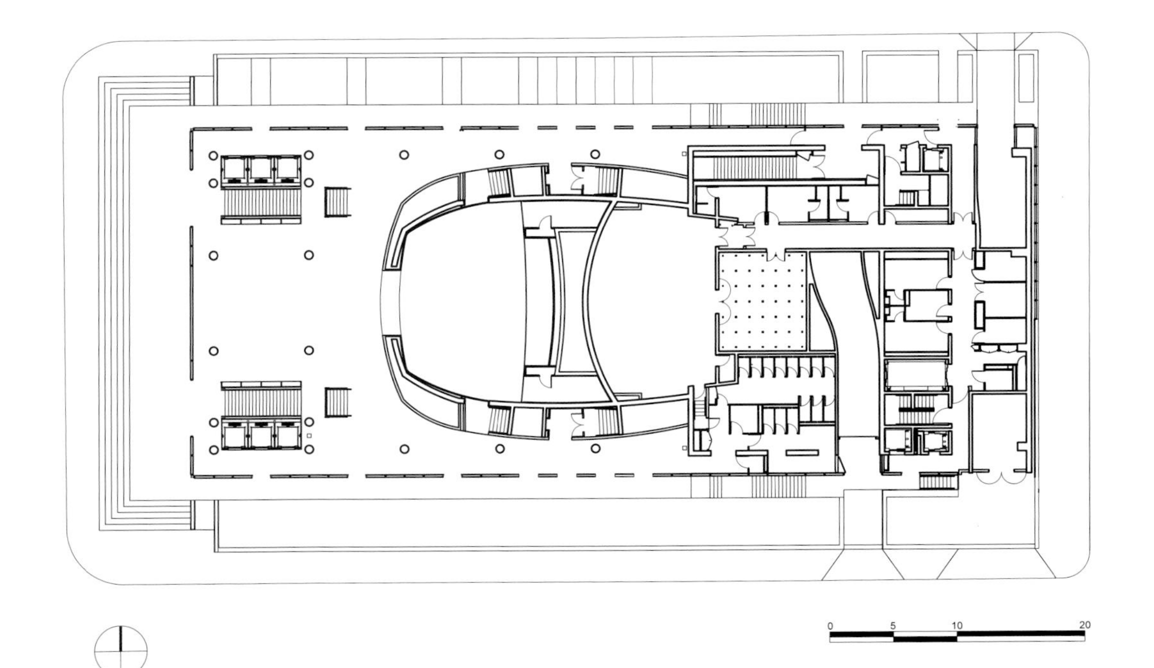

Ground floor plan.

一层平面图。

Central atrium.

中庭。

Aerial view of Athens: in the foreground the Onassis Cultural Center, in the background the Philopappos, Acropol and Lycabette hills.

雅典鸟瞰视野：近处的奥纳西斯文化中心，与远处的雅典卫城与吕卡维多斯山。

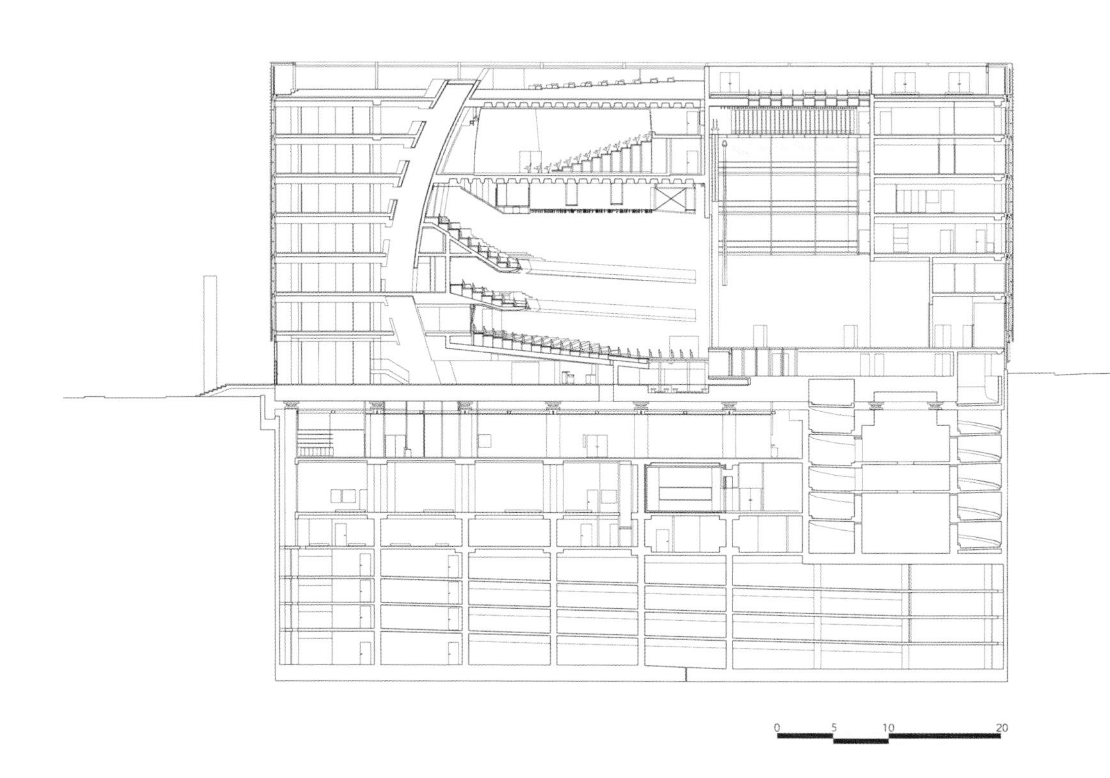

Longitudinal section of the building.

建筑剖面图。

AGS1

Le Quai Theatre
Angers, France
2003 / 2007
16 500 m²

河岸大剧院
法国，昂热
2003 / 2007
16 500 m²

The Le Quai Theater in Angers belongs to the category of amazing tools in the service of the common good. Its building is tailor-made for actors, dancers, directors and their audience, in the same way the digitalized French National Library (BNF) is custom-made for any reader or researcher eager to explore the world of print. However complex the make-up of a tool, its use must be simple. The deployment of certain elements of design – a singular setting, a coherent decoration – all emphasize the specific utility of each activity which would otherwise be considered as one entity under the regulations relative to the trade. A theater often looks unappealing when viewed from the back. Here, there is neither front nor back: space is adaptable. The theater of 971 seats can turn into a smaller theater of 400 seats. It is flexible in size and may be used for different workshops. The props that are usually stored away are visible here.
A drama school, a dance academy, a hall for theatrical art and a restaurant promotes closer ties with the city. The hall is big enough to have a circus installed in it and functions as the town's antechamber to cultural events and social exchange, visible from the exterior esplanade through the monumental doors leading into the enormous atrium. This dramatic theater-within-the-theater effect is served by a succession of galleries leading to the stage through a thick concrete veil pierced with apertures. Bright colors adorn this permanent theater set, sprightly as a scene in a Goldoni play, as happy as a sunny summer's day, even when, come winter, the neighboring Maine River has turned grey. H.L.

昂热河岸大剧院拥有强大的公共服务能力，它是专门为舞台剧演员、舞蹈家、戏剧导演和观众量身建造的建筑，就像数字化的法国国家图书馆是为所有渴望探索文字世界的读者或研究人员而设计的一样。但是无论设计如何复杂，其使用必须简单。建筑每个部分都遵循统一的设计概念，装饰相互呼应，并且根据相关行业规定所设计。通常剧院的幕后都是毫无吸引力的装饰，然而，这座剧院不分台前或幕后，就像一座艺术工坊，那些在别的剧院储藏在道具间的舞台道具在这里却是可视的。剧院有一个971座的大剧院厅，和一座可容纳400个座位的可变形剧院厅，强大的灵活性，让剧院能够满足多样的用途。集舞蹈学院、戏剧学院、表演艺术大堂和餐厅于一身的剧院建筑与城市保持紧密联系。大堂的大小足以容纳马戏团表演，也可以作为城市文化活动和社交交流活动的举办空间。透过透明的玻璃立面从外部广场即可看到大堂，剧院内部的剧院厅由布满大小不一窗口的混凝土“大幕布”所覆盖。明亮的色彩永恒地装饰着剧院，像戈尔多尼剧中的一个场景，与阳光明媚的夏天一样快乐，即便到了冬天，附近的曼恩河也变灰了，剧院的色彩依然明丽。H.L.

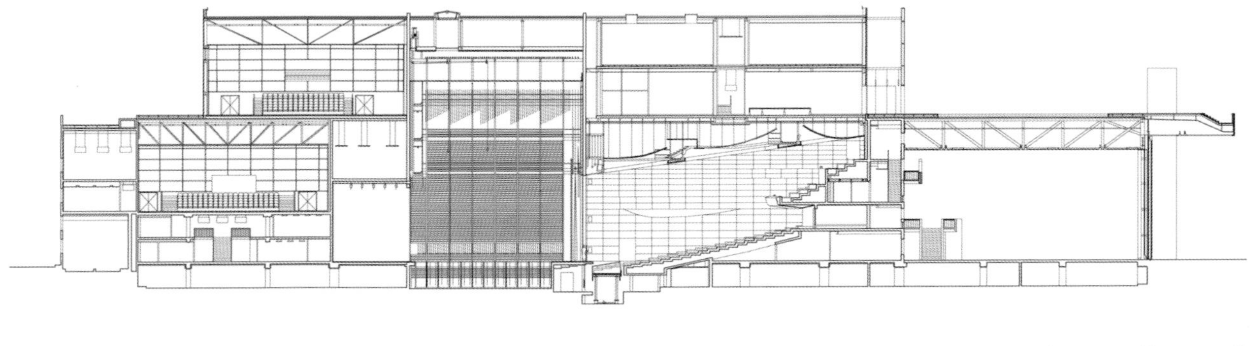

Longitudinal section.

纵剖面。

Façade on the Maine River.

面向曼恩河的建筑立面。

The forum, which hosts shows, workshops and exhibitions.

可举办演出、工作坊和展览的大堂。

Interior view of the theatre.

剧院厅的内视图。

Interior façade of the forum (detail).

大堂内部立面图（细节图）。

05

18-20432
MONITOR
#66
ZOOM > LIVING
FEVRIER 2006
CANSON
par
CASA CASUARINAS
ENTREVISTA
PATIO VISCONTI
ESPECIAL VERANO
ADVANCIA / NEGOCIA
UNE LANTERNE ROUGE
A MONTPARNASSE
CO
N-S TRU
C-TION
A- H
EAD
LE POINT DE LA SEMAINE
Pierres précieuses à Saint-Tropez

Créteil
Une cathédrale flambant nef
Culte Le premier édifice du genre érigé en France au XXIe siècle sera inauguré en septembre, dans le Val-de-Marne. Certains estiment que la mosquée ouverte en 2008 dans la ville a précipité le chantier
CRÉTEIL CATHÉDRALE+
UNE CATHÉDRALE DÉPLOYÉE
HOUT BLAD
Télérama.fr
A Créteil, la cathédrale du XXIe siècle
CATHÉDRALE DE CRÉTEIL
La cathédrale dans sa réalité urbaine
Baptême à Créteil
Pierres Vivantes
De l'église enfouie à la cathédrale déployée
Neubau der Kathedrale von Créteil bei Paris
Künstlerfenster von Udo Zembok
2016
archiSTORM
DOSSIER SPÉCIAL
ENSEMBLE, DÉPLOYONS LA CATHÉDRALE DANS LA CITÉ
LA CATHÉDRALE DÉPLOYÉE
5façades
LE SIGNE VISIBLE
de notre Église en Val-de-Marne
cathédrale
du XXIe siècle

05 BRINGING A PLACE TO LIFE

05 REINVENTER LE LIEU

建筑为周边带去生机

An obvious characteristic of the work of Architecture-Studio is its relation to place, to context, but also to the very life of a place. Since their early work in Paris and in other French cities, the partners have often referred back to their "Stimuli" concept, embodied in a single project, the Rue du Château-des-Rentiers apartment building (1986). Winners of a 1984 competition, the architects made reference to the so-called Grands Travaux that President François Mitterrand launched beginning in 1982 with such projects as the Louvre Pyramid (Grand Louvre, I.M. Pei) or the *Arche de la Défense* (Johann Otto von Spreckelsen). These costly urban landmarks led Architecture-Studio to study the numerous small, empty lots in Paris and to propose weaving together a series of small structures on those sites that collectively might indeed have had more impact than the iconic Mitterrand structures. The Château des Rentiers, a small, low-income housing project, showed their intentions by fitting closely into its difficult site, with one exterior wall entirely given over to a very large map of the area of Paris where the apartment building is situated. The point made with its partially exposed steel skeleton and the map was that the building was very much a part of the neighborhood, and not just a closed, modern building. Rather than simply filling an empty space with a building, they chose to reconnect the formerly empty site with the changing city.

As is often the case, the partners refer to historic examples to explain their thinking. In their interpretations of the Stimuli, an analogy can be drawn to the 62 toll barriers of the Paris Wall of the Farmers-General (1784–1791) designed by Claude-Nicolas Ledoux (1736–1806). The few remaining monuments by Ledoux from this period include the *Rotonde de la Villette* at the Place de Stalingrad, the *Barrière de la rotonde de Chartres* at the Parc Monceau (modified in 1861), the *Barrière du Trône* near the Place de la Nation and the *Barrière d'Enfer* at the Place Denfert-Rochereau. The neo-classical style of these monuments was clearly meant to unify the city through the erection of a series of relatively small projects. In keeping with the complexity and density of modern Paris, the architects point to the three new railway stations[1] that they are designing for the Greater Paris project – at Nanterre La Folie, Saint Cloud and Rueil-Suresnes "Mont-Valérien" – as a contemporary interpretation of the earlier Stimuli. These railway stations do not seek to match the encircling grandeur of Ledoux's toll barriers, but they signal, by the quality of their intent and their association with

法国AS建筑工作室设计作品的显著特点是它与周边环境、背景文化以及人们的生活之间是密切相关的。从他们在巴黎及其他法国城市中开展的早期创作开始，合伙人们就提出"Stimuli（激励）"概念，并将该理念应用到了Château-des-Rentiers街公寓楼建筑中（1986年）。从1984年开始，工作室的建筑师们运用"Stimuli"概念参与建筑竞赛，此概念是由1982年弗朗索瓦·密特朗总统提出的"大工程"项目而引发的反思。比如罗浮宫金字塔及拉德芳斯区新凯旋门等这些耗资巨大的地标性建筑均是那时开始启动的，这触动了法国AS建筑工作室开始统计和整理巴黎及其郊区的大量闲置小面积地块。他们希望通过改造和利用这些位于居民区的空白地块，让众多的小项目组成一个网络架构以带来巨大的社会效益，以此与被密特朗列为优先项目的地标性建筑的影响相媲美。Château-des-Rentiers街公寓是一个规模不大的廉租房项目，却显示了法国AS建筑工作室对建筑设计的执着。建筑巧妙地建设在一块狭小的地块上，其中一面外墙上刻画着一张巨大的地图。部分外露的金属建筑架构，以及刻画在外立面上的周边街区的地图，都表达出它是整个街区的一部分，而非一座封闭的现代建筑。建筑师们并不只是用建筑填补了空地，而是将原本的废弃地块与正在改变的城市重新联系起来。

合伙人经常举用历史上的例子以表述他们的想法。此次，他们将"Stimuli（激励）"概念与由克劳德·尼古拉斯·勒杜（1736年—1806年）设计并于1784年至1791年间建造的巴黎税收城墙的62个征税关卡做类比。勒杜的遗迹已现存无几，包括位于斯大林格勒广

The Abdali Gate complex is located in the heart of Amman, the capital of Jordan. The two towers are designed in such a way that they are immediately visible in the cityscape. Their silhouettes bend towards each other and form a gate to the new Abdali district on one of Amman's main axes.

Abdali Gate多功能建筑位于约旦首都安曼的中心，两座大楼在城市景观中十分显眼。互相向彼此轻微倾斜的两座大楼形成了一个进入Abdali区的大门，矗立于安曼的主轴上。

the enlargement and modernization of the city, that architects still very much have a place to play in the modern city. The stations can be considered as many types of "stimuli", as there are locations in the 25 new facilities planned in an operation run by Grand Paris Aménagement. The relationship between these stations and their contexts bring together not only transport but new uses for urban areas and, in fact, a new architectural morphology. Architecture-Studio points to its own project for the connected station at Créteil (which they did not win) as an example of how these stations can, and will, stimulate the developing city.

The partners have gone on from the example of the Château des Rentiers, not as much toward the concept of weaving together many small projects, but rather to insist on the rootedness of their work and its relationship, usually, to an urban environment. They have also developed original ideas of context that can be applied to "macro" (very large) urban structures, or to green field industrial or technological projects. A building, be it very large or quite small, thus participates in bringing its site to life. Such buildings integrate themselves into the city, or into a commercial and industrial complex that is taking form, as was the case with their 550,000 square meter Renault Technocentre proposal (Guyancourt, 1990), or the smaller Danone Vitapole (Palaiseau, 2002). Other "macro" projects include the European Parliament Building and the newer Jinan Cultural Center. The Technocentre refutes the usual industrial logic based on the idea that "form follows function" instead using a square plan with an orthogonal grid with an awning that is partially hollowed out for functional reasons but which has a façade designed to relate the main avenue. The result is a flexible, urban system. For the Danone Vitapole, the interface between the building and the natural setting is developed by buildings with planted facades and organically shaped gardens.

Roueïda Ayache explains how the logic of the Stimuli is applied to one of their current projects in Amman, Jordan – the Rotana Tower. "A tower can be seen as a form of urban stimulus," she states. "Our old idea about stimuli finds itself updated in this way. The concept at the time was to use empty sites to create new focal points, in a kind of acupuncture method. What we are referring to is points in the city that leave an impression and create a positive urban effect. A city such as Amman requires a stimulus on a larger scale than Paris for example. Amman is not like Paris with its rather clearly defined axes; it is in a way amorphous and

场的维莱特圆形大厅、位于蒙索公园（1861年曾改建）的沙尔特圆形大厅、位于民族广场附近的“御座”关卡以及丹佛 · 罗什洛广场的“地狱”关卡。这些新古典风格的征税关卡遗迹很明显用一系列相关的小型工程将整个城市统一并联结起来。面对当下巴黎城市的复杂性和高密度，工作室为大巴黎项目设计的三个新城际快线车站[1]：南泰尔—拉富丽车站、圣克鲁车站和勒伊—苏雷讷—瓦雷里昂山车站，是“Stimuli（激励）”精神在当今的展现。主导大巴黎项目的大巴黎公司旨在为大巴黎地区建设现代化和扩张性的交通网络，他们邀请建筑师进行车站规划设计说明了建筑师在巴黎都市圈发展中扮演着重要角色。

根据大巴黎项目的总体规划，将有68个新的城际快线车站被建设起来。法国AS建筑工作室与大巴黎公司一道对其中的25个车站进行了可行性研究。法国AS建筑工作室认为像车站这种大型公共设施不应只局限于承担交通运输的本职功能，而应成为周边城区发展的催化剂，事实上，这是一种新型的城市规划与建筑形态。合伙人用克雷泰伊雷沙（竞赛项目）作为典型案例，论证这些车站与车站周边的连接项目可以促进城市发展。

从“Stimuli（激励）”的经验说起，合伙人除了强调可以利用小工程将城市有机地联结起来的理念，又更多地谈论了建筑与环境的关系。他们曾落实过结合环境的建筑解决方法，比如在郊区工业园或技术园区内建设具有城市规划性的大型宏观建筑。不论一个建筑的规模大小，不管是建于城市中又或是仍在规划的地块上，关键在于能够促进建设地点的发展。例如位于基扬古尔占地550 000平方米的雷诺技术中心（1990年调研项目），以及较小型的帕莱索达能研究发展和质量中心（2002年）。雷诺技术中心的设计摒弃现代功能主义提倡的“功能决定形式”的常规理念，建筑师以正交概念设计广场，并预留了部分功能空间，凸起的玻璃立面跟随主干道的走向，这是一个复杂又灵活的城市规划系统。达能研究发展和质量中心在建筑和自然景观间采用了植物墙壁和有机形态的花园。其他“宏观”项目还有如欧洲议会中心，以及近期落成的山东省会文化艺术中心。

The two teaching units of the Jules Verne High School, in Cergy, are formed around a street and a central square, where a sculpture by Piotr Kowalski stands.

位于塞尔吉的儒勒 · 凡尔纳高中，两座教学楼围绕着一条街道和立有Piotr Kowalski制作的雕像的中心广场建筑。

The roof of this business center is treated as a cover, marking the periphery of a virtual dome. The meeting of two geometries, characterized by the curve of the cover and the vertical façades, generates a third geometry of unique logic that weaves relationships with the streets, axes and directions.

该商业中心的屋顶由一个如圆顶般的连贯顶盖构成。带有弧度的顶盖和垂直的立面相交，形成独特的几何结构，与街道、城市轴线和周边各个方向相互呼应。

our work there is a matter of creating context and defining an urban presence. This is the creation of a context beginning with an urban object. The tower here becomes a way of defining the city. There is a thought process about the city, about the pedestrian scale and about points that allow people to orient themselves, for example. Ideally, a network of points can be sewn together, changing the use of the city; in basic terms we can refer to icons, or landmarks. It is more a point of orientation." Again, in a form of reasoning that can also be related back to the work done by the firm at the CA'ASI, the point is not only to impose a piece of architecture on a city, but to participate actively in improving day-to-day life. "The idea is that even if a project is large, like the Rotana Tower in Amman," she continues, "it stimulates the city even more than its scale might imply. Amman has no points of orientation so this project in a sense addresses itself to the entire city. The idea in both cases (the small and the large) is to act in a stimulating way on the larger aspects of the city. The stimulus flows from the architectural object to the scale of the city or a pre-determined territory."

On a much smaller scale than the Rotana Tower, the architects worked intensively on the idea of a cubic church, which was the origin of their parish church, Our Lady of the Ark of the Covenant (1998). As Martin Robain explains, "this project was the basis for research on the future of churches in the range of 450 seats, it became a programmatic study of the needs of churches, in the context of post-Vatican II liturgy. We won the competition for the church in Rome in 1995 that Richard Meier finally built (Jubilee Church, Tor Tre Teste, Rome, 2003) and our proposal was cubic. The idea was to create an archetype or a focal point." The form of the Paris church, a cube symbolically supported by 12 pillars, is also the result of the constraints of the site. The church takes into account a public garden and generates the minimum possible impact on the site.

The relation between tradition and modernity, but also between this new church and its neighborhood, was very much in the mind of the architects when they began their work in Paris. "There is, of course, a relation between our ideas and those of the church of the past which often sought to be a focal point of towns and cities," explains Robain. "We succeeded in an area influenced by Haussmann, but also by residential towers, in creating a kind of keystone for the neighborhood. It brings diverse elements together." More specifically, Martin Robain explains the very traditional idea of the narthex, the entrance area usually located at

为说明工作室如何在其他当下的项目中应用“Stimuli”理念，罗伊达·阿亚斯举例约旦安曼的罗塔纳酒店。她说：“这座大楼被视作激励与促进城市发展的标杆，我们的‘Stimuli’传统理念通过这种形式进行了更新。当年，我们利用城市闲置地块建设出促进周边发展的建筑，就像在城市的关键脉络进行针灸疗法。

这些建于关键位置的建筑旨在给大众留下一个积极印象，并对城市产生积极影响。安曼这座城市与巴黎有着很大的区别，它不像巴黎一样有清晰的轴线，因此它需要一个更大体量的建筑作为标杆以促进城市发展。某种程度上来说，我们的作品是从一个标志性建筑出发，进而为城市打造市区环境。在安曼的案例中，罗塔纳酒店大楼成为改造城市的开端，它象征着整座城市对于规划反思的起点，又作为一个显著的地标，为城市里的每一个人指明方向。最理想的情况是能够创造一系列的地标性建筑以形成一个网络，随之提高城市功能。目前，罗塔纳酒店是安曼最高大楼，像城市灯塔一般具有明确的导向功能。“这种工作理念也是工作室运营CA'ASI艺术展览馆时所坚持的，建筑不应该强加于城市之上，而应该积极参与到提升城市日常生活质量中来。”罗伊达继续说道：“我们希望安曼罗塔纳酒店大楼这个规模巨大的项目，能够更加有效地激励与促进城市发展。安曼这座城市原本没有可以指引方向的建筑，因此本项目成为了全市的导向。不管从行人的尺度，还是从这个城市的尺度来说，罗塔纳酒店大楼都表现了单一建筑物对城市整体发展的促进作用。”

并不是所有的建筑项目都有罗塔纳酒店大楼如此宏伟的规模，工作室的建筑师们就曾经构思一个立方体的小教堂项目——巴黎约柜圣母教堂（1998年）。马丁·罗班说：“该项目不仅旨在建设能够容纳450名教徒的新型小教堂，也是根据第二次梵蒂冈大公会议后的礼拜仪式，对当前和未来的教堂建筑进行方案性研究。我们的立方体状小教堂的设计概念还曾于1995年中标过一个罗马的教堂设计竞赛，但最终是建筑师查理·麦尔完成了建筑（仁慈天父教堂，罗马，2003年）。”由于受到周边环境的限制，巴黎约柜圣母教堂的立方体结构由12根立柱支撑，整座教堂像是悬浮在公园中。

An architect's model of the Renault Technocentre. Between architecture and urbanism, this is macroarchitecture on an urban scale.

雷诺技术中心的建筑模型，介于建筑和城市规划之间，这是一座城市规模的宏观建筑。

the west end of the nave, opposite the main altar. "In the Middle Ages," he states, "non-believers could come into the narthex, a kind of intermediary space between interior and exterior. We transposed the idea of the narthex into a space between the city and the church. We put an external grid around the church that one can walk through, that is to say, an in-between space. It is around the church but it is also the entry into the city if you will. It is associated with the public garden and thus becomes a kind of shared space." The idea of "stimuli" and context is thus very obviously defined in a broad manner by Architecture-Studio. It has to do not only with neighboring buildings, streets and pedestrian passages, but also with the very roots of the church in this instance, and even beyond, as Martin Robain states: "We sought voluntarily to give new forms to ancient symbols of the church. You can find all the elements of a traditional church but transposed into a contemporary form and idiom. There is also a correspondence here between the Old and the New Testament, the Tables of the Law, the 12 Tribes of Israel, the 12 Apostles of Christ and of course the Virgin Mary. Stained glass windows by the noted artist Martial Raysse are also part of the project. This was quite new for a Catholic Church, but Cardinal Lustiger (Archbishop of Paris, 1981–2005) backed the project at the time."

Architecture-Studio systematically interprets issues of context as having to do not only with what exists, but also with the past, and surely the future. "The Ark of the Covenant Church," says Roueïda Ayache, "re-establishes a situation of hierarchy in the city, something that appears very early in the thinking of Architecture-Studio. Churches were disappearing when the Ark of the Covenant Church was designed. This was the first new church to be built in Paris since 1967. The project is a statement that a church is an element of the city that contributes to its structure. The design was a response to the desire of Cardinal Lustiger to render the church more visible in the city. It is dedicated to the community." Despite its relatively small size, the church has the ambition to tie a community together, and to build bridges between religious communities. This will together with the idea of making the church more visible in the city was also at work in the Créteil Cathedral project. Clearly this kind of weaving action must be conceived in a different way when large projects on new sites such as the Jinan Cultural Center, the Danone Vitapole or the Renault Technopole are concerned "Either we bring an existing context to life, or we create a new context as we did for Danone," says Ayache. "This has to do with

在开展项目研究之前，建筑师便有意将传统和现代的关系以及新教堂和周边环境的关系都考虑在内。罗班说："当然，我们所设计的新型教堂和以往的教堂是有关系的，教堂建筑往往都追求能够成为社区，乃至城市的焦点建筑。巴黎约柜圣母教堂所在的街区混合有奥斯曼风格和现代风格的建筑，并伴有许多住宅楼，因此我们设计的教堂就像一条将各种因素结合起来的重要纽带。"马丁·罗班更详细的解释了传统的教堂前廊经常设计于中殿西侧尽头，圣坛的对面。

他说道："在中世纪，初入教会的教理者可以进入前廊，这像是教堂内外部的过渡区域。我们将理念进行调换，在城市与教堂之间设置'前廊'空间，即在教堂外部的四周设置了三维的网格区域，人们可以直接通过，就像是个中间区域，既围绕教堂，又与公共花园相通，变成了一个共享区域。"法国AS建筑工作室所说的"Stimuli（激励）"理念和背景是具有开放性的，这个案例不仅影响了周边建筑和街道，同时还对教堂传统固有的组成和标志产生了影响。马丁·罗班继续补充道："在这个项目中，你可以从中找到所有传统教堂的元素，但在这里它们以当代的形式和表达方式展现。在建筑中，你也可以发现新约和旧约的各自对应：以色列十二支派和基督十二使徒是基督教堂的支柱，约柜还包含受圣神感召而怀孕的圣母玛利亚的元素。著名艺术家马歇尔·雷斯创作的花窗玻璃也属于建筑的一部分。值得强调的是当时的红衣主教Lustiger（巴黎大主教，1981年—2005年）也对这项工程给予了大力支持。"

对于法国AS建筑工作室来说，项目背景包括现存事物，还包括以往的历史和对将来的考虑。罗伊达·阿亚斯说："在项目构思的初期，法国AS建筑工作室便认为巴黎约柜圣母教堂将重塑教堂在城市中的形象。从二十世纪六十年代起，教堂似乎慢慢地消失在城市环境中。巴黎约柜圣母教堂是自1967年以来，巴黎新建的第一座教堂。红衣主教Lustiger希望重新赋予城市中的教堂更高的可见性，巴黎约柜圣母教

The Olympic Stadium of Canton stands as an urban landmark on a regional scale as well as the beginning of a new district. It aims to create a living place for the city which goes beyond sport events.

广州奥林匹克体育场为区域范围内的城市地标，也代表着新区的起点。该体育场不仅仅供体育赛事的使用，更为城市创建一个生活场所。

a strong will to intervene in a context. What one might call the "macro", or large projects outside of the urban environment, constitute a context in themselves. You might say that our work is about either the creation or the recreation of context," she concludes.

What might be called "macro" architecture in the oeuvre of Architecture-Studio includes the *Maison de la Radio* renovation in Paris, the European Parliament in Strasbourg. Rodo Tisnado states: "Men and civilization have developed in the urban environment. The architect's essential task is to intervene in the city. A large enough building has to be considered in a way as something like a piece of a city. Macro architecture is smaller than a city but bigger than a normal building." The partners explain that their task in the case of the *Maison de la Radio* (French national radio headquarters) was to deal with problems having to do with the original architecture and how it was used. Martin Robain says "We have given the building the routes it needs in order for the entire structure to behave a bit as a neighborhood in a city does."

Architecture-Studio has thus defined its work on the basis of certain early ideas, such as that of the Stimuli, where a large number of projects were intended to be woven together, creating an overall impact and form of coherence in the existing city. This concept is now carried over into individual buildings that the office designs – for example, the Rotana Tower in Amman and the railway stations for Grand Paris – where the creation of new places and urban markers proves decisive. Many of the ideas that they develop are, of course, part of the logic employed by many contemporary architects, but they obviously take a sense of deep rootedness further than their colleagues. Of course, an individual building must function as planned for its own users, but it is also seen by Architecture-Studio as being either an integral part of the city, or an element of a new development that must also have its coherence and its relation to the environment in the broadest sense of the word. The step up from the individual building to larger "macro" structures which may have their own internal "urban" aspect is clearly inscribed in the thinking of Architecture-Studio, providing a seamless transition from small interventions to very large ones.

1. For contractual and confidentiality reasons, no illustrations of these stations could be published in this book.

堂是根据他的这个要求所设计的，它承担起建立宗教团体间桥梁的重任。”同样以提高教堂在城市中的可见性为出发点的项目还有克雷泰伊大教堂的扩建。

很显然，新建筑与社会建立联系的方式，在各大型建筑项目中有不同的体现。罗多·蒂斯纳多说道：“人类和文明在城市中发展，建筑的首要任务是要融入城市。一座足够大的建筑就像城市的一角，宏伟的建筑与城市相比虽然规模较小，但却是城市的重要组成部分。”法国AS建筑工作室建造的宏伟建筑包括巴黎的法国国家广播电台总部、斯特拉斯堡的欧洲议会中心、山东省省会文化艺术中心以及马斯喀特文化中心。其中在法国国家广播电台总部改造的案例中，建筑师解释道他们必须通过改造解决原有建筑存在的问题，以及建筑在使用时产生的问题。马丁·罗班说：“我们在改造设计中不得不重新定义整个建筑的流线设置，让它变得与周边环境更和谐一些。”

法国AS建筑工作室一直秉承着这些早期理念进行建设，例如“Stimuli（激励）”理念，也就是主张将多个建筑项目联系起来形成网络架构，以对现有城市产生整体影响，并且与之合为一体。这种理念后来又被转化引入单独的建筑设计中，从安曼的罗塔纳酒店大楼，到大巴黎的城际快线车站，体现了新建筑和地标性建筑的重要性。他们所提倡的这些思路或许也是许多其他当代建筑师所遵循的，但法国AS建筑工作室对这些概念的理解却更为深刻。一座建筑不仅仅会对于其使用者带来影响，它作为城市的一部分，更需要与城市融合并为城市带来更多的活力，因此必须与其所在的环境协调发展。从单独的建筑到具有“城市规划性”的大型宏伟建筑，无论建筑尺度的大小，法国AS建筑工作室的介入都尽力保持最大的协调性。

1.因保密协议，大巴黎新城际快线车站的效果图不能展示于本书中。

Clermont-Ferrand Art School stands as a sculpture, a copper monolith.

克莱蒙费朗高等艺术学院的建筑就像一座大型的铜雕塑。

PAPR01

Our Lady of the Ark of the Convenant Church
Paris, France
1986/1998
1 600 m²

约柜圣母教堂
法国，巴黎
1986/1998
1 600 m²

The Ark of the Covenant was a chest made of very hard and dense acacia wood. It contained the Tablets of the Law given to Moses by God on Mount Sinai. Situated in Paris's 15th arrondissement, this church represents the Hebrews' older covenant with God. Fulfilling the wish of the archbishop of Paris, Cardinal Lustiger, to have reconciliation between the Jewish and Christian traditions, this edifice symbolizes the healing of old hurts. This parish church, built for 450 worshippers, has a rectory for four clergy at the mezzanine level. Its aesthetic is imbued with theology and spirituality. The volume is that of a perfect cube, an object chosen for its simplicity. The 3-D square's equal sides express the invisible potency of divine power. Familiar to everyone, this figure embodies stability and a sense of shared experience. It is enveloped in a skin of metal mesh, letting air and light pass through it, allowing the gaze to wander through the empty spaces over the building's surface. At garden-level, 12 pillars (the number of apostles) are distributed around a base that protrudes from the cube. The church, raised above the ground and free of its foundations, is a sign of revival in the middle of the city. H.L.

巴黎约柜圣母教堂由非常坚硬的高密度皂荚木制造而成，像是一个木制方盒。教堂内陈设着上帝在西奈山上授予摩西的“十诫”石板。这座教堂位于巴黎第十五区，代表了希伯来人与上帝的旧约。正如巴黎红衣主教Lustiger的愿望，这座教堂象征着世仇的消解，希望在犹太教和基督教传统之间实现和解。这个可容纳450名礼拜者的巴黎教区小教堂，在夹层阁楼为四名神职人员安排了住宿区。它的美学充满神学审美，外观是一个简约而严谨的立方体。等长的边长象征着上帝的无形力量，而在人类的日常生活中，正方体代表了稳定性和共享感。立方体建筑被悬空置于一个三维的金属网格所界定的体量内，气流在空间内自由流通，光线透过玻璃窗神秘变幻，路人可驻足欣赏通过镂空透出的建筑表面。矗立于花园地面上的十二根支柱（代表使徒的数量）分布在突起的中心地基周围，将教堂支撑，使其像是悬浮在地面之上，代表教堂在市中心的复兴。H.L.

The cross pattern printed all around the building stands as a strong symbol and echoes both the shape and the meaning of the place.

建筑中随处可见的十字架图案强有力地告诉人们这是教堂所在地。

Standing above a public garden on twelve pillars, Our Lady of the Arch of the Convenant Church was given the title of 20th Century Heritage in 2012.

被十二根支柱支撑于公共花园之上的约柜圣母教堂，在2012年被授予二十世纪遗产的称号。

The cubical volume of the church, surrounded by a metal grid, becomes part of the Parisian urban landscape.

教堂的立方体构造被三维的金属网环绕，成为巴黎城市景观的一部分。

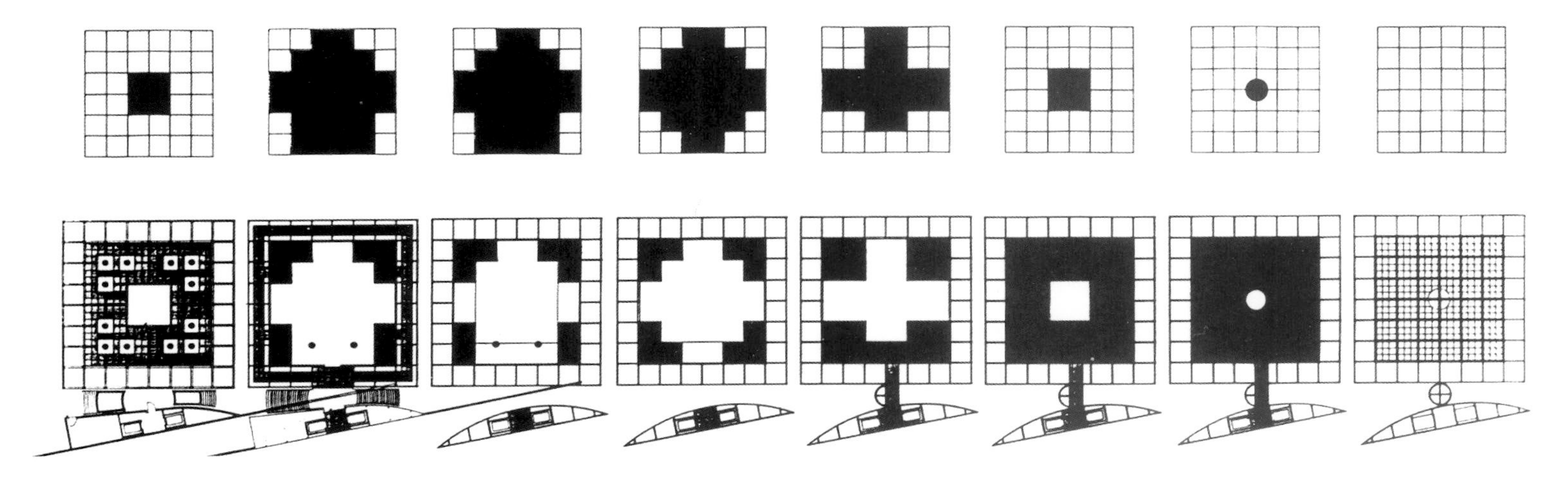

Horizontal sections of the cubic cupola and interior plans of the church.

教堂立方穹顶的横剖面和教堂内部空间平面图。

JNN3

Cultural Centre
Jinan, China
2010/2013
380 000 m²

山东省会文化艺术中心
中国，济南
2010/2013
380 000 m²

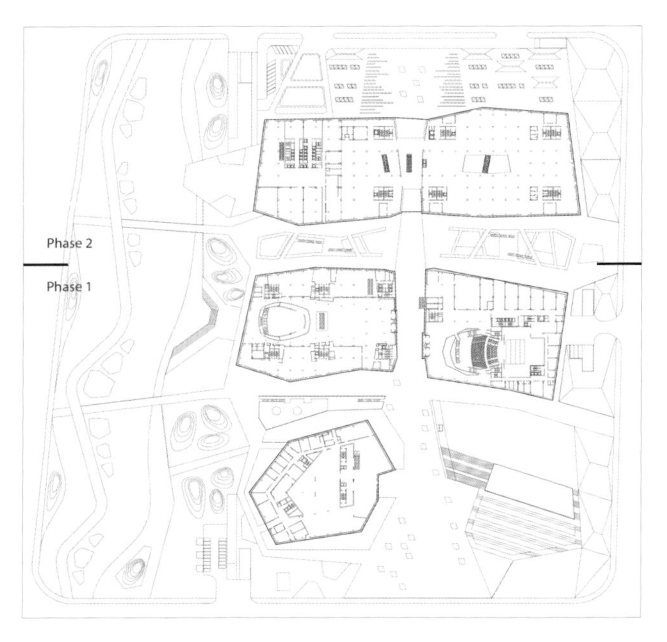

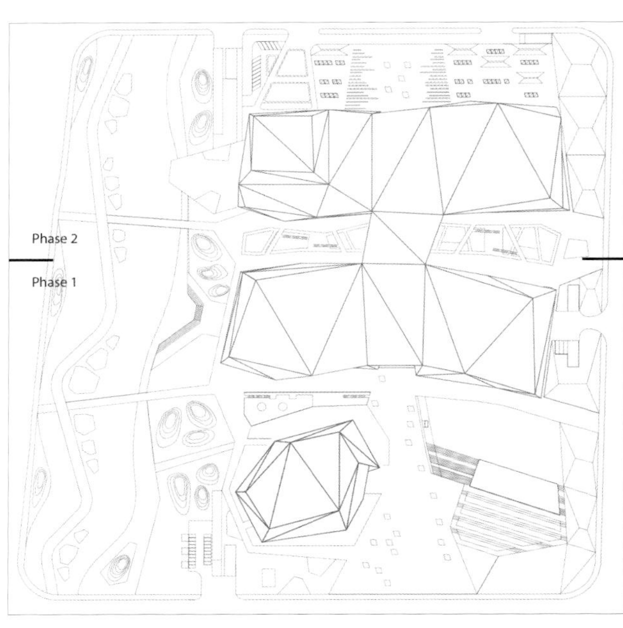

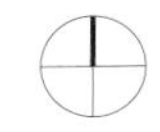

Erected in a new district near the high-speed train station, west of Shandong's capital, Jinan Cultural Center includes three major municipal buildings: the Museum of Fine Arts, the library and the dance and drama center. Areas around stations are often neglected and lack essential amenities, so this cultural venue, a veritable jewel of glass and marble, multifaceted like a diamond, really stands out. Its square, planted with elegant trees, welcomes the visitor and redefines luxury in environmental terms. With the ambition of becoming the hub of the district, a passageway leads to the different parts of the Center. In addition to the theater dedicated to the performing arts – with a 800 seat capacity and several studios for rehearsals – a shopping mall, a tower housing a hotel and a parking lot for 2 000 vehicles will complement this huge cultural venue. The main building has a double façade of perforated aluminum overlaid by a glass skin. This double-layered façade allows the sustainable distribution of light and energy inside the building. This building is a rare sight in a station area – surely this beautiful design deserves an entirely new district to be constructed around it. H.L.

山东省省会文化艺术中心位于济南西部新城核心区客站片区，主要由美术馆、图书馆和群众艺术馆三大文化建筑构成。火车站周边经常被视为没有文化生活，且缺乏必要便利设施的区域。而这个位于济南高铁西站附近的大型文化场所与周边广场的优质绿化交相辉映，形成一个优雅的环境迎接四方来宾。考虑到未来文化中心将成为该区域的中心，建筑在广场上预留出通道以方便公众出入。美术馆的外立面通过玻璃和大理石的交错运用，将建筑打造得宛如钻石般剔透夺目。群众艺术馆中则有800个座位的群星剧场以及多个排练用房。这座大型综合文化建筑还将建造一个购物中心、一座包含酒店及办公室的大厦和一个拥有2 000个停车位的大型地下停车场。主体建筑部分采用了双层幕墙系统，内幕墙部分采用浅灰色玻璃幕墙，外层立面由一个巨大的、不规则的镂空铝板构成，铝板镂空部分的大小与疏密与其对应的空间功能相呼应，在满足室内采光需求的同时也兼顾了保护的功能。在车站周边区域有这样的建筑是罕见的——当然，这样美丽的建筑也值得成为一个新区的中心。H.L.

Ground floor plan and ground plan: phase 1 was delivered in 2014, and phase 2 on construction.

底层平面图和总平面图：一期工程已于2014年交付，二期在建。

The main building, enclosing the library and cultural center, has a double skin – an inner, grey mirrored elevation and an outer elevation of perforated aluminum plates. The double façade perforations provide sustainable building management of daylight and internal temperature control.

主建筑连接图书馆和群众艺术馆，双层幕墙系统，内幕墙部分采用浅灰色玻璃幕墙，外层立面由一个巨大的、不规则的镂空铝板构成。双层立面系统使室内光线利用和温度控制效果达到最佳，是可持续发展建筑的特征。

Designed as a cultural district, Jinan Cultural Center is intersected by water features and green public spaces.

山东省会文化艺术中心被设计成一个文化区，水景设计和绿色景观设计穿插其中。

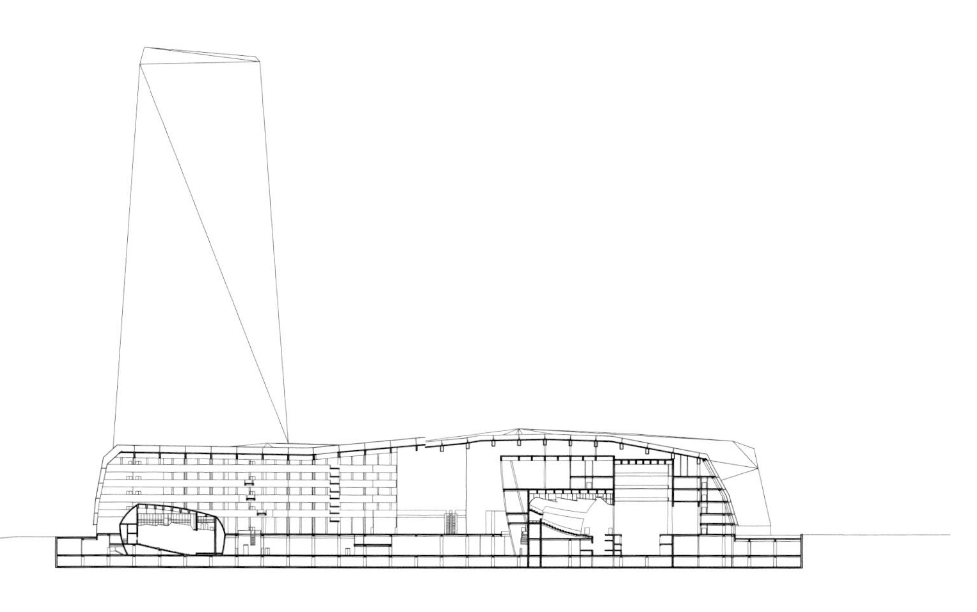

Cross section.

横剖面。

Interior view of the lobby of the Cultural Center.　文化中心大堂的内景。

Interior view of the library.　图书馆的内视图。

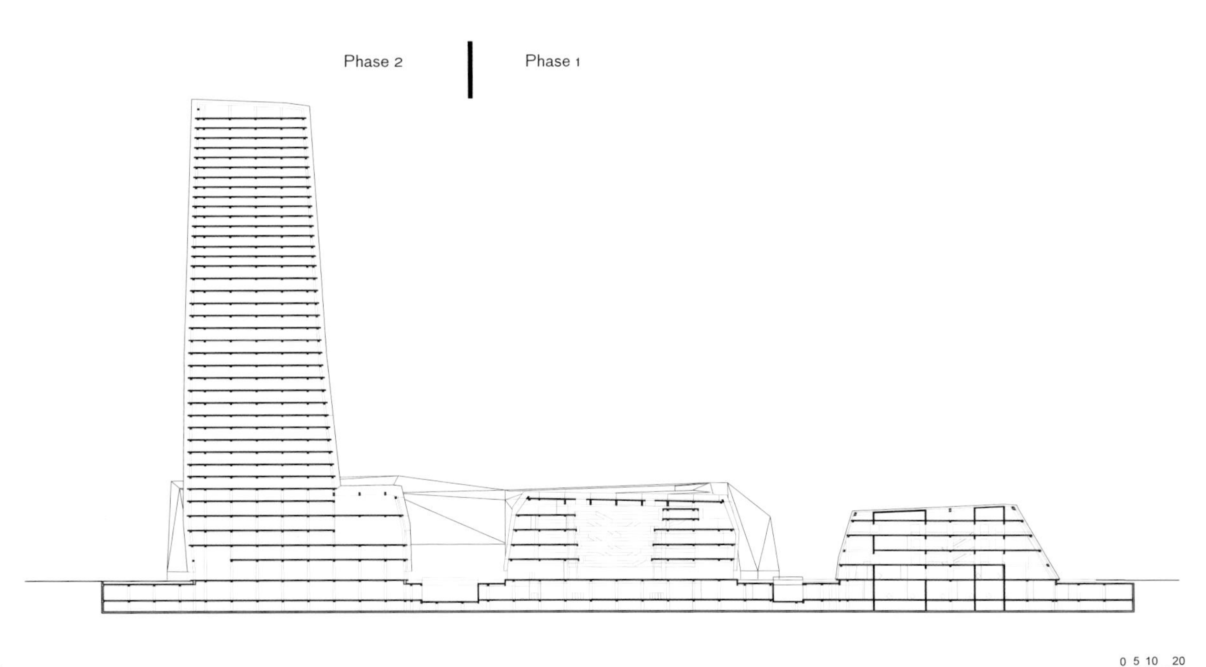

Longitudinal section.

纵剖面。

AMN1

Amman Rotana Hotel
Amman, Jordan
2005/2016
45 265 m²

安曼罗塔纳酒店
约旦，安曼
2005/2016
45 265 m²

The Rotana Hotel is one of the three towers of the Abdali district, Amman's new downtown. This building has become a landmark in the Jordanian capital. Consistent and coherent with the remaining office blocks to be built to complete this urban ensemble, the Rotana Tower has a distinctive silhouette, a voluminous cylinder, voluptuously rounded, as elegant as a pencil dress. This tower welcomes visitors with an alluring gap in the square's amphitheater like an invitation to enter. The hotel has generous amenities: 400 rooms, an atrium, a café, restaurants, a ball-room, a swimming-pool, a spa, gym facilities, a VIP lounge with a panoramic view and a conference center. The glass tower stands as a transparent voluminous block whose height ties in with the roofline of the lower buildings in the neighborhood. From the hotel's salons, cafés and restaurants that open on to the atrium, the tower offers different points of view on the new district. On the sixth floor, next to the pool and the lounge, customers can enjoy after-dinner drinks and the night's cooler temperatures in the agreeable and refined atmosphere of a terrace with a beautiful vista. This urban edifice's façade with its curtain-wall protected by aluminum sunshades expresses a high-tech and eco-friendly hospitality. The structure is composed of a central core covered with an exterior metal mesh allowing more space inside and greater circulation within the loadbearing elements. Emblematic of Amman's redevelopment, the tower stimulates public spaces, invigorates pedestrian pathways, structures the immediate environment and gives perspective to the faraway gaze. H.L.

罗塔纳酒店大楼是安曼新市中心Abdali区的三座摩天大楼之一，这座建筑已经成为约旦首都的标志性建筑。罗塔纳酒店大楼与将在不久后完工的另外两座大楼保持着共容性和连贯性，但其独特的轮廓又使它尤其出众，线条优美的圆柱体如铅笔裙那样优雅。缓和的弧度朝着剧院广场方向开敞呈现出酒店热情接待各方来宾的姿态。酒店设施完备，包括400间客房、中庭、咖啡厅、多家餐厅、舞厅、游泳池、水疗中心、健身设施、全景贵宾休息室和会议中心。建筑底部透明化的外立面构成室内外空间对话的桥梁，透明底座的高度与相邻街区的低层建筑完美融合、交相呼应。酒店的沙龙、咖啡厅和餐厅一面向中庭敞开，一面可以通过玻璃幕墙欣赏新区的城区景观。位于七楼的露天全景平台设有游泳池和环绕在精致花园中的休息室，客人在享用餐后饮品的同时，可以感受凉爽的清风和悠闲的氛围。由铝制遮阳板和玻璃幕墙构成的外立面既体现了高科技的特征，又完美兼顾了周边的地理环境。整幢大楼的结构设计是由内部中央核心筒和外部网状支撑立面构成的，创造建筑内部开敞的空间效果。罗塔纳酒店大楼能够促进公共空间发展，象征着安曼的重新发展，透明的底座使经过大楼底部的路人不会产生距离感，而从远处观赏大楼时，却又为它的高度和曲线感到震撼。H.L.

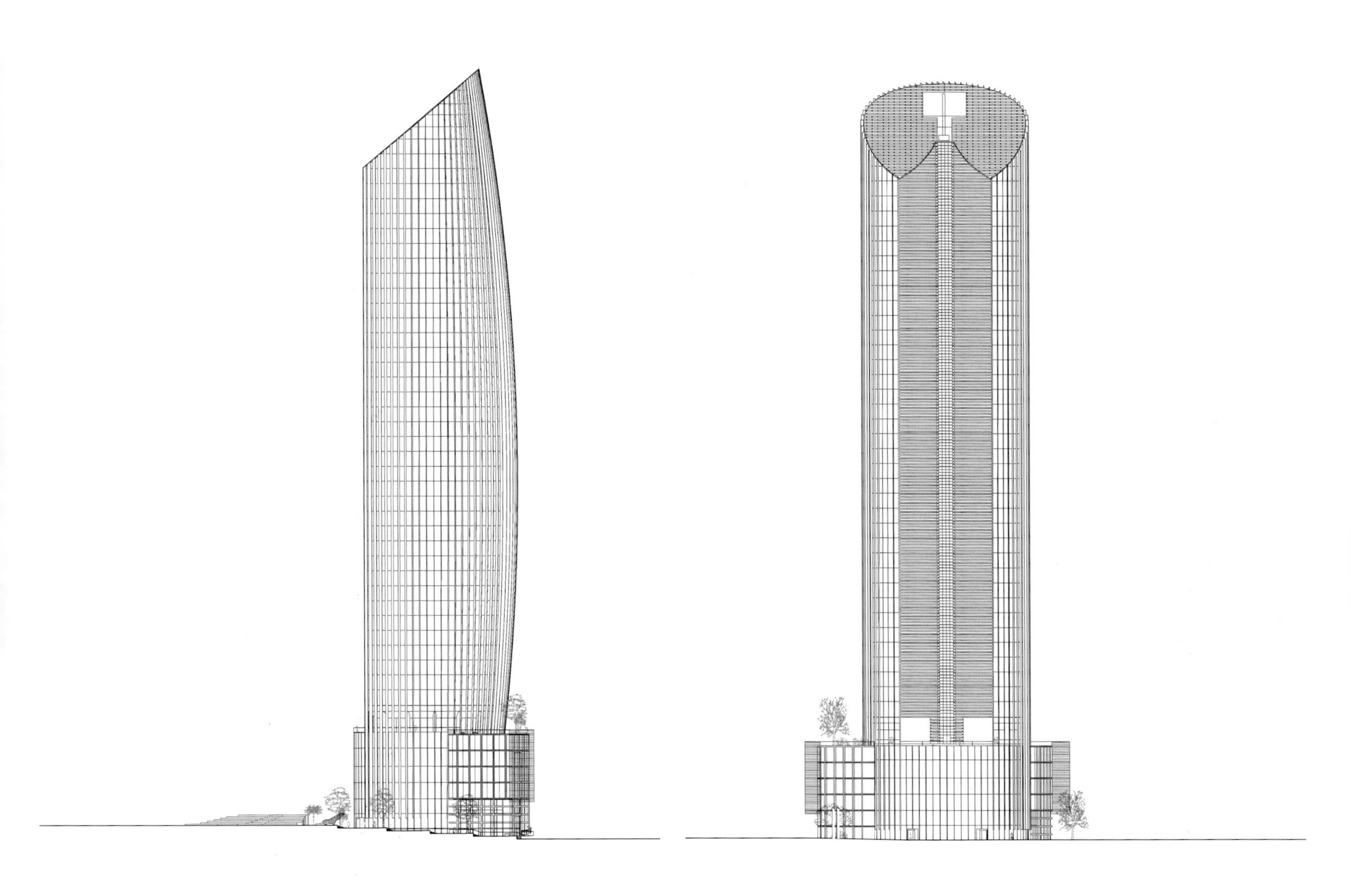

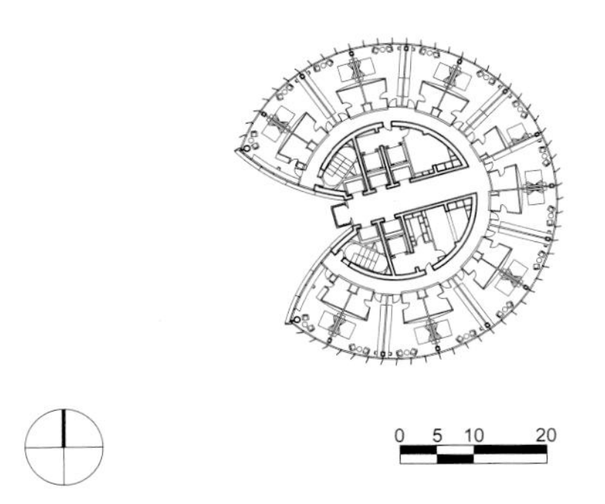

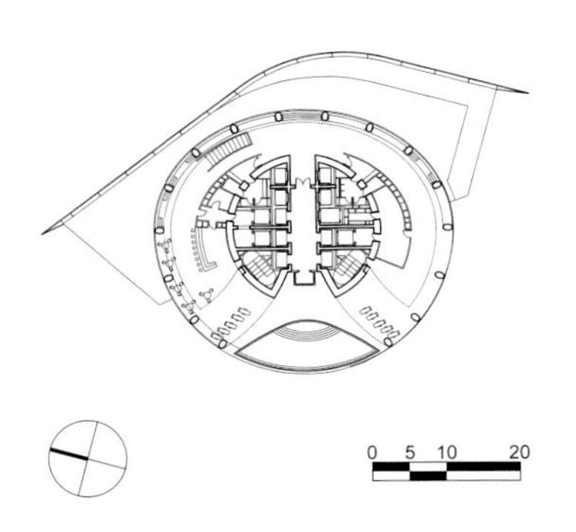

South elevation (top) and twenty-first floor plan (below).

南立面图（上图），二十二层平面图（下图）

West elevation (top) and sixth floor plan (below).

西立面图（上图），七层平面图（下图）

Galaxy S6 edge

Grand Paris Railway Stations and related projects
Paris, France
2014 / 2025

大巴黎火车站联结项目
法国，巴黎
2014/2025

The project of Grand Paris deals with strategic issues for the Île-de-France region. Its ambition is to extend the metro lines by an extra 200 kilometers of rail-track in order to have every part of the metropolis connected. This is a historic opportunity to promote public transport, which is a cleaner means of transportation than individual vehicles, as well as to develop the oft-neglected neighborhood around stations and include business activity and housing in these areas. Further projects related to the development of stations entail the upgrading of the districts through which the Grand Paris Express will run. Illustrating its taste for design research and innovation, Architecture-Studio is responding to this dynamic with sustainable answers and improved interconnected functionalities. In 2015, the firm won competitions for the design of three stations – Saint-Cloud, Rueil-Suresnes – Mont-Valérien, Nanterre-La Folie – a program related to Issy-les-Moulineaux's urban planning and a commission for the study of the renovation of the zone between the suburbs of Créteil and Maisons-Alfort, called the Triangle de l'Échat. Stations are used in all sorts of ways and have the potential for growth. So Architecture-Studio's role is manifold – promoting the use of the areas surrounding them; building in existing pockets of derelict land; reinventing underground routes through new lighting and increasing sensorial experience. The firm's feasibility survey for 25 stations in the Île-de-France, Grand Paris region has set the tone for these renovation projects. H.L.

大巴黎项目是打造法兰西岛地区都市群的战略性发展项目，目标是延长和建造200公里铁轨，完善大巴黎的轨道交通网络。这是促进公共交通发展的历史性机遇，不仅要发展更清洁、更环保的公共交通，同时还要对车站周边区域进行具有前瞻性的规划，大力发展商业和住宅区域。车站联结项目旨在重新定义大巴黎快线车站站点的周边街区，对其进行城市规划，法国AS建筑工作室乐于为建设可持续发展、智能化、功能完善的新型街区提供创新并可行的设计方案。2015年，工作室中标了大巴黎快线其中三个车站的设计竞赛，分别是：圣克鲁车站、勒伊—苏雷讷—瓦雷里昂山车站和南泰尔—拉富丽车站；另外还有伊西莱穆利诺的车站连接项目，以及位于克雷泰伊和迈松阿尔福之间，与车站相连的雷沙三角区域的城市规划。车站的利用方式多种多样，在考虑如何增加旅客的感官体验的同时，也必须为长远发展考虑，如加强车站周边的公共服务能力、优化景观设计、改造地下空间环境和照明等。工作室曾为大巴黎地区的25个车站进行过可行性调查，为这些改造项目提供了基础。H.L.

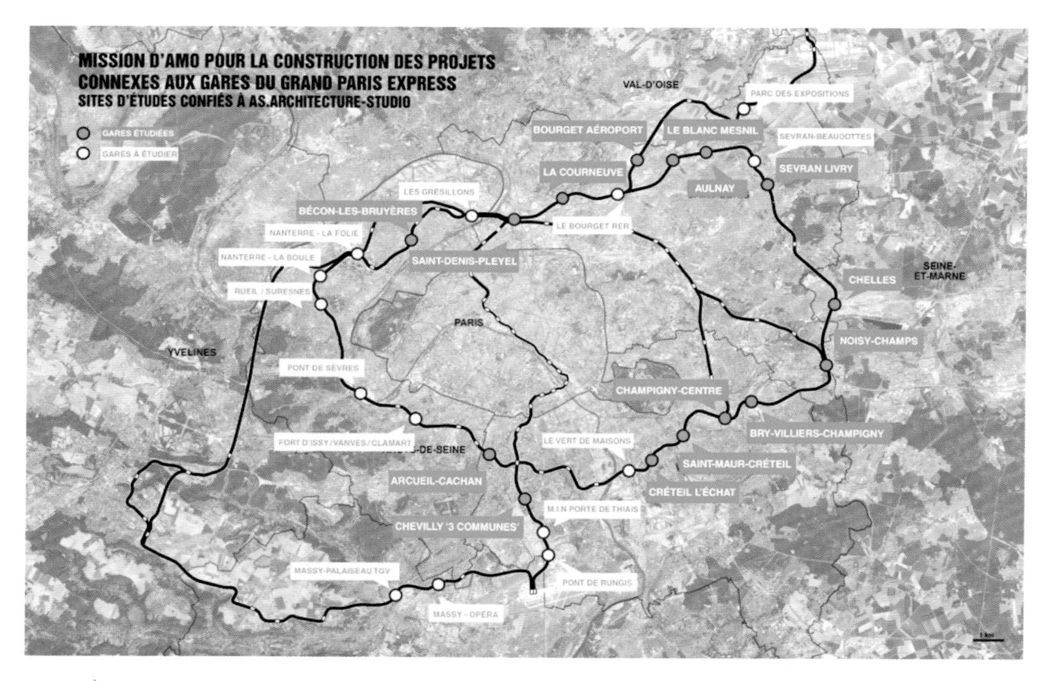

Related project combining housing and offices (Architecture-Studio) above the station of Issy-les-Moulineaux (Brunet-Saunier).

连接伊西莱穆利诺Brunet-Saunier车站的住房和办公综合建筑项目由法国AS建筑工作室设计。

Gare d'Iss

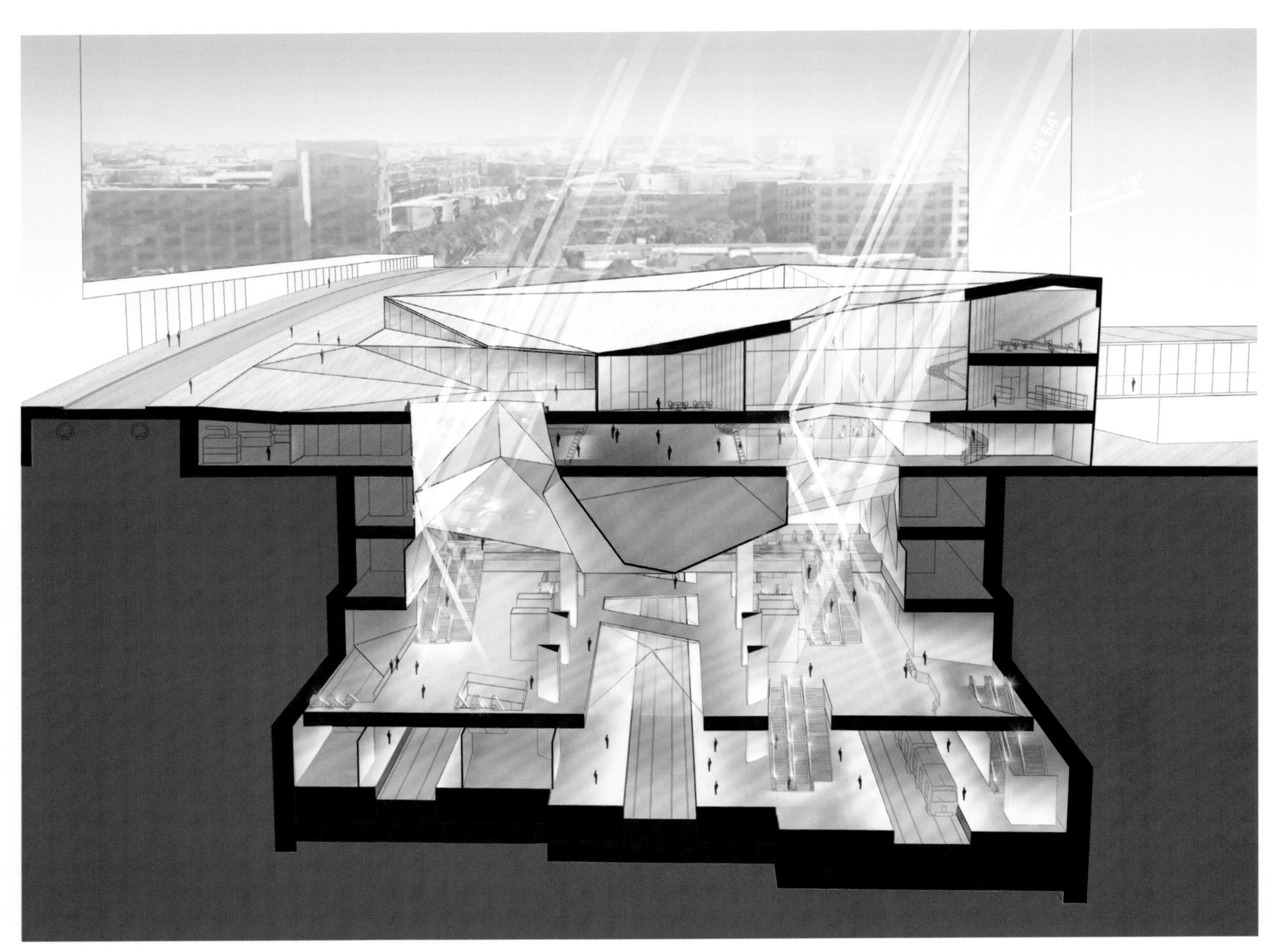

The emblematic train station of Saint Denis Pleyel, Grand Paris.

大巴黎圣丹尼斯・普莱耶尔车站是最具象征性的火车站。

next spread:
The related project of Créteil l'Échat.

下页：
克雷泰伊—雷沙的相关项目

La Grande Passerelle Cultural Hub
Saint-Malo, France
2009/2015
6 500 m²

圣马洛文化中心
法国，圣马洛
2009/2015
6 500 m²

Beyond the historic city-center Saint-Malo discards some of its cultural significance at the harbor's landmark fortress and becomes much less design-friendly. The urban space here looks like architectural "non-space". However, a good cultural project can always reboot the city's social fabric and be a fresh start for the whole of the community. In order to renovate the neighboring rail station's esplanade, the ground level has been raised through the dual uplifting of the ground and the buildings, whose increased size is incorporated into this revitalized environment. "La Grande Passerelle," "The Great Bridge," is the emblematic name given to this open-minded cultural venue. This high-tech multimedia library houses both pre-Gutenberg editions as well as the latest material on new media and technology and organizes cultural events of urban and regional scope. It was designed in a spirit of open-mindedness. Beyond and above cultural content or utilitarian purposes, this project fulfills the aesthetic demands of a work of art. A nation-wide audience that regularly comes in numbers to the local book festival "Étonnants Voyageurs", justifies the municipality running an ambitious cultural program alongside this venue. Inside the building is a big atrium, three art-house cinemas and rooms dedicated to multimedia. Outside there is an amphitheater and a new esplanade. A brand new urban landscape is alive! H.L.

圣马洛因抵御海盗的古城墙和海上堡垒而闻名，但却因为在规划上摒弃了有文化价值的建筑，逐渐使城市变得缺乏人文气息。不过，一个好的文化中心项目可以重新构建城市的结构，成为新市中心发展的开端。为了给毗邻的火车站广场注入一缕动感，文化中心由两座呈起伏的波浪形的建筑组成，即是建筑又是广场的文化中心完美地融入周边环境中。其法语名“La Grande Passerelle”，可译为宏大的桥梁，象征着这个文化场所的开放性。建筑包含多样的文化功能和一个多媒体图书馆，建筑整体似乎拥有艺术品的触感。多媒体图书馆装备了最新一代的系统，这里既保存有在古登堡印刷机之前的出版物，又包含新媒体和高新技术，还承担起了组织城市和地区性文化活动的角色。圣马洛举办的“Étonnants Voyageurs”图书节吸引了全国各地大量读者的到来，说明该市政活动的成功和文化中心的吸引力。除了多媒体图书馆，文化中心内还包括三个艺术电影与试映电影厅、休息厅、多媒体展厅等；而室外城市空间包含一个露天剧场的大型市民广场。全新的景观让城市展现出勃勃生机！H.L.

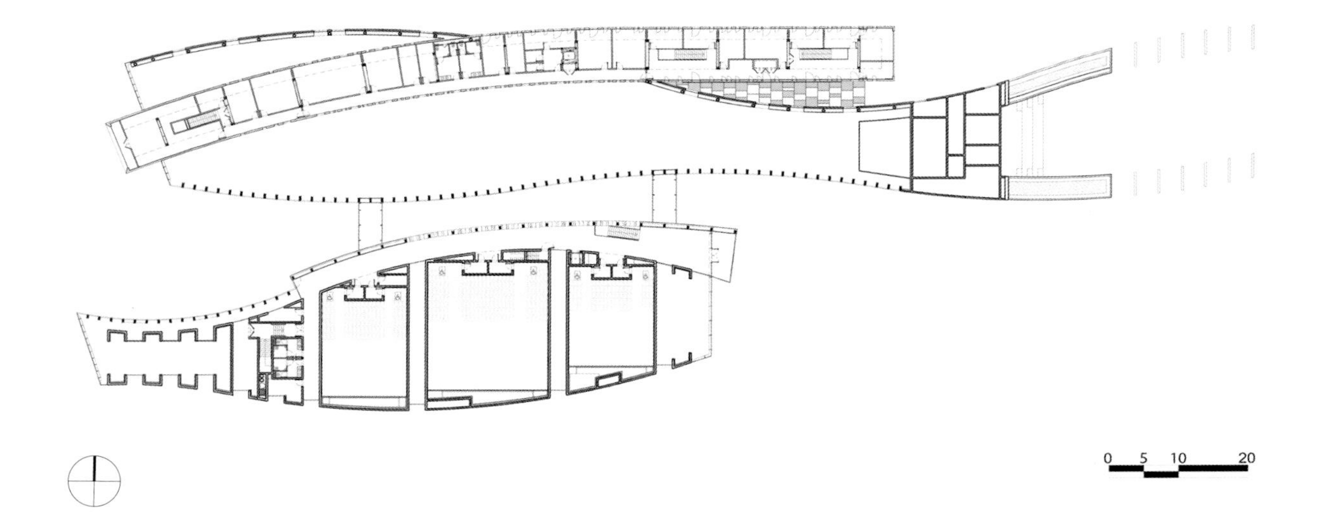

Ground floor plan.

一层平面图。

Aerial view of the Cultural Hub from the station to downtown. The Saint-Malo axis is indicated by a blue LED light on the floor.

从火车站面向文化中心和城市的鸟瞰图。广场上的蓝色LED灯表示的是圣马洛的城市中轴线。

View from the esplanade.

广场视图。

Interior view of the multi-media library.

多媒体图书馆的内景图。

Nouveautés

06
LE MONITEUR
Les cinq points qui font débat
CAMPUS UNIVERSITAIRE
Paris Jussieu achève sa métamorphose
FINAL CONCEPT
ARCHITECTURE - STUDIO
PARIS
IDEAT
ege

ΕΚΠΑΙΔΕΥΤΙΚΑ
ΕΡΓΑΣΤΗΡΙΑ
ONASSIS
CULTURAL
CENTRE
ΠΑΝΟΡΑΜΑ
ΕΛΛΗΝΙΚΗΣ
JAZZ
NEW
GENERATIO
2
sgt.gr
ΣΤΕΓΗ
ΓΡΑΜΜΑΤΩΝ ΚΑΙ ΤΕΧΝΩΝ
ΩΝΑΣΣΕΙΟΥ ΙΔΡΥΜΑΤΟΣ
ΣΤΗΝ ΑΘΗΝΑ
ΠΡΟΣΒΑΣΗ
ΑΝΕΥ ΟΡΙΩΝ
UNLIMITED ACCESS
BACH
ΚΑΜΕΡΑΤΑ

06 ARCHITECTURAL HERITAGE RENEWED

06 RENOUVELER LE PATRIMOINE ARCHITECTURAL

建筑遗产的修复

Unlike many architectural firms, Architecture-Studio takes a significant interest in the idea of heritage, defined in the broadest sense. This is to say that several of their projects have made use of existing buildings, sometimes completely redefining them in the process. Whether these buildings date from the late 19th century or the 1970s, Architecture-Studio approaches them as elements of architectural heritage. The definition of what constitutes heritage has clearly evolved over the years, with many modern buildings now either listed or considered as untouchable in many respects. When the partners of Architecture-Studio speak of architectural heritage, either based on what existed before they came on the scene, or in reference to what they hope to accomplish with their own new projects – their point is to create identifiable buildings that become an integral part of their neighborhood and city. The partners rightly point out that buildings or groups of buildings that come to be considered as part of architectural heritage achieve that status more because of how cities and users appropriate them than because of any desire of the architect to design an iconic building. Nonetheless, their careful consideration of issues of how buildings are used and how they fit into a neighborhood or a complex such as a university campus makes it more likely that their work will become part of architectural heritage.

Alain Bretagnolle sets out three forms of patrimony that interest Architecture-Studio: "We speak of the older kind of architectural heritage, or of more modern heritage on which we are beginning to work in terms of restoration, and finally of contemporary heritage which is what we are hoping to make." Admittedly, such an approach requires a certain amount of flexibility in terms of local culture. In China, where the firm has worked a good deal, the idea of preserving their heritage is relatively recent. In the Middle East, the approach to heritage is different. There is a chronological issue (how old is a building or a neighborhood), but there is also the matter of how respective cultures approach such heritage. Each context has its own problems and solutions. Sometimes countries or cities do not see their old buildings or heritage as being worth preserving, and, in those instances, Architecture-Studio has sought to make opinions evolve. One of the partners states: "In China the process is relatively simple since there is usually only one decider – so it is necessary to convince the mayor or the governor. We have to convince them in, some instances, to restrict the right to build so that elements in the general interest can be preserved or developed. We have also tried to explain that interweaving what is old and what is to be built is a way of creating a new heritage. We seek to avoid the *tabula rasa*. New directives have been issued to China relative to the manner in which cities should be built. The idea of creating very large avenues is being questioned, for example. We have, to some extent, anticipated these changes in attempting to bring the best of European approach to these heritage and urban issues." Clearly public space and attractiveness is playing an increasing role in the developing competition between cities. The question of just how dense city areas must be to be at once inviting and efficient is at the forefront of the thinking of the office. This was particularly the case for the development of the master plan for the heart of the city of Tirana in Albania (2003–2004).

For the partners of Architecture-Studio, particularly when they work in China or the Middle East, there are two extremes to be avoided – the *tabula rasa*, and, at the other end of the spectrum, "Disneyland", or the will to rebuild a pastiche of what existed, perhaps seeking to draw in tourists. This happens in China, but the firm was recently involved in a project in Tahiti (Tahiti Mahana Beach, 2014) that had exactly the same issues. There was a Chinese team that was proposing the *tabula rasa* approach, Hawaiians who wanted to create a kind of Disneyland and their proposal that made reference to local culture and engaged in neither extreme.

Although the concepts of architectural heritage most often bring to mind the more distant past, Architecture-Studio clearly applies

法国AS建筑工作室对建筑遗产有着浓厚兴趣，这里指的是最广泛意义上的建筑遗产。工作室所完成的几个与原有建筑相互融合，或对整座建筑进行修整的项目已充分证明了这点。不管是19世纪末期的建筑还是20世纪70年代的建筑，法国AS建筑工作室都把它们视为建筑遗产。过去数年间，建筑遗产的定义不断演化，如今，许多近期落成的现代建筑也都被列为或视为受保护的杰出成果。法国AS建筑工作室合伙人在谈及建筑遗产时，不仅谈到原本已存在的建筑及其历史意义，还谈到他们给建筑带去新活力的改建方案。他们的目的在于让建筑在其环境中变得更有意义，并使其成为整个城市必不可少的组成部分。合伙人特别指出，建筑物或建筑群之所以被视为建筑遗产的组成部分，并不是建筑师本身所能决定的，而是在于建筑物的使用者以及城市居民如何使用这些建筑资源。根据法国AS建筑工作室设计作品对使用者的重视程度和对城市环境的融合程度，将来或许有不少的工作室作品也将被称为建筑遗产，比如已建成的巴黎第六大学Jussieu校区东区改造项目。

阿兰·布勒塔尼奥勒列出法国AS建筑工作室感兴趣的三种建筑遗产类型："历史建筑遗产和现代建筑遗产，是近年以来需要整修的对象，还有当代建筑遗产，则是我们希望在当前或将来能够实现的。"毫无疑问，根据不同的地方文化，对建筑遗产的处理则需要一定程度的灵活性。在工作室已有许多项目落成的中国，其建筑遗产保护的理念是相较近些年才开始的。而在中东地区，建筑遗产的定义又有所不同。建筑遗产的价值可以体现在建筑学上又或者在历史发展之上，与建筑或街区的建造年代相关，同时又与政治因素、社会和文化因素密切相关，即在不同的文化背景下看待建筑遗产均有所差异，不同的环境背景有各自的问题和相应的解决方案。有些国家或城市并不认为其古老建筑物值得保护，并且忽视了一些公共建筑遗产。这种情况下，法国AS建筑工作室必须尽力去更新当地相关人士的观念。勒内-亨利·阿诺说道："在中国，这个过程相对简单，最重要的是通过必要的事实与最终的决策者（市长或者省长）进行沟通交流。某些情况下，我们必须劝说他们限制并严控在历史遗址建设的授权，以保护或者进一步发展公共利益。我们也尝试着去解释，只有将古老建筑遗产的历史元素与当代新元素相结合，才能使其成为未来的建筑遗产。我们不支持抹掉所有历史痕迹的做法。对于城市建设的方式，中国已发布新的相关规范指令。地方政府也开始反思一些建筑与规划所存在的问题，比如建造大型街道是否合理等。在一定程度上，我们早已预见到这些变化，即尝试将欧洲值得学习的关于建筑遗产和城市规划方面的经验和政策介绍给中国。"城市公共空间和公共服务能力所构成的城市魅力，在全球激烈的城市竞赛中扮演着越来越重要的角色。应当如何建设和改造城市中心区域为最大化城市吸引力和提高发展效率等问题一直是工作室所思考的，例如阿尔巴尼亚提拉那市的中心区总规划项目（2003年至2004年）。

对于法国AS建筑工作室合伙人而言，在他们工作时，尤其是在中国或中东区域工作期间，需要避免两种极端——白板理念，即彻底抹掉历史痕迹的做法；以及迪士尼化理念，即对某些已存在的建筑进行模仿

the same basic logic to new construction as to renovation and extension work. "We do our best to create contemporary heritage architecture," says one partner, "but basically we don't decide; it is history and society that make those decisions. We already have one building that has been recognized as being a 20th century religious heritage structure – Our Lady of the Ark of the Convenant Church in Paris."

When asked to cite examples of modern buildings that have clearly participated in the definition of a local culture in contemporary terms, Architecture-Studio readily evokes the Middle East with, for example, the Burj al Arab hotel in Dubai (Tom Wright, Atkins, 1999) or Asia with the Petronas Towers by Cesar Pelli (Kuala Lumpur, 1999). For Roueïda Ayache, "in some locations we are confronted with the question of contemporary cultural identity. Curiously a number of countries turn to foreign architects to design and build elements of what is to be their own contemporary cultural identity. No country wants to be exclusively identified with a nostalgic, backward looking version of its own culture. Post-Modernism survived even longer in the Middle East than elsewhere, until the 1990s, where pastiche ruled or nostalgia for past periods was dominant, and finally, on occasion the most radical modernity came forward as well. It is now felt that each country is an actor in the development of modernity. Our task, as the National Theatre of Bahrain and the Rotana Tower may demonstrate, is intended to be very contemporary, even as a building helps to define a cultural identity that is associated with this city or this place. A building that becomes famous surely participates in this kind of effort. Take the example of a building that is not ours, the Burj al Arab, that, in a sense, put Dubai on the map. In some respects, this building is the sign of a contemporary identity. This building encouraged a number of nearby countries to leave behind a nostalgic attitude in their architecture and to affirm a desire to look forward. The Burj al Arab appeared as a drawing on Dubai license plates, a proof that it had in some sense become a symbol of the Emirate, or even the United Arab Emirates that it belongs to."

The firm has often been asked, in competition briefs, to refer to culture, be it local or broader, as was the case for the Arab World Institute or the European Parliament in Strasbourg: the program for the Grande Passerelle Cultural Hub in Saint Malo specified that the authorities wanted a new cultural icon for the city; for the Onassis Cultural Center in Athens, the architects were asked to refer to Hellenic culture. Clearly, in Europe

并且不断复制，最终达到吸引游客的目的。虽然以上现象在中国也曾遇到过，而工作室在2014年参加的大溪地马哈娜沙滩规划竞赛中也遇到了相似的情境，三个团队在同样的竞赛要求下进行各自的创意设计，中国团队的设计作品极具自主创意但忽视了当地的原有文化，夏威夷团队的设计作品类似一个波利尼西亚的主题乐园，而法国AS建筑工作室的设计作品在规划上主张城市和滨海空间的共享，并由当地传统文化获取灵感来设计当代的建筑。

在通常情况下，建筑遗产会让人想到遥远的过去，然而法国AS建筑工作室却给予它们更广阔的定义，通过重新设计，或进行翻修和扩建为建筑遗产带去新的元素。马丁·罗班确认说："我们尽最大可能创造出当代建筑遗产，但建筑是否具有建筑遗产的价值是由社会和历史发展决定的。目前，我们有一处建筑已经被评为20世纪宗教遗产——巴黎约柜圣母教堂。"

当被要求罗列出对定义某地的当代文化做出贡献的现代化建筑时，法国AS建筑工作室马上想到了位于中东地区的迪拜帆船酒店（汤姆·赖特，阿特金斯集团，1999年），以及位于亚洲的吉隆坡双子星塔（西萨·佩里，1999年）。对罗伊达·阿亚斯而言，"在某些地区，我们需要面对当代文化认同的问题。一些国家会邀请外国建筑师设计并建造将作为本地当代文化标志的建筑，没有哪一个国家愿意让自己的历史和文化传达出落后、破旧的信号。中东地区受后现代主义建筑学的影响要大于其他地区，模仿和怀旧的建筑风潮一直持续至20世纪90年代，直至最为激进的现代思潮得以爆发。在这样的环境下，每个国家都希望自己成为现代化的重要角色。正如巴林国家大剧院以及约旦罗

Close to the Place de l'Étoile, the restoration of No. 38 Kléber pays discreet homage to the original architecture and the modern spirit of the 1930s. The creation of an open canopy on top floor offers an exceptional view of Paris.

毗邻凯旋门所在的星形广场，克勒贝尔街第38号的整修工程向原有建筑和20世纪30年代的现代精神致敬。设置在顶楼的玻璃棚为观赏巴黎景观提供独特视角。

or a country such as Greece, references to culture can be lively and rich. In a place like Bahrain, Architecture-Studio looked to the landscape itself for inspiration. For one of the partners, "Bahrain was an effort to respond to the identity of the landscape; it can be seen as an attempt to create a new cultural identity, but associated with the forms of the seascape. The horizon and the sea are sources of inspiration. A horizontal canopy responds to the horizontality of the landscape; it was also imagined as a focal point in the city where highways are very present. The theatre itself is imagined as a kind of waterfront magnet. The main auditorium interior was inspired by the forms, techniques and materials of fishing or pearling boats. The canopy may recall the roofs of traditional local houses, but these elements are secondary as compared to the horizontality of the landscape as a source of inspiration and identity. We have also taken care to integrate pedestrian spaces in connection with the National Museum and the waterfront so that not only the scale perception from the highway is integratednto the design, but that of the individual is as well."

The Novancia Business School in Paris is indicative of the approach of Architecture-Studio where the integration of older buildings into a renewed complex is concerned. They decided to retain a 1908 building on the site, but to demolish another structure from the 1950s to make way for new spaces. The partners judged that the 1950s building had no great architectural interest, but also that, above all, it did not function well as a school building. Distancing the firm from outdated visions of the architecture of the past, Mariano Efron says, "as opposed to Post-Modernism that imitated the past we made no effort to imitate the 1908 structure. In the opposition of two periods there is a sense of what may constitute a contemporary heritage as a kind of mirror of an older building that is already recognized as such. With the new building, we consciously tried to take into account the urban rules that have been applied to the broader area, and the oldest building in particular. But from that point we go on to see what else can be read (and written) into the rules. Starting with Haussmann period rules, we inscribed the project in the larger context of the city. One might imagine that Paris architecture is usually white like the stone that was used, but in this area, which was almost at the limit of the city when the first building of the school was erected, there was more color. We looked at the local color schemes and the forms of neighboring buildings to generate a completely contemporary

塔纳酒店大楼两个例子所示，我们的任务旨在实现当代化，并为城市或地区的当代文化认同做出贡献。比如帆船酒店这个著名的当代建筑，它让迪拜在世界范围内一举成名，本身也成为迪拜当代文化的标志，甚至鼓励众多周边国家抛弃对建筑物的保守态度，坚定展望未来的信念。迪拜帆船酒店形象出现在迪拜各类执照牌上，这在一定程度上代表着酒店已经成为阿拉伯联合酋长国的文化标示。"

在建筑竞赛的方案中，评审总是期待建筑能有与文化相关的内容，不管是当地的文化还是广义上的文化，比如巴黎阿拉伯世界研究中心或斯特拉斯堡的欧洲议会中心。在圣马洛文化中心的竞赛要求中，当地政府表明希望建筑成为城市文化新地标和城市的中心。而雅典的奥纳西斯文化中心的业主希望建筑师的设计能够反映出古希腊文化。在欧洲或者说像希腊这样的国家，可以参考借鉴的文化元素是多种多样的。在巴林的项目中，法国AS建筑工作室则优先从海洋风景中寻求灵感。其中的一个合伙人表示，"巴林国家大剧院项目是对当地风景的一种认同和回应，也可以被视为一种文化认同的创新型尝试，因为这里的文化恰恰是来源于海天之间。大海和水平面是项目灵感的来源，建筑水平顶棚像是海平面的延续，位于海岸的大剧院以其魅力让海滨景色更具吸引力。剧院大厅内部的形状、及技术和材料的使用都受到当地特色渔船和捕珍珠小船的启发；顶棚则参考了当地传统房屋常用的棕榈编制屋顶，然而这些均属于次要因素，是随着水平风景线这一首要灵感而来的。同时，我们还注意整合了大剧院与国家博物馆，以及海滨之间的人行通道，让人们不仅可以从周边的公路接近建筑，还可以在附近的人行通道上观赏建筑全景。"

The Arenas High School in Toulouse retains memories of the site as it follows the circular structure of the ancient Arenas of the Golden Sun.

图卢兹阿雷纳高中的造型灵感来源于黄金太阳古竞技场的环状构造，保留当地的记忆。

The restoration of the historical district of Taiyuan (Shanxi, China) offers a green strip composed of squares and public spaces. While connecting the Zhonglou and Gulou belvederes, the project creates a genuine harmony between tradition and modernity.

中国山西省太原的历史城区改造方案中，规划了包括广场和公共空间在内的绿化带，连接着钟楼和鼓楼瞭望台，该项目使传统和现代实现了真正的和谐。

element that is nonetheless rooted in the city and its history. It is at once contemporary and in context."

Two very large projects have recently allowed Architecture-Studio to enrich its approach to architectural heritage, starting with relatively modern buildings. These are the *Maison de la Radio* (completed in 1963) and the Jussieu university campus whose construction process was halted in 1972. Speaking for the firm, Gaspard Joly says, "I would say that we have not rehabilitated these buildings, but that we have redone the architecture. And we rededicated the buildings to their real functions. In both cases, we have maintained the original aesthetic aspects of the architecture. Both the *Maison de la Radio* and Jussieu are powerful structures, and we did not want to take away from their original nature, the identity that dates from the times when they were built. Our task was to transform these buildings in keeping with the changing use of the structures that has come with time."

The cylindrical form of the *Maison de la Radio* made it apparent that substantial changes in the external appearance of the building were not desirable. It is rather inside of the building that the changes were made. The architects sought to clear away the progressive occupation of spaces that had rendered the transparency and logic of the building difficult to understand. New digital forms of information storage were seized upon, for example, to liberate former archive space in the inner tower of the complex. Double-height spaces were also added to the tower to alleviate a kind of generalized claustrophobic feeling in the building and Architecture-Studio also sought to find ways to avoid having as many locked doors and offices as had existed until the date of their intervention. Another of their significant ideas was to create a transverse axis in the building, a kind of internal street, because users had explained that in the existing layout they wasted a great deal of time taking circuitous routes from place to place. Two former studios (102 and 103) were converted by the architects into a Grand Auditorium that is a new national symphony hall in Paris. An esplanade and parking area were added on the north side of the building and the architects also sought to reopen perspectives in the direction of the nearby Seine River and the Pont de Grenelle. Roueïda Ayache says, "the architectural renovation of the *Maison de la Radio* is a kind of acupuncture – we changed things that are small but that profoundly modify the use, the comprehension and finally the quality of the place."

位于巴黎的诺凡西亚高等商学院是法国AS建筑工作室将古老建筑与改造整体相融合的典型案例。团队一致决定保留下1908年建设的建筑部分，但同时拆除建于20世纪50年代的已不再适合使用的建筑部分，以创造新空间。工作室从古老建筑的过时审美中抽离，马里亚诺·艾翁说："与后现代主义的模仿恰恰相反，我们不会花心思去模仿这栋1908年的建筑。两种美学的强烈对比，相当于一幢让新老建筑相呼应的当代建筑遗产。建筑的新构造部分尽量将城市规则统筹考虑在内，然而，以这一出发点来说，有哪些规则需要被考虑到呢。在巴黎的整体大环境之下，我们又将项目归入其更具体的城市区域中。巴黎建筑常使用白色的石头，因此人们印象中的巴黎建筑通常都是白色的，然而，项目所在街区的建筑实际上是由多种色彩的砖头所建成的。我们仔细研究周边建筑的颜色使用范围和建筑形式，为的是创造能够植根于城市及其历史的当代建筑，让整修建筑既融入了环境中又有具有当代性。"

近期，两个大型工程项目让法国AS建筑工作室进一步积累了建筑遗产的处理经验。这两处现代建筑分别是于1963年建成的法国广播电台大厦，以及于1972年停工的巴黎第六大学Jussieu校区。贾斯帕·朱利说道："我更偏向于说，我们不仅是对建筑进行了翻新重建，还加强了建筑的辨识度，并且重新定义了建筑的实际功能。在这两个案例中，建筑物最初的美学要素均被保留了下来。法国广播电台大厦和巴黎第六大学Jussieu校区的原有建筑结构各具特色，我们不希望改变能代表其建筑年代的原始结构特色。我们的任务是跟随时代的变化确保建筑能够满足不同的使用需求。"

在法国广播电台大厦改造项目中，出于保留亨利·贝尔纳于1952年至1963年间设计的圆柱体原始结构的意愿，真正大幅度改造的是建筑的内部空间。建筑师将那些逐渐变成无用空间的多余部分清除，还以建

The Mahana Beach Art and Cultural Resort aims to highlight Polynesian culture by respecting local resources and knowledge, and the identity of Tahiti. Here, the Paul Gauguin Oceania Museum.

马哈娜海滩艺术和文化旅游胜地的设计重视本地资源和文化技术，旨在宣传波利尼西亚文化。图为保罗·高更大洋洲博物馆。

The restoration of a town house located on Place Vendôme for AXA Private Equity Headquarters redefines working spaces in a heritage building.

位于旺多姆广场的法国安盛私募股权公司总部的改造项目，为这座具有建筑遗产特质的府邸创建当代办公环境。

For Architecture-Studio, a first element in the analysis of a large project such as that of the renovation of the eastern section of the Jussieu university complex in Paris is to enter into a kind of dialogue with the original architect, in this case Edouard Albert (1910–1968). The overall renovation of the campus, which involved removing a good deal of asbestos, took a total of 15 years, with the western area handled by the noted Paris firm Reichen & Robert. Thus, aside from interpreting and indeed improving on the original design, Architecture-Studio also engaged in a fruitful dialogue with Reichen & Robert, developing further some of their concepts in order to create an overall feeling of continuity in the university. With respect to the collaboration with their Paris colleagues, Martin Robain states "we have a collective approach, we like working with other architects and this case was no exception. We tried to create a synthesis between their ideas and ours so that a sense of continuity could emerge. A certain humility is not a bad thing in this kind of project."

The intervention of Architecture-Studio was put into place during a three-year period ending in 2015. Marie-Caroline Piot sums up the intervention of the firm succinctly: "We worked on spaces, communication in the space and contact with the city. We connected the grid of the university with the city. We brought vegetation into the ground-level areas and light into the underground spaces that had been dark. There was a rethinking of the entire space. We covered an internal courtyard with an inflatable cover that added usable space to the complex. We closed off one of the car access routes and created a pedestrian passage that was harmonized with the western part of the campus. This is like an intervention on the scale of a city, or rather a kind of macro architecture. The real ground level is below the platform and had been treated architecturally speaking like a basement. Today, the connection with the outside world has been re-established – there are large glass facades."

It seems clear that there are interesting links between the approach of the firm in such apparently different projects as the Novancia Business School, the *Maison de la Radio*, Jussieu and the *Université de la Citadelle* in Dunkirk. In each case an urban structure or group of structures already existed. Respecting the urban rules concerned in each case, the firm sought to establish points of reference that give a new meaning to the buildings. These points of reference might be works of art or new small structures that would allow visitors to orient themselves and

筑透明度和流畅度。比如，利用将档案信息化后腾出的建筑内原本的档案储藏空间，打造成两层通高的内部空间使建筑内部更开阔，还有最大限度的将办公空间和录音室透明化和开放化。另外一个了不起的理念是在建筑轴线上打通一条内部街道，供建筑使用者能够方便快捷的在建筑内穿梭，减少绕路所浪费的时间。原有的102号和103号录音室均被拆除，为1500座的大音乐厅留出足够的空间，如今，改造完成的大音乐厅已经成为国家级别的交响音乐厅。露天平台和停车场增设在建筑物的北部，从而使建筑朝向塞纳河和格勒纳勒桥一面的视野被重新打开。像贾斯帕·朱利说的那样："法国广播电台大厦的改造工程像是一种针灸疗法——建筑内部进行了深度调整，但建筑的标志外形却几乎没变化。"

关于巴黎第六大学Jussieu校区东区大型改造项目，首先要考虑的是与原建筑师爱德华·阿尔贝（1910年–1968年）所设计的建筑形成对话，同时也要考虑到以往的多次改造，以及由赖兴&罗贝尔建筑事务所主导的校园西区改造，以通过改造提高整个校区的连贯性。在谈到与同行的合作时，马丁·罗班说道："合作是工作室的基本理念，我们乐于与其他建筑师合作，Jussieu校区的改造项目并不是特例。我们尽力整合双方的建筑理念，在这种建筑工程中适当体现更多的人文要素并不是一件坏事。"

法国AS建筑工作室对Jussieu校区的改造工程持续了三年，于2015年完工。玛丽卡·碧欧简要地总结道："我们的改造致力于空间的交流互通，以及与城市之间的联系。大学的结构框架与整个城市景观重新融合，其中包括在地面区域引入绿色植被和为原本阴暗的地下区域引入光线。与景观设计师米歇尔·戴斯威纳合作下，诺大的花园被安排在扎曼斯基大楼底层周边，为整个校区增添绿意。我们重新思考了整个空间使用，这样的介入是城市规模的宏观改造。"

The rebuilding of Liévin Regional Covered Stadium provides new functions and capacity: it changes the image of this high level sports complex.

列万地区带顶体育场经改造，发挥新功能和承办更多样的赛事，给予该体育中心全新的形象。

to create new uses for space. Mariano Efron explains: "On the occasion of a renovation or rehabilitation, we create spaces that may not have specific uses at the outset but that users define over time. At Novancia, between the new building and the old one, we stretched a Teflon covering that has become a place of passage but also of communication between users. The inflatable covering that we created in a courtyard space at Jussieu has been appropriated by the students and was not part of the original program." Roueïda Ayache suggests that there is a relationship between this tendency to create new, undefined spaces and the layout of the offices of the firm: "There is a kind of nave in our offices," she states, "that serves no specific purpose but it is where we meet and communicate. Maybe this space has had an unconscious influence on our approach. The Agora space we created at the *Maison de la Radio* functions in the same way." Another partner underlines the fact that, even in their new buildings, Architecture-Studio has frequently created undefined intermediate spaces that are then appropriated by the users in any way they please.

Another, smaller project that the firm worked on was the extension of the Créteil Cathedral, originally designed by Charles-Gustave Stoskopf (1907–2004). At the time of the original project (opened as a parish church in 1978), the Church sought a low profile, but, on the contrary, since that period, the Church as an institution has sought more visibility. The architects took on the challenge of opening the Cathedral to the city, creating two spherical shells "deployed to the sky like two clasped hands above the altar". This change in the appearance and function of a cathedral is surely a matter of adapting the building to a contemporary approach to heritage, but also to changing ideas about the Church itself.

In two other recent cases, Architecture-Studio has engaged in the creation of an idea of heritage that finds other sources than existing buildings or local architectural history. The Xie Zhiliu and Chen Peiqiu Art Gallery (2015), located in a new construction area designed by GMP (Lingang New City), is on a site that one partner describes as an "isolated no-man's land area." For this project, the firm looked to the specific poetry of the paintings exhibited and their presentation. This project is thus more about contents than context, and illustrates how a building can create a landscape and be inspired by heritage that is not specifically architectural.

很显然，诺凡西亚高等商学院、法国广播电台大厦、巴黎第六大学Jussieu校区以及敦刻尔克城堡大学这几个改造工程虽是如此的不同，但使用的建筑方法却又存在千丝万缕的关系。在每个项目中，工作室都对建筑整体结构或是周边城市环境进行了介入。在充分尊重城市建设法则的同时，通过结构或细节的改造让建筑使用者更方便地使用建筑空间，赋予建筑新的意义。马里亚诺·艾翁具体解释道："在翻新或改造的过程中，我们会创造出未被定义的新空间，刚开始或许没有明确的用处，但随着时间的推移，使用者会赋予其新的用法。例如在诺凡西亚高等商学院改造中，我们用特氟龙材料作为顶棚，覆盖新老建筑之间的空间，为其遮风挡雨并保温，该空间不仅可以作为通道，更能够成为交流的空间。而在巴黎第六大学Jussieu校区东区的改造中，充气式顶棚所覆盖的广场已经被学生充分使用。"罗伊达·阿亚斯认为设计未定义使用功能的空间之趋势与法国AS建筑工作室办公空间的整体布局有着密切的联系："我们办公室有一个无指定用途的中庭，但可以作为我们集会和交流的场所。这种空间或许对我们的建筑设计方法产生潜在影响。我们在法国广播电台大厦内设置的内部街道也具备类似的功能。"

另外一个规模相对较小的改造项目是克雷泰伊大教堂的扩建工程，该教堂于1978年向教区开放，原始设计师为Charles-Gustave Stoskopf（1907年–2004年）。在原始工程时期，教堂追求隐于闹市的风格，而如今，教堂寻求在城市中有更高的可见性。在确保教堂容量翻倍的情况下，法国AS建筑工作室建议让教堂面向城市开放。教堂钟楼独立矗立于广场一角，教堂主体的两个半球型外壳朝向天空，犹如"祈祷的双手汇聚在圣坛之上"。克雷泰伊大教堂的扩建是对现代教堂建筑遗产的当代改造。

法国AS建筑工作室在近期参与的其他两个与建筑遗产相关的案例，却与已存在的建筑和当地建筑史无关。谢稚柳与陈佩秋美术馆（2015

The design for the new city center of Tirana aims to create a reference point and consistent policy for the long term urban renewal of the city. The project involves all levels of the city, it determines the new densities, provides for historic preservation and ultimately allows for the redevelopment of public spaces.

提拉那市新市中心的规划设计，是在政策的规范之下为城市的长期可持续发展提供模板。规划项目囊括了城市的各个等级和城市密度，为历史建筑保护提供预案，并且对城市公共空间进行细致的安排。

Seddiqi & Sons headquarters. The exterior façades are made up of light-filtering metallic facets, reflecting sunlight and protecting interior spaces. It has become an urban landmark on Sheikh Zayed Road in Dubaï.

Seddiqi & Sons总部建筑的外立面由铝板构成，反射日光的同时保护内部空间，该建筑将成为迪拜Sheikh Zayed大道上的标志性建筑。

Another case requiring a kind of ex-nihilo creation of heritage is that of the projects being carried forward by the firm for the Lusail district of Doha, a place that until quite recently was undifferentiated desert space. They worked with the French landscape architect, Michel Desvigne, to define the identity of the landscape. Local ingredients such as an eco-park thus create a matrix for the development of this new city area. Roueïda Ayache states "we have used the salient points of the existing Qatari landscape as a source of information; this is a creation of heritage based on landscape. The idea is to stimulate the life of the city even before its buildings are built." For a master plan of the so-called Boulevard Commercial District at Lusail (2006), the firm envisaged using elements such as whiteness and wood, which is frequently used in the Gulf to define future architecture, rather than referencing European boulevards, which was inscribed at the outset in the program. Here, quality public space is developed with large shaded and generously planted sidewalk areas, ensuring that movement occurs at a human pace, allowing pedestrians to feel fully integrated into the movement of the city and not left behind by a parade of fast vehicles.

The broad conceptualization of architectural heritage brought forward by Architecture-Studio is very much in keeping with their analysis of context and patterns of use as a basis for the success of a building, rather than any insistence on a specific style or iconic form. In their analysis, the very nature of architectural heritage has evolved substantially now taking into account modern forms, even industrial ones that never would have been considered to be viable parts of urban or national heritage in the past. Alain Bretagnolle concludes, "we think it is possible to discover new uses for old buildings, doing things that in fact create a patrimonial value that would not have existed otherwise, this is true for our work at Jussieu or Dunkirk, surely. The heritage of the future is also made by creating new uses for existing buildings."

年）位于由德国GMP建筑事务所规划的临港新城，项目开始于新城的发展初期，而那时新城的环境正如合伙人描述的“荒无人烟”一般。因此针对此项目，建筑师团队从将在此展出的传统国画的诗意中获取灵感，可以说本项目更注重的是内容而非环境，但建筑却与环境完美的融合。这是一个从与建筑无关的文化遗产中获得灵感，并通过一座建筑造就一处风景的绝佳诠释。

另一个需要实现“从无到有”的设计是多哈卢塞尔新城公共空间景观规划项目，该区域直至最近几年还是一片沙漠无人地带。工作室与景观设计师米歇尔 · 戴斯威纳合作重新定义该区域的景观。当地某些景观的特征成为景观规划的要素，比如生态公园成为了城市规划的中心。罗伊达 · 阿亚斯解释道：“我们利用卡塔尔原有风景的显著特征作为创意的灵感来源，这是一种基于风景的建筑遗产和文化认同的创作。其理念在于，建筑落成之前便可通过景观促进城市的发展。”在卢塞尔新城商业街道总体规划项目中（2006年），工作室提出混合使用白色材料和木材，让步行街更具波斯湾特色，而不是单纯的像项目要求那样模仿欧洲的商业大道。在规划中，高质量的公共空间由林荫大道组成，确保行人出行的舒适度，能够让行人完全融入到城市的运转当中。

这种与建筑遗产相关的构思方法，和法国AS建筑工作室对于建筑环境和建筑使用性的分析方法是一致的，即不固守某一特定风格或表现形式。合伙人认为，建筑遗产的定义在不断的发展，现代建筑与工业建筑已逐渐被纳为建筑遗产，而在过去，它们的文化价值、使用价值以及遗产价值是不被认可的。阿兰 · 布勒塔尼奥勒总结道：“赋予历史建筑新的使用功能可以让其建筑遗产价值更加升华，正如我们对巴黎第六大学Jussieu校区东区的改造和对敦刻尔克城堡大学的改造工程。而未来的建筑遗产亦是通过对现有建筑增加新功能而形成的。”

Due to its size and ambition, the Muscat Cultural Centre has created a brand new cultural district for the city, where the present meets the past.

马斯喀特文化中心的宏大规模和雄心规划，为阿曼苏丹国的首都创造一个全新的文化区域，历史与当代在这里相遇。

PAM011

Extension and restructuring of Novancia Business School
Paris, France
2006/2010
22 360 m²

诺凡西亚高等商学院
法国，巴黎
2006/2010
22 360 m²

This project, on behalf of the Parisian Chamber of Commerce and Industry, deals with the extension and restoration of an existing school that was built in 1908. The immediate surroundings of the building have the warm color of brick and earthenware, which is in stark contrast with the city's typical cream limestone, often turned grey. The greater the knowledge as to what is already there, the more inventive renovation can be.
A reminder of the fire-colored bricks varying from red to yellow, a series of vertical sunshades light up the façade. The volume and proportions tying it in with the Parisian Haussmannian style have been respected but are set to new architectonic rhythms.
This selective business school, with its 1 555 students, wanted a new visual identity. The restoration of the building aims at underscoring the beauty of the original design while adding the urban dimension which it lacked with an extra atrium and buffer zones.
This part of the city is a junction on the city's boundary. Alongside the classrooms, the project includes three theaters, a 300-seat auditorium, a film-set and a video control system, offices, catering areas. Heritage and contemporary design work hand in hand. Born out of collaborative harmony, this place for learning and understanding will be an opportunity for many generations to come. The automated glass sunshades regulate the ingress of sunlight. They treat light and color like modeling clay, reshaping the building from dawn to sunset. H.L.

巴黎工商会希望整修并扩建这座建于1908年的学校建筑。与巴黎建筑最常见的灰色或奶白色石灰石墙面不同，该建筑周边街区的红砖建筑外墙显得尤为突出。随着对周边环境的深入了解，本项目的改造便越具有创新的精神。与周边的红砖相呼应，红色与黄色的垂直遮阳板外立面将建筑包裹，像一团火焰一般夺目。改造后，这座容纳1555名优秀学生的高等商学院建筑既保留了巴黎奥斯曼式建筑的特点，又带来了别样的风情和更高的辨识度。为建筑增添美感的同时，更为建筑设置了原本缺失的中庭和缓冲区等社交空间，此处就像是建筑与外部城市的交界处。除了教室外，建筑内还包括三个阶梯教室、一个能容纳300位观众的礼堂、一套摄影棚系统、办公室和餐饮区。建筑遗产与当代化设计在该改造项目上完美融合，为师生间和校友间的一代又一代传承提供场所。彩色玻璃遮阳板随着阳光自动调节，灵动的色彩让建筑不断变幻、充满生机。H.L.

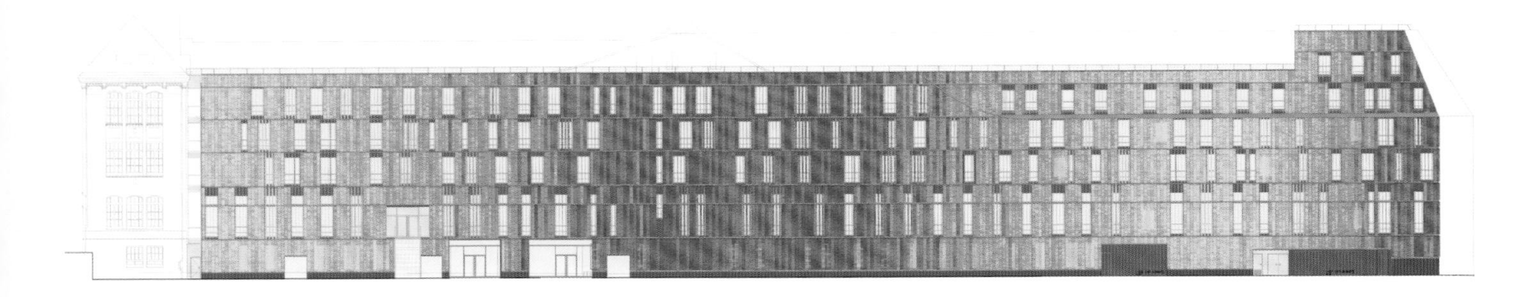

Façades on the courtyard, on Rue Antoine Bourdelle (top) and on Rue Armand Moisant (below).

面向安托万 · 布德尔街和庭院的立面设计（上图），和面向阿尔曼 · 莫桑街的立面设计（下图）。

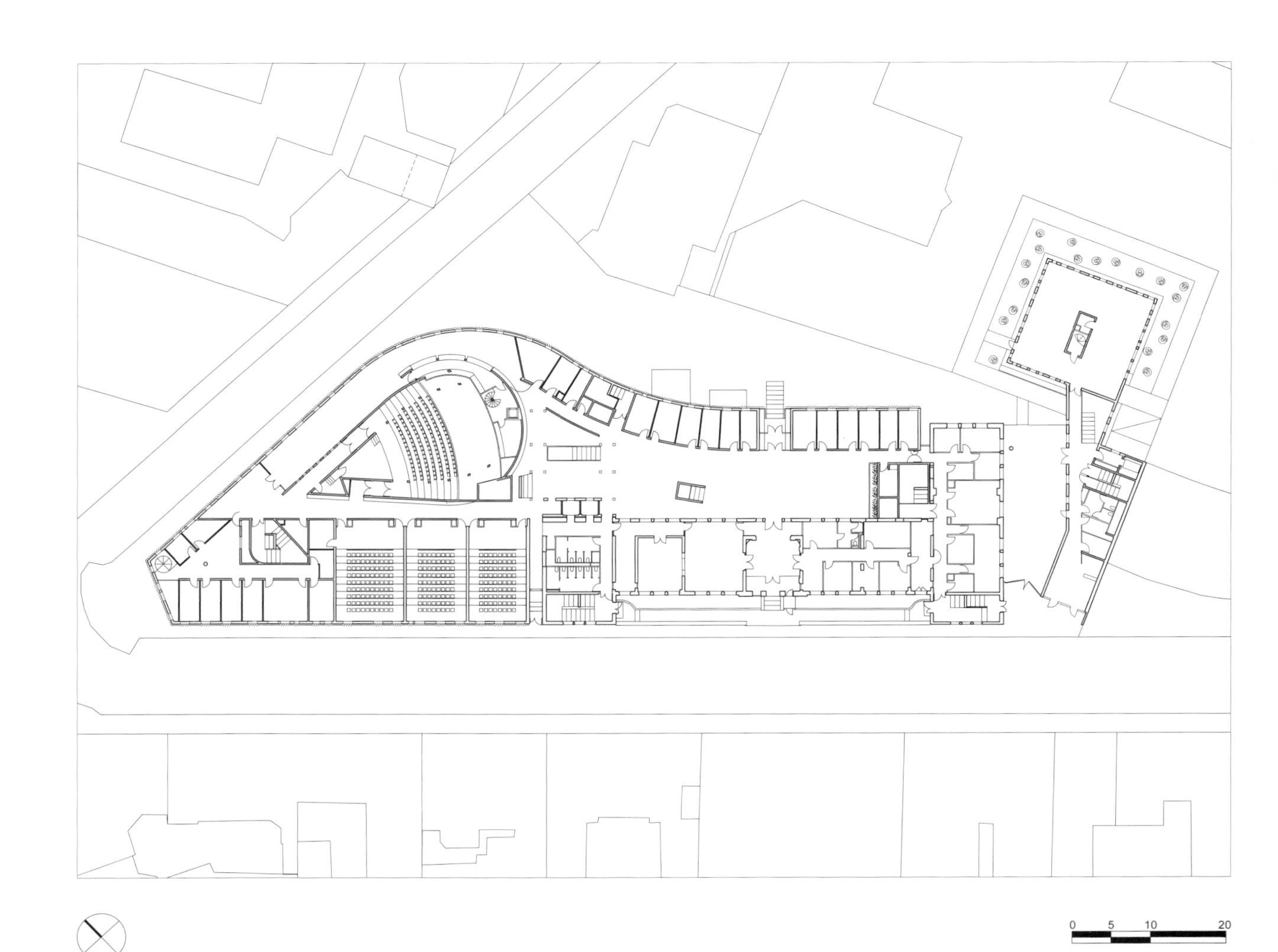

First floor plan.

二层平面图。

The atrium, the link between the old and new buildings.

旧建筑与扩建新建筑之间的连接中庭。

The view from Rue Armand Moisant.

阿尔曼 · 莫桑街的视角。

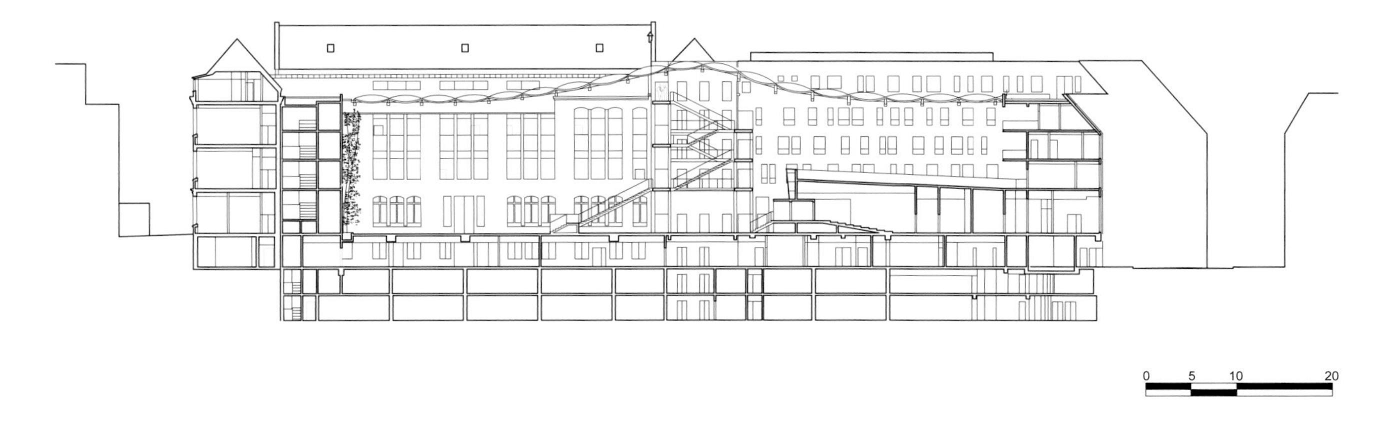

Longitudinal section of the atrium.

中庭纵剖面。

BHR1

National Theater of Bahrain
Al Manama, Bahrain
2003/2012
12 000 m²

巴林国家大剧院
巴林，麦纳麦
2003/2012
12 000 m²

Bahrain, in Arabic, means “two seas.” This name reflects the geographic situation of the island country, composed of an archipelago of the western shores of the Persian Gulf. Its longitudinal shape has inspired the project. A 12 000-square-meter theater has been oriented toward the lagoon so as to draw your gaze onto the glorious view. The translucent openness of the edifice is in harmony with the National Museum, which shares with it a common space uplifted by the new cultural dynamism. The kingdom has put this project to the fore and has invested in it to promote its international image. Built between land and water, the National Theater of Bahrain has a distinctive plaited aluminum-mesh canopy, which is inspired by local wicker roofs. This large

巴林在阿拉伯语中意为“两个海洋间的国家”，这个名字反映了位于波斯湾的该岛国之地理特征，这里海天一线的景观为项目带来灵感。面积为一万两千平方米的大剧院立于无限水平延伸的潟湖之上，景观壮丽。大剧院透明的外立面与不远处的国家博物馆相呼应，与之共享公共空间，并共同为未来创造文化活力。巴林十分重视这个项目，希望借此机会宣传国际化形象。受当地传统的棕榈编织屋顶启发，大剧院巨大的铝带编织遮棚不仅具有特色，而且能够过滤阳光，在这炎热的国度，为大剧院内部和外圈前廊提供

umbrella, that filters the blinding sunlight and moderates the high temperatures, covers a long ambulatory. Embedded in the canopy, this metallic shell houses a 1 001-seat auditorium, a number referring to the Arabian Nights. A counterpoint to the translucence of the main hall, the golden-colored case inside is self-contained, as if to say: here drama is, literally speaking, contained in a nutshell. Its organic beech-wood curves resemble the hull of a dhow, the pearl fishers' boat. Emblematic of the country's traditional trade, the pearl now symbolizes its commitment to fully engage with the global community. Culture and aspiration have merged. A new Arab soft power is born. H.L.

一丝阴凉。含有1 001个座位的剧院厅被包裹在一个镶嵌于铝制顶棚的巨型金属外壳之中，象征着一千零一夜。与大堂透明玻璃立面相对应的是闪耀着金属光泽的“珍宝盒”，似乎在传达着：金属壳的内部才是上演精彩剧目的剧院厅。剧院厅内部由榆木打造，勾勒出的形状宛如当地传统出海捕珍珠的渔船。珍珠代表巴林的传统贸易，尽管如今珍珠已不再是巴林与外界交流的唯一因素，它依然象征着巴林与国际社会充分接触的决心。项目实现了文化与梦想的融合，代表并促进着阿拉伯的软实力。H.L.

View of the lobby.

大堂视图。

View of the 1 001-seat auditorium.

拥有1001个坐席的剧院厅。

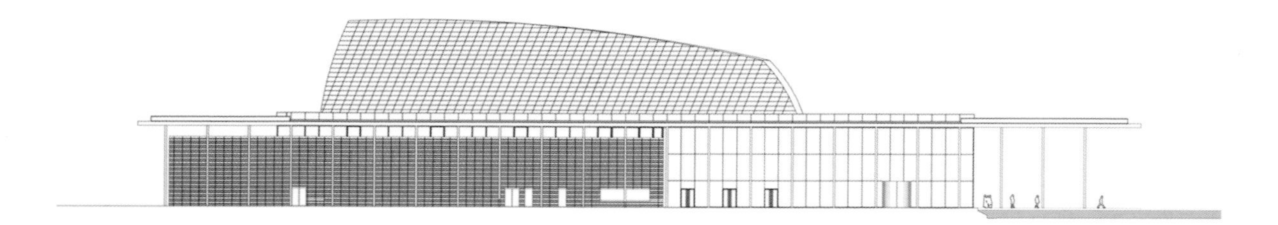

East façade.

东向立面。

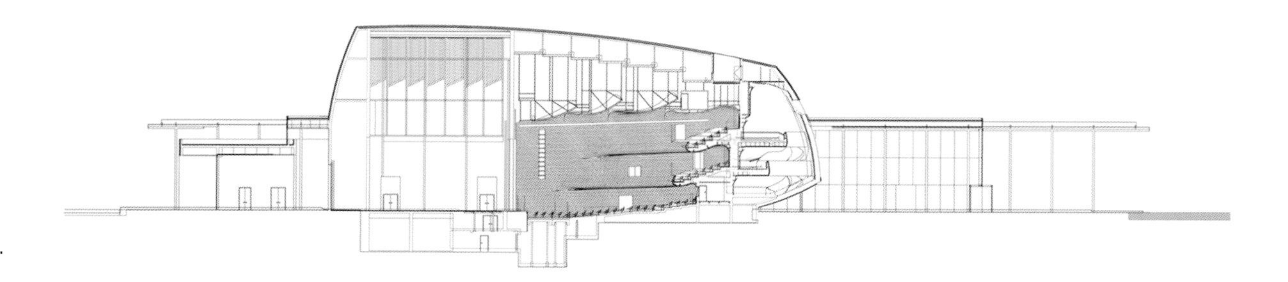

Longitudinal section.

纵剖面。

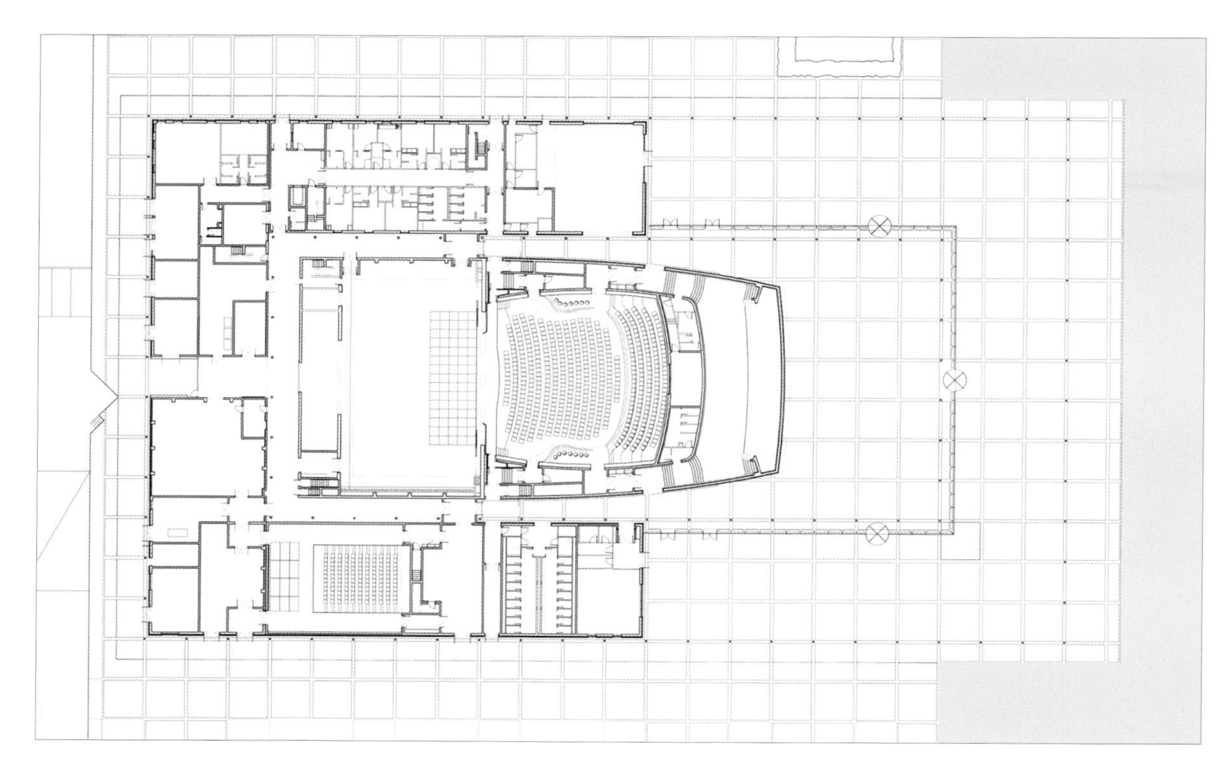

Ground floor plan.

一层平面图。

View of the *foyers*.

休息室视图。

PAPRK2

Restructuring of the Maison de la Radio
Paris, France
2005/2014
110 000 m

法国广播电台大厦改建工程
法国，巴黎
2005/2014
110 000 m²

Increasingly, architectural design is characterized by mental plasticity uniting optimism and engineering for the creation of forms. No building can stand on the leg of design alone and treat urban planning as some inferior limb. An ambitious rehabilitation plan demands even more pragmatic wisdom than it does technical know-how. You need intuition, which can only be acquired through a deep understanding of the location and respect for it. Restructuring the French Broadcasting House, Maison de la Radio, was a *tour de force* because it happened in a working environment. To avoid getting in the way of employees and staff carrying on with their jobs, new ways of navigating inside the building had to be devised, with extra corridors. The plan also comprised an agora, an auditorium with a 1 461-seat capacity, an underground parking lot for 700 vehicles and new green areas outside. Architecture-Studio has kept the original cylindrical design by Henri Bernard. However, the circulation had to be reorganized. Linking the core building to its circular periphery with an agora made sense of the edifice's various functions. Four new 32-meter long glass gangways bridges the outer ring to the smaller ring on the 5th floor level. Having the parking lot situated underground liberated more green space so as to set off the design of the building. The entrance and park were co-created with landscape designer Michel Desvigne. The library of Radio France's archives has been turned into a fluid office space. The production and recording studios that used to be oriented toward the inside of the building now look out onto the city's panorama. The famous Studio 104 has become a concert hall where all types of music will be played. This is a national treasure reborn.
The Grand Auditorium embodies the saying of end-of-eighteenth century French architect Claude-Nicolas Ledoux: "Architecture should draw the spectator into the alluring arms of wonder." To keep this sense of awe, the distance between the stage and audience does not exceed 17 meters. This all-embracing space, that was produced with incredible economies of scale and a baroque touch (the wainscot paneling looks like sculpture) combines both majesty and intimacy.
The acoustics have been carefully tailored in collaboration with Nagata Acoustics.
The Auditorium's stage – 22-meter wide and 15-meter deep – is equal to the greatest in the world. Like drapes that have been slightly pulled, the wood-panels behind the seats above the choir let you catch a glimpse of the monumental organ designed by Gerhard Grenzing. With its spacious glass atriums overlooking the River Seine, the renovated Maison de la Radio symbolizes an ever more welcoming attitude toward the public. It is an open-mindedness that has made Radio France renowned worldwide.
H.L.

建筑的概念设计越来越需要积极向上的态度和造型创造的巧思，以将既定的环境与经验、能力和专业性等更好地融合，为创建可持续发展城市做出贡献。建筑设计不能够独立存在，而需要与城市规划相配合。负责一个大规模改造项目，务实的智慧显得比技术专业更加重要：只有通过深入了解建筑本身以及所处的环境，才能获得利于改造的灵感与直觉。法国广播电台大厦的改建工程是个费力的工程，不但需要保证电台的正常运作。而且需要在不太过多打搅员工工作的情况下，重塑建筑内部的通道流线、打造一条内部街道和一个广场、完成含1461座的大音乐厅、设置含700停车位的地下停车场，并整合户外的绿化空间。法国AS建筑工作室的改造项目保留了亨利 · 贝尔纳设计的原始建筑外形，但重新规划了建筑内部的流线，以提高空间使用的便捷性。中心广场将环形建筑的所有功能区域紧密相连。新增的四条32米长的玻璃通道设置在第六层，将内环建筑和外环建筑相连。地下车库的设置为地面绿化提供了更大的空间，建筑被环绕在内、外花园之中，这些绿化景观既是花园又是公共公园，由景观设计师米歇尔 · 戴斯威纳完成。曾作为法国广播电台档案储存空间的大楼如今被改造为宽敞明亮的工作空间。过去朝向建筑内部的录音棚和监控室，经改造后均拥有朝向城市的视野。著名的104录音室则成为了一个多功能音乐厅，可承办多种风格的音乐会。这是一个国家级建筑遗产的重生。
法国广播电台内的大音乐厅完美的体现了十八世纪法国建筑师克劳德 · 尼古拉 · 勒杜的名言："建筑应将观众环抱在极具魅力的美妙之中"，其舞台和环绕在四周的观众席之间最远的距离不超过17米。大音乐厅的震撼之处不仅在于观众与中心乐池的超近距离，还在于如同雕塑般的木质内壁营造的巴洛克风格。声学部分由法国AS建筑工作室与Nagata Acoustics公司共同研究，以保证大音乐厅的优秀声学效果。长22米，宽15米的舞台规模可以与世上现存最大的舞台相媲美。在大音乐厅中轴线上，唱诗班阶梯上方，木质的墙壁微微开启，展现出由Gerhard Grenzing设计的巨型管风琴。大音乐厅的接待空间位于面向塞纳河美景的落地窗长廊之中，透明化处理象征着改造后的法国广播电台对公众更加开放的态度，而这种开放的态度正是法国广播电台闻名于全球的原因。H.L.

View from the agora.

从中心广场仰望大楼。

View of the inner thoroughfare.

内部街道视图。

Longitudinal section.

纵剖面。

Aerial view, on the Seine
River in western Paris.

巴黎塞纳河右岸的鸟瞰图。

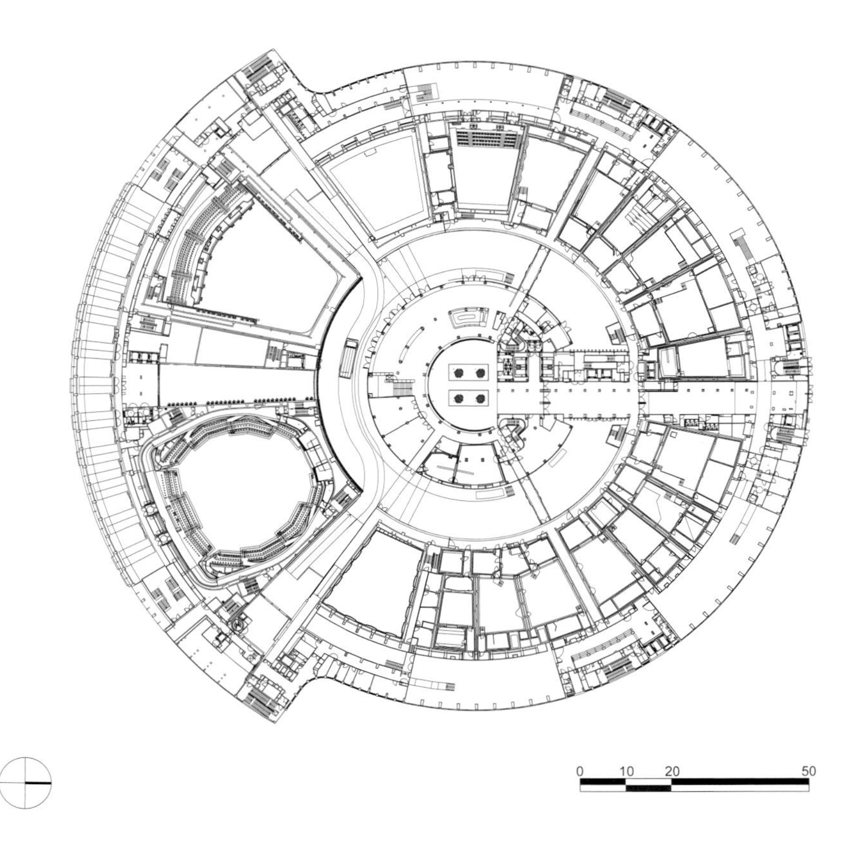

First floor plan.

二层平面图。

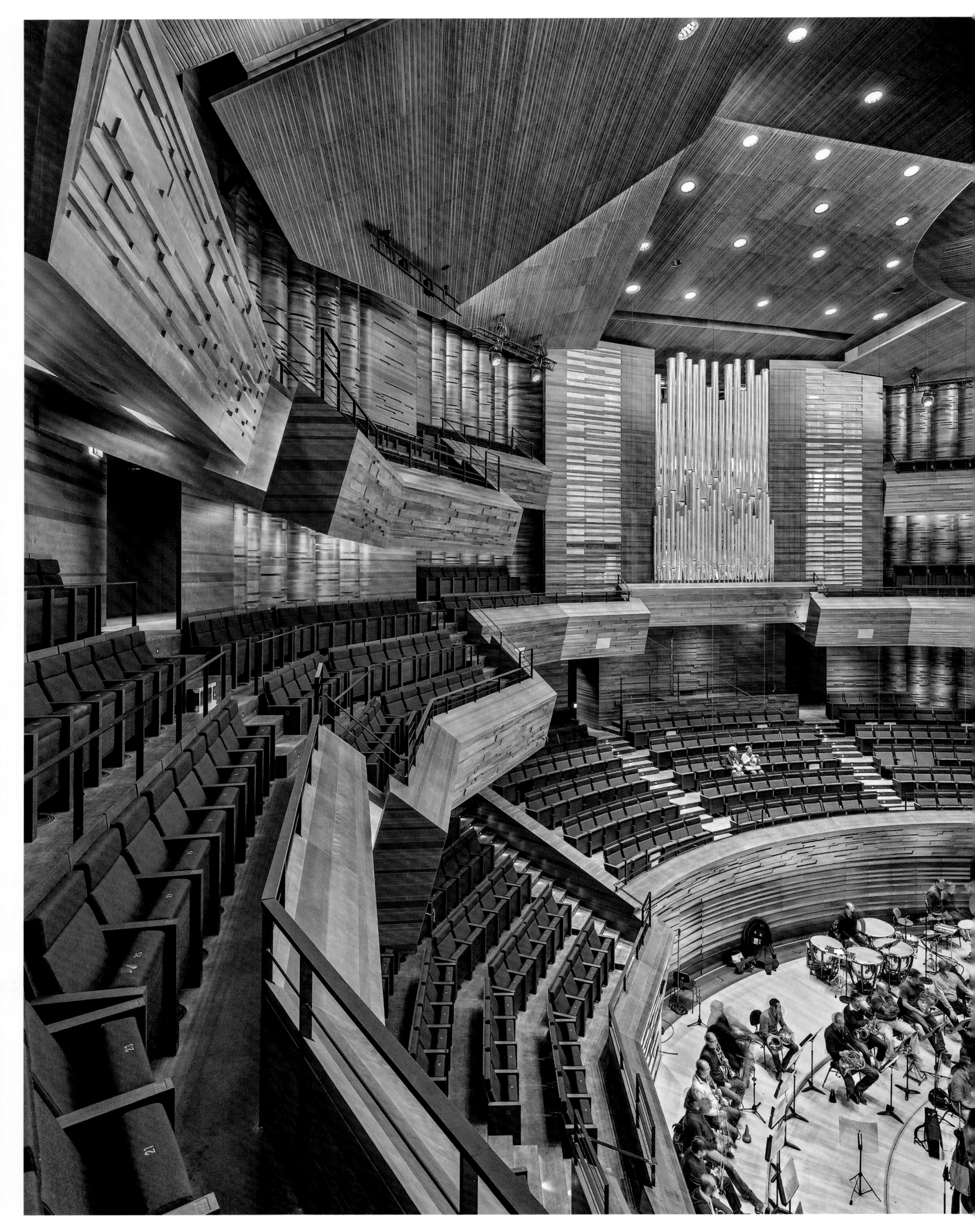

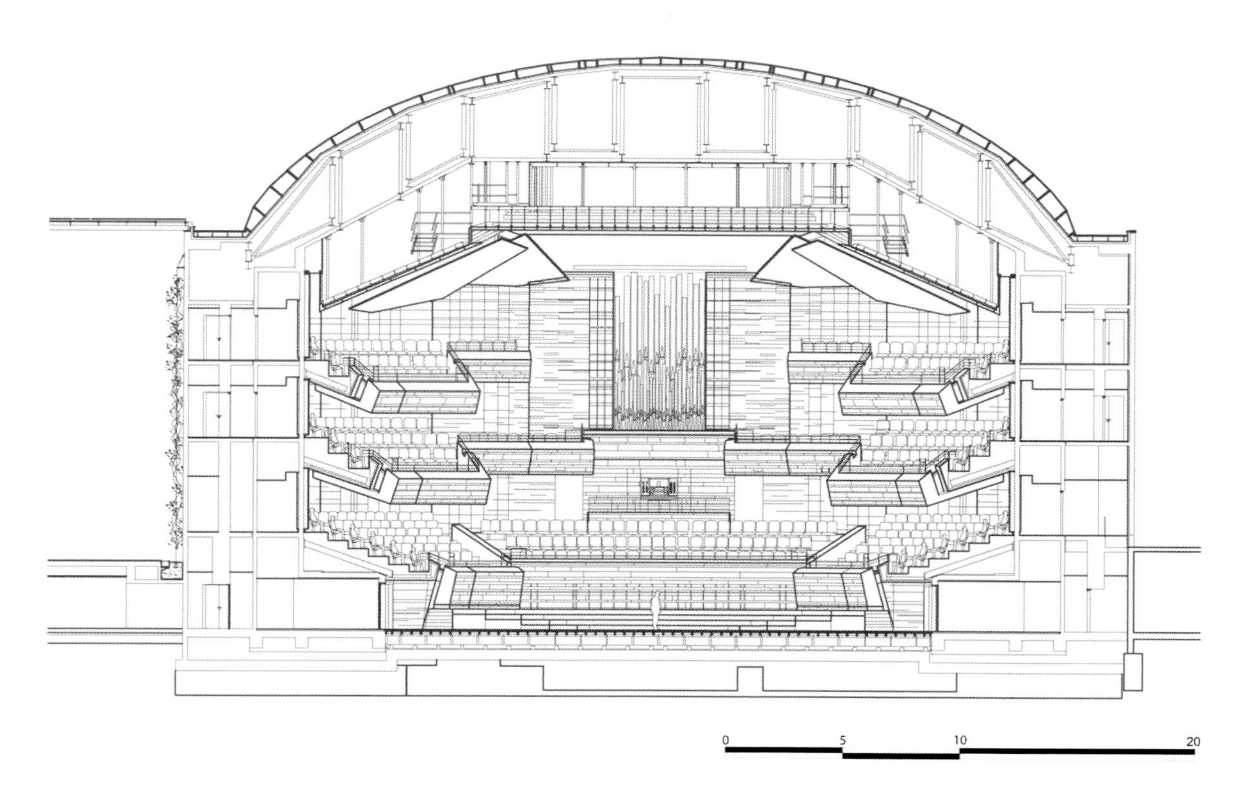

Cross section of the Grand Auditorium.

大音乐厅横剖面。

Previous page: The Grand Auditorium of Radio France.

前页：法国广播电台大音乐厅。

PAJUS5

Restructuring of the Eastern Sector of the Jussieu Campus
Paris, France
2008/2015
100 000 m²

巴黎第六大学Jussieu校区东区改造工程
法国，巴黎
2008/2015
100 000 m²

The restructuration of University of Paris VI-UPMC (Pierre and Marie Curie) campus began after all the asbestos was removed, and was made on an occupied site.
It comprises the research laboratories of the physics, biology and chemistry departments, conference and seminar rooms, a 500-seat lecture hall and the university library. Due to the quality of its academic team and its sheer size, this campus is unique in its kind in France, let alone Europe.
The scale and rigorous, if not dry, aesthetic are the imprint of Seventies architecture, like the works of art of the same period which have been kept on the site. Architecture-Studio's mission was to refurbish the campus's surroundings and rehabilitate its global visual identity, making it a landmark in the eastern part of the capital where it is situated. Édouard Albert's original design was thus put back into perspective and highlighted in its urban context. This site has amazing clarity because of the use of transparent signboards made of glass. Plants and greenery counterbalance the overall mineral atmosphere. A poetic network of pathways keeps pedestrian navigation fluid. The campus and the esplanade were designed with the same idea of fluidity in order to facilitate access to the neighborhood and extend the urban axes. Your gaze is drawn towards the open perspective, the trees of the Jardin des Plantes and the façade of the Institut du Monde Arabe. The campus and its surroundings fully embrace this urban context. Open to the city, Architecture-Studio was able to capture its dynamic and rephrase it into the style of the building. H.L.

巴黎第六大学（皮埃尔和玛丽·居里大学）的改造在拆除了所有石棉后的原址上进行。改造内容包括物理、生物和化学系的研究实验室、教学空间、学生生活中心、科研会议中心、一个500座的阶梯教室和大学图书馆。凭借高质量的教学，以及超大的占地规模，巴黎第六大学Jussieu校区是法国乃至欧洲最大的大学科研中心。它的规模及严谨的几何体量体现的是整个街区的尺度，它的整体设计理念是70年代的典型建筑象征，就像保留在现场的怀旧艺术作品。法国AS建筑工作室对于Jussieu校区东区的改造完工意味着整个校区改造工程的结束，也意味着此处将重新成为巴黎东部的地理标志。在这次新的构思中，法国AS建筑工作室将爱德华·阿尔伯特的原始设计与城市环境有了更深层的融合。玻璃材料被大量运用于区间划分，让项目整体拥有高度透明度；随着绿化带的嵌入，校区内钢筋水泥建筑的冰冷感被削弱，显得更加柔和；充满诗意的通道网络让行人穿梭自如。校区内部与校园对面广场的统一规划，让Jussieu校区与周边街区的联系更加密切，城市轴线得以延伸。而水平面的视野经过改造也变得更加开阔，能够看见巴黎植物园内的植被和阿拉伯世界研究中心的外立面，校园及周边环境充分体现了这个城市的文脉。法国AS建筑工作室以对城市开放的改造方式，捕捉城市动态，并将其更好的融入建筑风格中。H.L.

View of the License Library.

本科学生图书馆视图。

View of the covered patio.

带顶篷内院视图。

View of the court.

庭院视图。

SUD 43

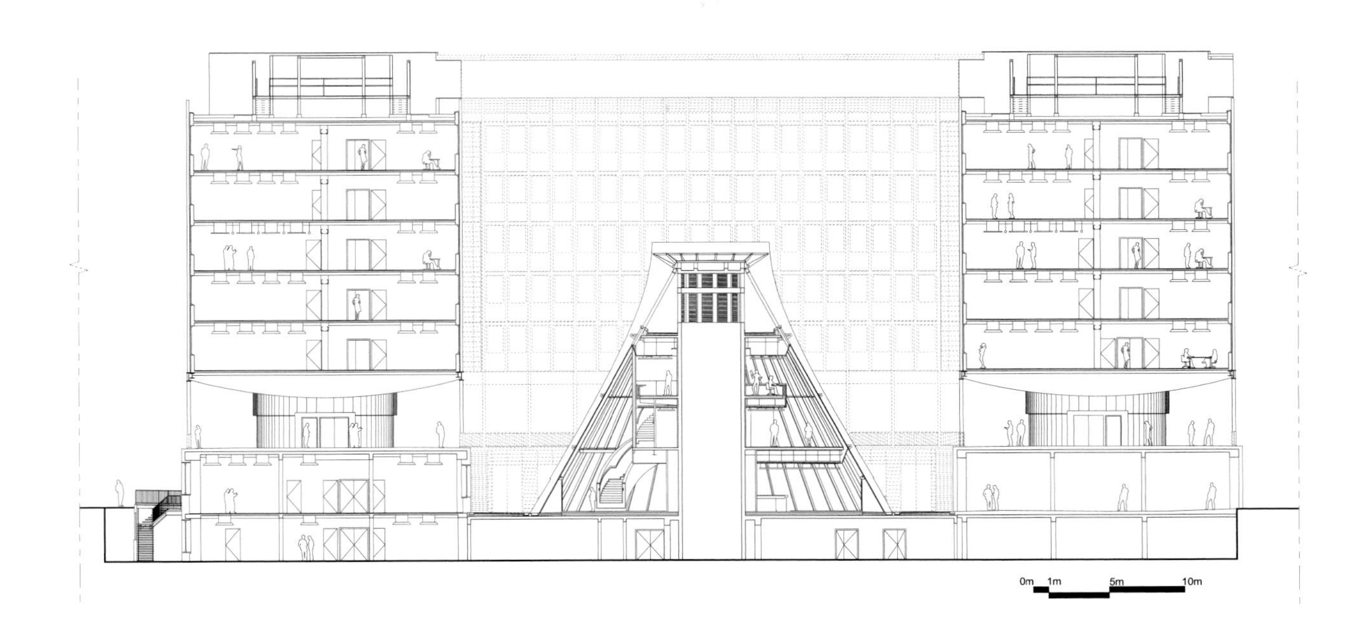

Section and view of the Tipi.

Tipi的剖面和视图。

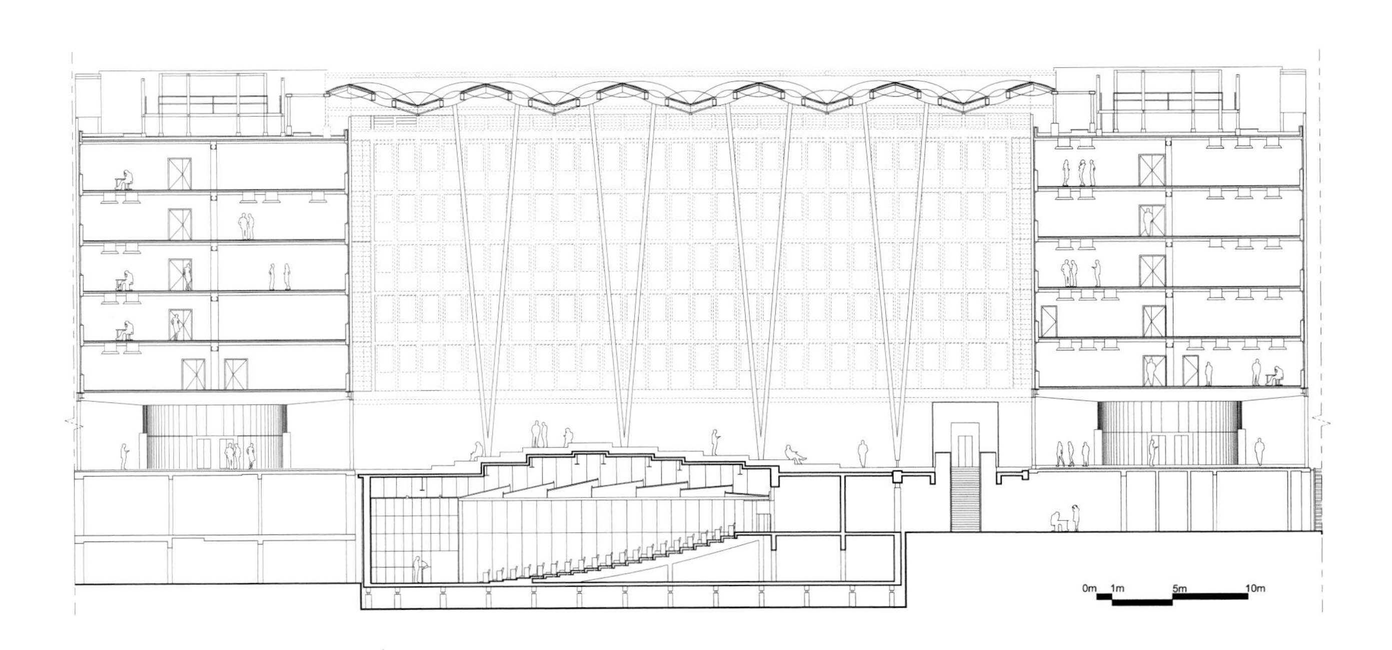

Section and view
of the covered patio.

带顶篷内院的剖面和视图。

CRT2

Extension of Créteil's Cathedral
Créteil, France
2008/2015
1 400 m²

克雷泰伊大教堂扩建
法国，克雷泰伊
2008/2015
1 400 m²

The extension of Créteil's Cathedral was designed to hold a congregation of 350 people and intended to become a landmark in the city. When extending this building, the original design of Charles Gustave Stoskopf, emblematic of Seventies theology – a church buried in the city – was on Architecture-Studio's mind. Both architectural concepts had to be able to respond to each other and, though they differed, needed to be geometrically coherent. The main entrance, designed on a human scale, ties in well with the monumental volume of the new project. Today's dome is respectful of the one drafted in the original plan.
The church's roof is composed of two semi-spheres joined together like hands in prayer but leaving an interstice of pastel blue that recalls the sky, hanging above the altar. Seating levels have been distributed around the church's interior space, thus enlarging the size of the cathedral and allowing more seating capacity. The choir has been lowered, tracing a neat curve into the ground, with the seats forming a hemicycle. Detached from the main building, the steeple stands at the corner of the church entrance, thrusting its tall and elegant silhouette into the heights as if sending a message of spiritual elevation. With its urban scale it integrates the cathedral into its surroundings. The dome in the shape of an ogive restores the awe-inspiring presence of a place of worship. The cathedral is the new "face" of this once ignored popular district made of non-descript apartment blocks. The newly designed entrance and adjoining square are like a breath of fresh air in the neighborhood, revitalizing parish life.
H.L.

克雷泰伊大教堂的扩建工程旨在将教堂容量扩充为350座，并将其打造为城市标志性建筑。法国AS建筑工作室希望保留Charles-Gustave Stoskopf于70年代所设计大教堂的原有基础——隐于闹市的教堂风格。教堂原有建筑与扩建部分通过既各具特色又和谐共处的几何建筑语言构建对话。具有人性化尺度的原有入口处轮廓，与新项目以中殿为中心的宏大规模相连接。穹顶的形状遵循原有建筑平面图的轨迹，两个半球型外壳朝向天空犹如祈祷的双手汇聚在圣坛之上。新添加的看台大幅提高了教堂的可容纳人数，原有祭坛被保留下来，并以其为中心设计了半弧形座席。教堂钟楼独立矗立于广场一角，修长的外形凸显着教堂入口。钟楼顶端指向天空的设计不仅代表了教堂的精神象征，而且让克雷泰伊大教堂融入到城区中。拱形圆顶赋予了大教堂新面貌，并使其在由高层住宅组成的街区中凸显出来。教堂前的广场和新绿化区域使教堂周围的空间更宽阔，重新定义了室外空间，并打造富有活力的教区生活空间。H.L.

View from Avenue Charles de Gaulle.
夏尔·戴高乐大道的视角。

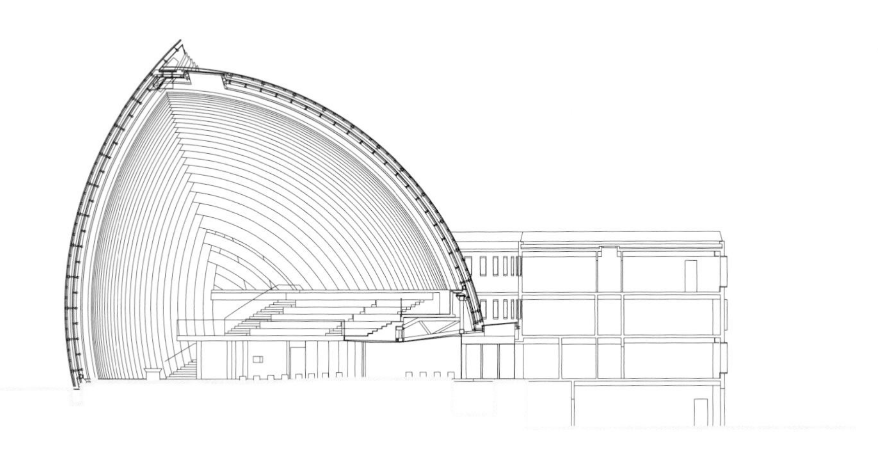

Transversal section.

横剖面。

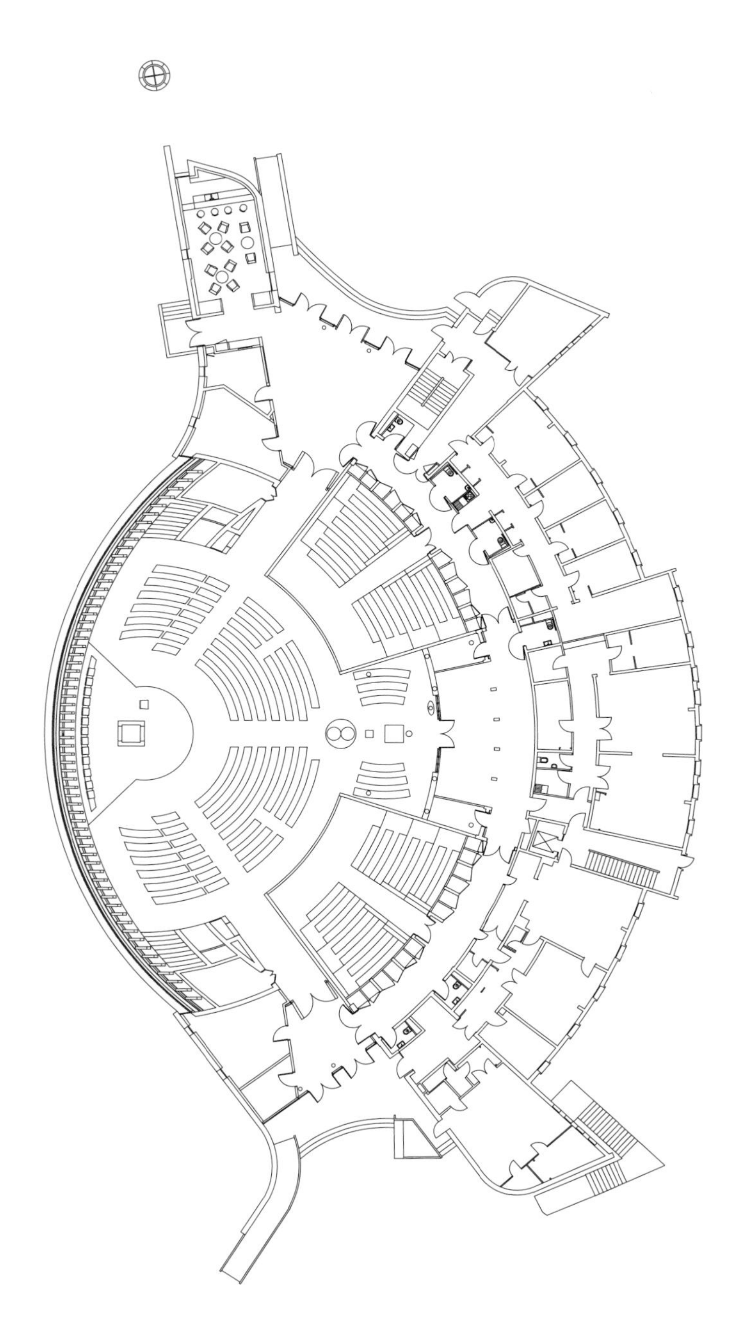

Ground floor plan.

一层平面图。

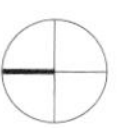

0 5 10 20

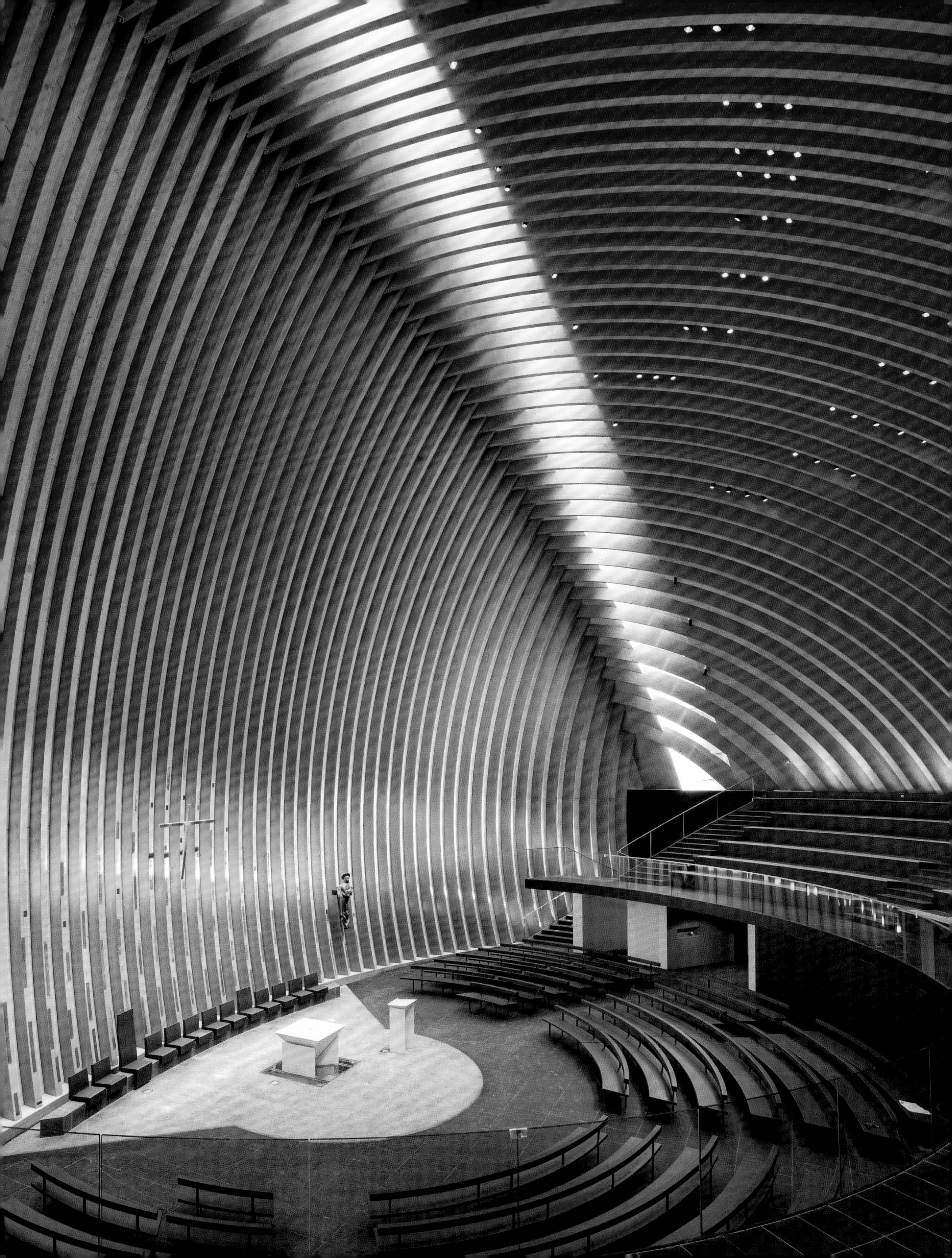

SHPG30

Xie Zhiliu and Chen Peiqiu Art Gallery
Shanghai, China
2012/2016
11 000 m²

谢稚柳陈佩秋艺术馆
中国，上海
2012/2016
11 000 m²

This famous Shanghai art gallery owes its name to two Chinese landscape artists. Xie Zhiliu and Chen Peiqiu practice traditional painting, which is deeply rooted in China's collective imagination and strikes a chord in the Chinese psyche. This art venue, designed using the same principles, is itself a painting: elegant design, simple layout, vivid colors, harmonious views... Visitors navigate through this space as though they were walking through a painted landscape. Exterior and interior arches open up to high ceilings: your gaze is drawn toward the faraway horizon. You feel an amazing sense of space as soon as you enter. This project is the fruit of Architecture-Studio's research on traditional Chinese painting and its free interpretation of it. According to the principles of this noble art, every inlay of ink is not just an isolated mark but actively contributes to the whole picture by intermingling and mixing constantly in order to liven up the landscape. Likewise, each part of the project is not merely present but constitutes an integrated element of the organic whole. Much as they differ, both approaches end up with the same conclusion. Whatever the material used, the work prolongs the natural space contemplated by the painter. Here the architect has transposed the fantasy of the brush onto stone. H.L.

这座位于上海的艺术馆得名于两位中国当代国画艺术家代表——谢稚柳和陈佩秋，他们不仅在绘画艺术上有着极高的造诣，同时在对中国文化的推广上有着突出的贡献。传统的国画致力于意境的表达，写意往往大于求形似。艺术馆的设计运用了相同的逻辑：优美的线条勾勒出简单的轮廓，明快的颜色搭配简洁的布局和清幽的景色，已然一种人在画中游的意境。拱形的建筑底部，像一座座开启的门洞，让人们有机会看到后方下一层的建筑，却又因为每一层的拱形大小不一，位置不一，营造出"犹抱琵琶半遮面"的效果，吸引参观者进入，穿梭其间，感受空间的改变。建筑布局的灵感来自于法国AS建筑工作室对传统绘画技法的研究以及对中国古典审美的理解。传统的绘画技法需要墨色的层层叠加，才能形成一副完整的画作，每一层互相渗透，互相影响，缺一不可。建筑布局异曲同工，每一部分都是一"层"，且逐"层"递进。每一层都有自己独立的形态和功能，但却又相互联系，组成一幅完整的图画。建筑师运用建筑材料，将如国画般极具意境的优美建筑呈现在自然景观之中。H.L.

Ground floor plan.

一层平面图。

0 5 10 20

View from Lake Dishui.

位于滴水湖畔的艺术馆。

PACAB4

Renovation of the Sainte-Anne Hospital: Psychiatric Clinic
Paris, France
2004-2012
6 000 m²

圣安娜医院：精神治疗中心翻修
法国，巴黎
2004-2012
6 000 m²

Since its founding in 1870, the Sainte-Anne Hospital has known a considerable number of changes due to progress in psychiatry – from prevention of mental disorders to the latest developments in neuroscience. Sainte-Anne, located on the historical 14-hectare site in the 14th arrondissement of Paris, has preserved much of the original architecture designed by Charles-Auguste Questel. The building had lost coherence and harmony through successive additions and piecemeal changes. Despite changing attitudes towards the mentally ill, this hospital was redolent with madhouse imagery which had been romanticized by the presence of such patients as the great poets Antonin Artaud and Paul Celan (committed respectively to the institution in 1937 and in 1967). A new architectural scheme was therefore been launched in order to convey a greater synergy with the urban context. Being in the public eye, it had to reflect the concerns of both the hospital and the city. Over the next 15 years, this project will provide a museum of psychiatry, a conference hall dedicated to research and cultural and public health equipment. The approach is progressive so as not prevent Sainte-Anne from delivering its public service. This prospective project keeps options open for future development. At the same time, the client expected Architecture-Studio to build an up-to-date clinic specializing in mental health and brain-related issues, as befits the latest medical advances. So the hospital will be provided with a quantitative psychopathology unit and a therapeutic mediation unit. Psychiatric care per se will be divided into three sections with a total capacity of 77 beds. There are both surgeries and classrooms adapted to both university teaching and compatible with hospital care. Section 3 allows for such a combination: 26 beds, 10 open wards, accessible to an enclosed garden. The design of a new clinic illustrates Sainte-Anne's ambition of renewing its image and showcasing a strong vision for the future: the highest standards for the accommodation of patients, better working conditions for medical and non-medical staff through a well-defined circulation separating the different services. While helping people find their way round, the project enables them to appreciate the hospital's original building and garden. The new building integrates the arc of the historic composition and the symmetric design by Questel. In addition, much has been improved between the hospital and its neighborhood. This clinic does not inspire fear. Nothing in its style is awe-inspiring; on the contrary, it exudes warmth and humanity. Letting sunshine come in is a way of showing openness and shaking off the social stigma attached to mental illness, as if to say: all is calm. H.L.

自从1870年成立以来，圣安娜医院随着精神病学的发展——从过去单纯的预防功能到最新应用的神经系统科学等，使得医院发生了相当多的变化，进而历经了数次改造。位于巴黎第十四区的圣安娜医院占地面积14公顷，建筑师Charles-Auguste Questel设计的原始建筑大部分被保留了下来。由于不断增加的设施和零碎的改动，整体建筑群已经失去了一致性与和谐性。尽管人们对精神病患者的态度有所转变，或许因伟大的诗人安托南·阿尔托和保罗·策兰分别曾于1937年和1967年住院的经历，为精神病院增加了一丝浪漫的气息，但人们对于精神病院的传统印象似乎没有多大改变。作为一个公共项目，圣安娜医院全新的总体规划需要医院与城市共同商议完成，以实现和城市形成更好的协同作用。该总体规划将作为未来十五年发展的指导，其中包括在不防碍医院正常运作的情况下，渐进式地建设一所精神病学博物馆、一间专门用于科研讨论的会议厅、以及公共健康文化普及设备。前瞻的规划为未来发展提供了开放性的选择。
同时，圣安娜医院希望法国AS建筑工作室设计建造一个针对心理疾病和脑神经疾病的现代化治疗中心，以适应最新的医疗发展。医院配备一个咨询服务门诊、一个精神病理分析部门、一个治疗观察部门、医疗专科学院、以及共计77个床位的三个住院部。与之相连的三号区包括拥有26个床位的病房、十个位置的开放治疗区和一个封闭型花园。治疗中心的整体建筑设计和规划充分体现了新一代医院的设计特色，并且考虑到其今后可持续发展的功能：优化患者的住院康复环境、提高医务人员工作的舒适度、促进合理化的分流效果。位于圣安娜医院的建筑群主轴线上的治疗中心建筑，不仅为建筑群和花园景观增色，还具有导向功能。拱形建筑作为圆形景观天际线的一部分，遵从了Charles-Auguste Questel原始设计的对称性特点，成为圣安娜医院历史遗迹中心和周边街区之间联结的桥梁。人性化的建筑设计让治疗中心更加平易近人，不再使人恐惧。自然地光线在治疗中心被最大化利用，象征着社会给予精神疾病患者更多的包容，好像在向人们表达：一切都很平静。H.L.

View from Rue de la Santé.

健康街的视角。

DOH-

Landscape Design of Lusail Public Spaces
Doha, Qatar
2008/2022
1,300 000 m²
Landscape design: Michel Desvigne

卢塞尔公共空间景观改造
卡塔尔，多哈
2008/2022
1 300 000 m²
景观设计师：米歇尔·戴斯威纳

Lusail is a new city. Facing the sea, it is situated a few kilometers north of Doha, Qatar's capital. Once construction is completed, it will boast a surface of 35 million square meters, a true transformation that will have changed the desert environment into a modern urban one. In the master production schedule, Architecture-Studio was asked to design a commercial artery in order to upgrade a roadway and turn it into a lively boulevard with high standard apartment buildings, shopping malls and recreational areas for tourists.
The urban plan for buildings and public gardens on the coastline comprises a marina and promenades with a view of the sea, streets, squares and boulevards, playgrounds and sports grounds, showcase buildings and an eco-park – Wadi Park. The vision for Lusail has been one of a Garden City offering, much like an oasis, genuine comfort within a micro-climate where green space is usually present in the form of vegetal pockets in the middle of desert land. The Marina District constitutes the first phase of the project: a marina, a four-kilometer seafront with palm-trees, tree-lined boulevards making up an 18-kilometer network of promenades. The plan unfolds in a pattern of light and shadow, alternating piazzas and parks, cooled by the shade of trees and fountains. The orthogonal design underscores the project's formal unity. These public spaces are built with a particular care for finish, enhancing the aesthetic enjoyment of pedestrians, cyclists and customers of public transport alike.
Wadi Park is a 2.5-kilometer strip of green running between sand and sea. "Wadi" refers to the local valley where fresh water, usually scarce in Qatar, is found. The river irrigates the oases and mixes with seawater in the mangroves neighboring the shores. Wadi Park embodies the transition from dry land to sea with the three ecosystems forming the Qatari landscape. This sustainable park, though tourist-friendly, retains its natural integrity thanks to an educational farm and information desks along the paths. This park epitomizes public spaces respectful of their environment's identity and is a good example of how to design the sustainable city of the future. H.L.

卢塞尔是位于卡塔尔首都多哈以北数公里处的海滨新城，占地3 500万平方米，新城的建立改变了原有的荒漠状态，创造了一个现代生活的场所。由法国AS建筑工作室负责的总体规划包含商业主街道规划，以强化新城的主轴线，将其打造为高品质的林荫大道，设置高标准的公寓楼、舒适的购物区和形式多样的游客休闲区。卢塞尔新城的城市规划和公共景观改造包括滨海区、沿海大道、林荫大道、街道、广场、儿童娱乐区、体育活动场、标志性建筑和生态公园——Wadi公园。围绕着"城市-公园"这一理念，口袋状的大小绿化带有效的缓解沙漠气候带来的高温，打造舒适的微气候生活条件。首期海滨区项目包括一个游艇码头、一条4公里长的海滨大道、以及由棕榈小径和林荫大道组成的长约18公里的景观走廊。由公园、广场和藤架覆盖的人行小道等形成的网状结构更是提升了街区的景观。高质量的公共空间促进了公共交通、自行车和步行等绿色出行方式的使用。Wadi生态公园在沙漠和大海间的2.5公里半径范围内展开，Wadi的原意为溪谷，是卡塔尔稀少的由雨水侵蚀而形成的沟壑。淡水灌溉着绿洲，与海水的交界处形成了红树群落。公园内的三个生态系统展示着卡塔尔景观的特点，体现了沙漠到海洋的过渡。园内设置了方向路标、信号亭和教育农场以便于游客出行，同时又保留了自然的完整性。"城市-公园"的设计理念强调共享公共空间和尊重景观特征的重要性，勾画出一个未来的可持续发展的城市。H.L.

Wadi Eco Park extends over 2.5 km between the desert and the sea. *Wadi* means the small valley through which flows rare rainwater in Qatar. The Park travels through the three eco-systems that define the Qatari landscape. Visitors can discover Wadi Park thanks to marked out courses, information kiosks and an educational farm.

Wadi生态公园在沙漠和大海间的2.5公里半径范围内展开。Wadi的原意为溪谷，是卡塔尔稀少的由雨水侵蚀而形成的沟壑。公园内多样的生态系统展现了卡塔尔的地貌特征。游客可通过方向路标、信号亭和教育农场深度了解Wadi生态公园。

Eastern restaurant on Lusail Waterfront.

卢塞尔海滨区东岸饭店。

07
LA MAISON DE LA RADIO
La partition d'AS Architecture-Studio
ARCHITECTURE-STUDIO
NOUS NE LACHERONS RIEN !
la maison de la radio
Parti architectural
ORCHESTRE NATIONAL DE FRANCE
N° Y. 1601
GROUPEMENT DES ENTREPRISES
R.T.F.
MAISON DE LA RADIO
ARCHITECTES
Elévation Mur
MODIFICATIONS
COMMUNICATION AS
Studios Moyen
ORCHESTRE NATIONAL DE FRANCE
Musique, on ouvre !
MAISON DE LA RADIO
DE CHOSES A VOIR A LA RADIO.
WEEK-END
Z VOIR ADIO
PATRIMOINE
LA NOUVELLE MAISON DE LA RADIO
LE GENERAL DE GAULLE INAUG MAISON DE LA RADIO.
Le Général de Gaulle a ina hier soir la Maison de la Paris.
NIM - Le général de Gaulle le discours d'inauguration M. Alain PEYREFITTE (mini de l' information), et pré général, M. BORDAZ, direct général de la R.T.F.
69 858 AFP Photo 15.12
Une verrière sous tension
116, avenue

18-20432
L'ÉCOLE COMMERCIALE
DE LA CHAMBRE DE COMMERCE
ET D'INDUSTRIE DE PARIS
ECCIP
MONITOR
#66
ZOOM > LIVING
CO
N-S TRU
C-TION
A- H
EAD
中国建筑
装饰装修
INTERIOR
ARCHITECTURE OF CHINA
LE POINT DE LA SEMAINE
Pierres précieuses à Saint-Tropez

07 MATERIAL CULTURE

07 UNE CULTURE DES MATÉRIAUX

建筑材料文化

Aluminum canopy of the National Theater of Bahrain, Al Manama (detail).

位于麦纳麦的巴林国家大剧院的铝制顶棚（细节图）。

The Arab World Institute, Paris.

巴黎阿拉伯世界研究中心。

The intellectual process of Architecture-Studio forms a continuum that leads logically from office organization to finished buildings. The shared, open method chosen by the partners is focused on creating works of architecture that are fully consistent, from their initial concept to the final details. This approach amounts to the "style" of the office, a style that is not expressed in the repetition of specific materials and details from project to project, but rather in a way of looking at architecture itself. "We try to be as singular as we are in the development of a building in its details. We also try to develop details that are specific to projects. The genesis of these details is of course related to the concept of the architecture. There is also a will to work on meaning, emotions and the physical coherence of the materials," explains Alain Bretagnolle.

A closer look at the details and surfaces of many contemporary buildings yields a feeling that materials are chosen for purely aesthetic reasons, or because the architect is known for a certain appearance in his or her work. This is clearly not the case of the work of Architecture-Studio. They insist that their choice of materials and surfaces is not a formal procedure but one that gives meanings to volumes and expression to concepts. For some buildings such as their Onassis Foundation (Athens, Greece, 2010), the choice of materials imposed itself as one of the basic ideas of the project. White marble was chosen and was treated in bands that allow the building to filter the light coming in. The entire concept of the project is based on the idea of a certain purity,

The façade of the Museum of Science of Tibet, Lhasa (detail).

位于拉萨的西藏自治区自然科学博物馆的立面（细节图）。

法国AS建筑工作室从内部运作到建筑项目管理，其构思和反思的过程是统一连贯的。为了保持建筑从最初的概念构思到最终细节敲定的一致性和协调性，合伙人选择分享共担和开放的工作方式。这是工作室的工作“风格”，然而这种风格不是指在不同的项目中重复使用同样的材料和细节处理的方法，而是从建筑本身出发，做出相应选择的一种坚持。“每一个建筑细节都必须与建筑的总体概念相符，因此我们只能根据每一个建筑项目的特点做出相应的细节设计。建筑材料的运用所带来的感受、情感和物理反应都是我们要考虑的，细节都是由建筑的特征自然发展而来的”，阿兰·布勒塔尼奥勒解释道。

近距离观察当代建筑的立面和细节时，经常会让人感觉选择这些材料纯粹是出于美学原因，又或者是因为建筑师作品需要某种既定的特点，但这显然不是法国AS建筑工作室的做法。他们一致强调，建筑立面的材料选择没有一个完全固定的流程，而要思考建筑空间需要传递的建筑概念和带去的感观体验来决定。有一些项目的材料选择便是项目最初构思的一部分，例如奥纳西斯文化中心（希腊雅典，2010年）

Thassos marble on the façade of the Onassis Cultural Center, Athens.

雅典奥纳西斯文化中心的立面使用达索大理石。

expressed also in its geometry, for example. The marble is characterized by an alternation of voids and full volumes of marble. This alternation allows sufficient daylight to come in during the day and makes the building glow from within and reveal hints of its activity at night. The skin is an interface between the building itself and its context on a busy avenue. Here reference is also made to the volumes of ancient Greece and its own materials, but reinterpreted in a contemporary manner. Thus the surface, or skin, sometimes considered only to be the final layer of a building, becomes, in the hands of Architecture-Studio, the expression of an idea, of a project. This skin made of bands of white marble announces the presence of the Onassis Foundation, makes it stand out from its urban environment, connects it back to the ancient history of Greece and also serves to protect its interior from the hot sun.

In Athens, as in many if not all of the projects of the office, there is an expression of an economy of means, but also of architectural "gestures" and a sense that the repetition of details generates the overall image of the project. There is a focus on overall texture and materiality that develops out of an essential detail. There is often an almost tactile dimension in these details. "We avoid being too verbose in our architecture," continues Bretagnolle. "We seek to select materials that are coherent. Working often with a single material, the visibility of the entire project is increased. We do not seek to compose a façade out of windows, for example, but to create a kind of abstraction in which

Interior façade of Saint-Malo Cultural Center made of wood (detail).

圣马洛文化中心的木制内墙和天花板（细节图）。

Façade detail of the terra-cotta façade of Sainte Anne Hospital, Paris.

巴黎圣安娜医院的陶瓦立面细节。

选择条状的白色大理石组成建筑的纯白遮帘，为建筑内部过滤掉一部分光线，简洁的材料选择与建筑整体的简单几何结构相呼应。

光滑的条状大理石与透明的玻璃体相间组成间隙，日间阳光透过间隙为室内带去舒适的照明，夜幕降临时，建筑内部的灯光透出间隙，让建筑夺目，也象征着基金会开放的运作特征。镂空外立面既能够保证建筑内部免受烈日灼伤，又够促进建筑内部与周边繁忙街道环境的相互交流。文化中心的体量和选材均参考了古希腊建筑，但以当代方式重新诠释，法国AS建筑工作室善于通过建筑立面表达项目的理念。纯白无瑕的大理石外表面让奥纳西斯文化中心从城市环境中凸显出来，并与古希腊文明紧密相连。

雅典奥纳西思文化中心与法国AS建筑工作室的许多其他项目一样，从经济意义和建筑表现方式来讲，均选用简洁的材料和正确的搭配方式，并通过单一细节的不断重复，让建筑的整体形象得以更好的呈

The mobile vertical colored – silkscreen – shutters of the Novancia Business School, Paris.

巴黎诺凡西亚高等商学院的灵动彩色百叶窗。

the windows are included." The abstraction of their façades is born of an analysis of the requirements of the project and its context in the largest sense. For the Novancia Business School (Paris, 2011), they worked with a well-known Paris urban form but made color into one of the constituent elements of the design. In this instance, the goal was not to make the materials used identifiable but to create a kind of abstraction that makes the overall project fits into its city environment and meets the requirements of the client. According to René-Henri Arnaud, "there is also an opposition between autonomy and context that we work on extensively. The character of a building can in a sense reactivate the context rather than contradicting or ignoring it." In other cases, Architecture-Studio has willfully characterized materiality as they did with the Onassis Foundation.

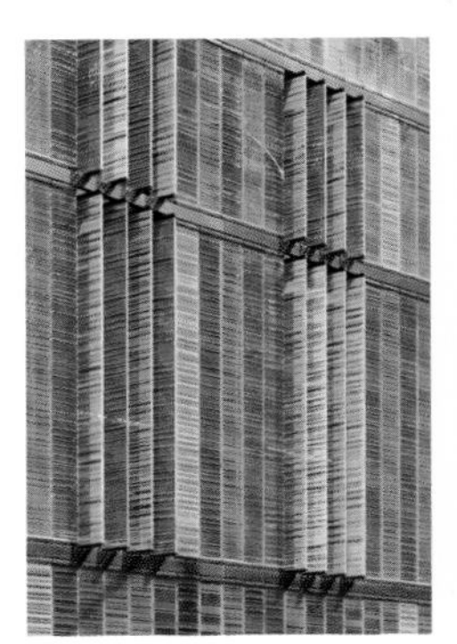

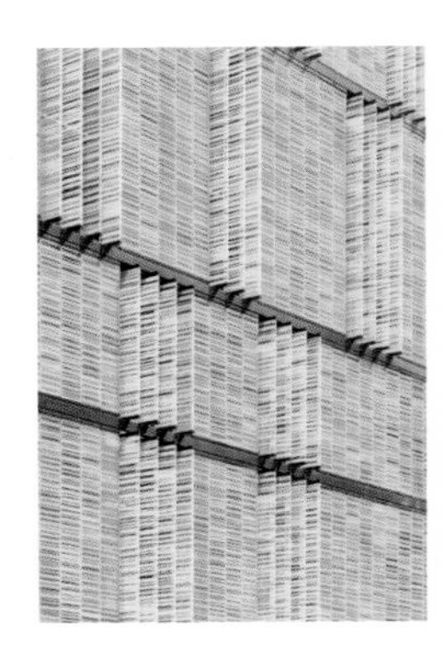

Materials and surfaces are used by the architects in very different ways, according to the circumstances and the nature of the building. In the Our Lady of the Ark of the Covenant Church (Paris, 1998) there are questions asked by the different layers of the skin of the building. The outer metallic grid creates a certain distance between the object itself and its environment. From a distance, there is the appearance of an abstract cube but, when one comes closer, there is something more – there is a kinetic element in many of the surfaces. It is only in approaching the actual façade of the church that it becomes apparent that the words of the Hail Mary (*Je vous salue, Marie, pleine de grâce, le Seigneur est avec vous...*) "Hail Mary, full of grace, the Lord

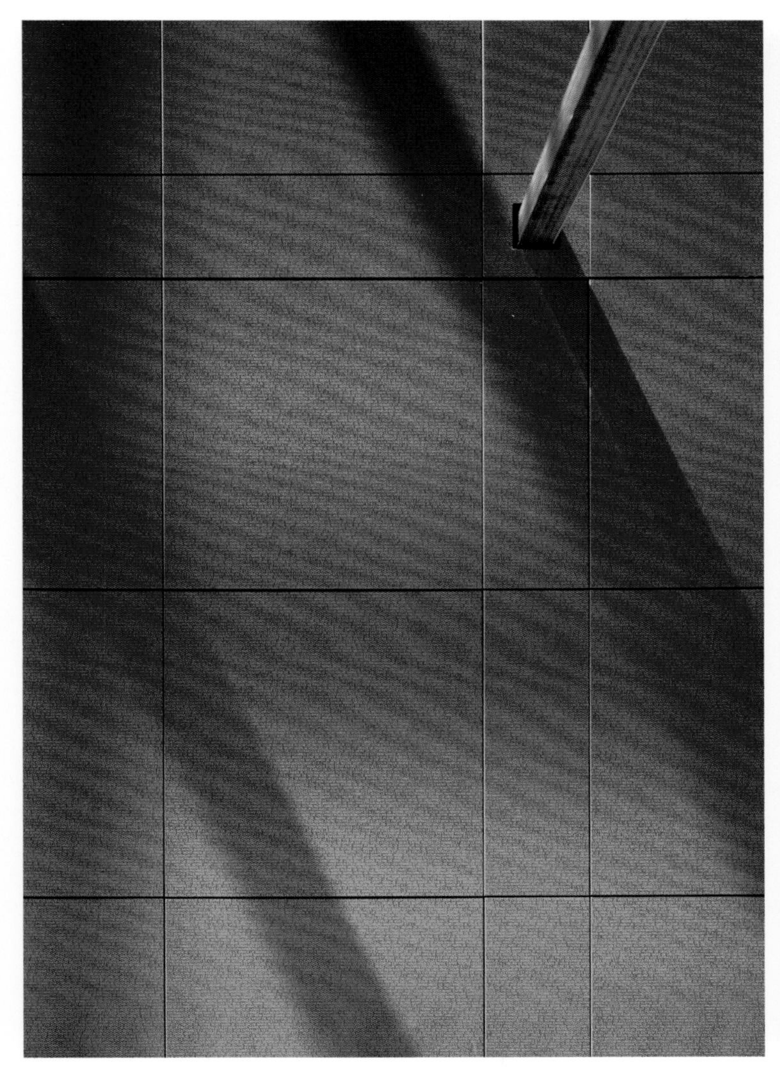

Silkscreened façade of the church of Our Lady of the Ark of the Convenant in Paris.

巴黎约柜圣母教堂的丝印外墙细节。

现。建筑的某一个重要细节足以决定建筑的整体质感和重要性，而这种重要细节经常与触感相关。“为避免过于繁琐的堆砌”，阿兰·布勒塔尼奥勒继续说道，“我们尽量选择与设计概念相符的材料，使用同一种材料能够增大建筑整体的可视性。我们不会在外立面设置无意义的抽象窗户，但会将窗户融入到建筑的抽象设计中。”立面设计的灵感来源于对每个项目背景和要求的具体分析。诺凡西亚高等商学院（法国巴黎，2011年）改造项目尊重周边街区原有的城市形态，建筑外立面采用附近建筑砖墙的常见颜色，色彩因而作为背景因素参与建筑整体设计。由此可见，项目的目标并不是简单地使用高辨识度的材料，而是选用能够与老建筑产生抽象式对话，并能融入到周围环境中的材料。勒内-亨利·阿诺说道：“自主设计和尊重项目背景之间的关系是我们设计时需要把控的。这样，建筑才能够更好的融入并激活环境，而不是忽视环境甚至与环境相悖。”其他例子比如奥纳西思文化中心，工作室经过各项考虑，选择了足够有特点的材料。

法国AS建筑工作室根据建筑的性质及其环境，给出不同的材料运用和立面设计。约柜圣母教堂（法国巴黎，1998年）外立面根据观看距离

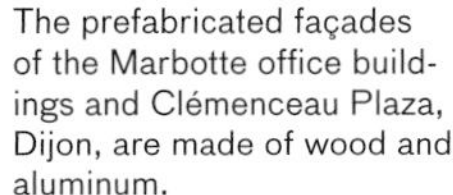

The prefabricated façades of the Marbotte office buildings and Clémenceau Plaza, Dijon, are made of wood and aluminum.

第戎Marbotte办公楼和克雷门索广场的预制立面板，由木材和铝材制成。

Interior façade of Le Quai Theatre, Angers (detail).

昂热河岸大剧院内部立面（细节图）。

is with thee…" are engraved continuously on the wood fiber panel surface of the building.

The architects obtain variety in their buildings depending on points of view with such methods as the superimposition of materials, or folding and curving. The columns at St Malo (Grande Passerelle Cultural Hub, 2015) are set at a 2.5-meter interval but on a curve – thus their transparency depends on the placement of the viewer. The Rotana Tower (Amman, Jordan, 2016) has large vertical *brises-soleil* that modulate light and have a curved form that plays on light as well. Reacting to context in this instance has to do with reacting to light. Again, there is a kinetic element that makes the appearance of the building depend on the location of the viewer. The Tipi – the student space at Jussieu (Renovation of the Jussieu University Campus, Paris, 2015) also has this effect – the structure is a simple glass cone that has an exterior skin of aluminum tubes which are spaced and twisted so that their geometry is different to that of the main cone – from the interior it thus becomes apparent that there are two different geometries that are not superimposed. **As Alain Bretagnolle says, with a smile, "You can say that this is our interpretation of Le Corbusier's *promenade architecturale*." The *promenade architecturale* was a central element of Le Corbusier's architectural and city-planning work – it is the sequence of images that unfolds as a visitor gradually advances through space and through the structure, expressed for example in the Villa Savoye (Poissy, France, 1928).**

The brick façade and aluminum canopy of the Engineering School of Mines, Albi (detail).

阿尔比矿业工程学院的红砖立面和铝制遮棚（细节图）。

远近有不同的效果，建筑外部金属网架为建筑及其环境之间隔出了恰当的距离。从远处看，教堂像一个抽象的立方体，走近了才能看到细节，雕刻在木质纤维立面上的连续文字变得清晰“万福玛利亚，充满恩宠者，上主与你同在……”。

立面的设计和材料的选择是多种多样的，根据角度的变化，采用玻璃丝印、折叠、弯曲等方法为建筑带来多样性。比如圣马洛文化中心（2015年）的每根木质支柱之间相距约2.5米，呈曲线分布，因此建筑的透明度取决于观赏者的位置。罗塔纳酒店大楼（约旦安曼，2016年）带有优美曲线的外立面安装了大型垂直铝制遮阳板来调节室内取光。各具特色的建筑外立面设计与光线玩着多变的光影游戏，而光影效果则取决于观察者的位置。Jussieu校区（巴黎第六大学校园改造项目，2015年）内的Tipi学生空间也有相似的效果，它是一个外立面由铝管环绕所形成的玻璃锥体结构，相互间隔的扭曲铝管与中心的玻璃几何体是隔开的。因此，从内部往外看Tipi的双层外立面并非相互重叠，而是两个分开的不同几何结构，形成动态效果。阿利安·布莱托诺尔微笑着说，“你可以说这是我们对勒柯布西耶“建筑的散步”的重新诠释”。”散步的建筑是勒柯布西耶建筑和城市规划工作的中心元素。正如他在1928年法国普瓦西萨伏伊山庄中，用空间和结构进行的表达，彷佛在一位游客面前，随着他前行的脚步，逐渐打开了一系列建筑图景。

Double façade in lace honeycomb aluminum of Jinan Cultural Center (detail).

山东省会文化艺术中心的蜂巢状镂空铝板双层立面（细节图）。

The smooth façade of the Wison Headquarters, Shanghai.

上海惠生集团总部的光滑立面。

The variety of surface treatments and intentions of which the firm is capable might well be expressed in projects erected in China in a relatively short space of time (2012–2013). In the case of the Jinan Cultural Center (China, 2013), they worked on a parametric design for the metal façade made of aluminum and perforated like a honeycomb, that takes into account views and light. For the Wison Headquarters (Shanghai, China, 2013) there is a kind of tension between different aesthetic intentions – what they call meaningful oppositions (*oppositions signifiantes*) – that reinforce the identity of each element through opposition. Here red *brise-soleil* contrast with whiteness. Aside from its fundamental opposition to (sun shades) white, the red chosen can refer to China itself, or more prosaically to the corporate logo of Wison – it thus has a meaning that goes beyond that of pure color. The concept of the *opposition signifiante*, or meaningful opposition, appears prominently in the development of French structuralism, particularly in the work of Roland Barthes and Jacques Lacan. Although they do not often couch their project descriptions in the language of structuralism, such intellectual currents do run through the work of Architecture-Studio. In the Zhonghua Tobacco Factory (Shanghai, China, 2012), which is voluminous vis-à-vis its urban context, they sought to treat the project as a kind of landscape. The mirrored stainless steel elements of the façade reflect the ground-level garden – in a sense, the building thus disappears into its own landscape. And the appearance of the building also changes with the seasons. Though mirrored façades might seem

The epoxy strips of the façade of the Shanghai tobacco factory.

由镜面不锈钢节状物构成的上海烟草集团建筑外立面。

法国AS建筑工作室对建筑外立面的细节处理在中国近期落成的建筑中得到了充分的体现。比如山东省会文化艺术中心（中国济南，2013年），工作室通过参数化设计的蜂巢状镂空铝板立面，使建筑内部光线得以统一。而上海惠生集团总部建筑（中国上海，2013年）的立面设计是不同美学之间的对立和融合，红色百叶窗与白色墙面的对比使各元素得以凸显，而红色不仅是中国的象征色彩，也是惠生商标的代表色彩——这超越了颜色的单纯含义。对立意义的概念，或者说有意义的对立在法国结构主义的发展过程中尤为显著，例如罗兰·巴特和雅各·拉岡的作品。尽管建筑师们并不使用结构主义来形容他们的项目，但这种思想潮流却一直贯穿在法国AS建筑工作室的作品中。为减少规模大型的中华卷烟厂（中国上海，2012年）建筑坐落在市区的突兀感，工作室将建筑设计成景观的一部分。建筑外立面采用不锈钢镜面元素反射地面花园——如此，建筑完美地融入到城市风景中，而建筑外观也会随着季节而发生变化。尽管镜面的立面看起来更像是20世

The patinaed copper roof of the Exhibition Park, Paris Villepinte.

巴黎北部维勒班特展览馆的青铜顶棚。

more related to architecture of the 1970s, this particular use is both original and coherent with the architecture and the context.

Each of these projects changes in appearance depending on what angle or position it is viewed from. The Onassis Foundation is either opaque or transparent depending on the hour or the angle of view. This opposition also applies to exterior and interior – seen from the outside during the day, the Foundation often appears to be opaque, but, from within, it seems totally transparent. This opposition is reversed at night. This kind of variation is also visible in Albi (*École des Mines*, Albi-Carmaux, France, 1995) whose façade is made of triangular columns of brick. While completely transparent from certain angles it also offers, from other angles, a compactness worthy of the Cathedral of Sainte-Cécile in Albi. On entering, surprisingly, it becomes apparent to the visitor that the facade is fully glazed. "We are also seeking a certain amount of surprise with this work on the skin or façades of our buildings," says Marc Lehmann. Surprise of this kind is a way for the architects to interact with users.

There is often an immaterial aspect that informs the details and choice of materials in the projects of Architecture-Studio. In the Radio France Auditorium – such as the acoustics in the Radio France auditorium – the forms are neither superficial nor accidental; they are purposeful and thought through entirely. This space obliged the architects to be concerned with acoustics and that constraint was the point of departure for their design. There is thus a particular aesthetic sense that emerges

The copper skin of Clermont-Ferrand School of Fine Arts (detail).

克莱蒙费朗高等艺术学院的铜质外立面（细节图）。

纪70年代的建筑特色，但在这个项目上的特殊使用却让建筑与环境融为一体，这就是它的创新之处。

每个建筑的外观随着观看位置和视觉角度的不同而变化。奥纳西斯文化中心呈现出的透明或不透明的效果，完全取决于观看时间或角度，而建筑的外部和内部也形成对立——白天，从外往里看，建筑呈不透明状，但若由内往外看，则有通透的视野；到了夜晚，情况则相反。这种变幻在阿尔比矿业工程学院（法国阿尔比，1995年）也有展现，其外立面由红砖和玻璃相间组成，并与立面墙体呈三角的立柱支撑起透明遮棚。从某些特定的角度看建筑呈透明状，而从另外的角度看，建筑又让人联想起阿尔比圣塞西尔教堂的全红砖结构。实际上，只需要稍微靠近建筑便会发现建筑立面和遮棚都是透明的。“我们总是希望在建筑表面或立面上制造一些惊喜”，马克·莱曼说道，这种惊喜是建筑师与用户相互交流的一种方式。

法国AS建筑工作室对建筑细节处理和材料选择也体现在非物质方面。例如法国国家广播电台大音乐厅所采用的圆厅设计并不是巧合，而是经过设计师深思熟虑后的结果。“以舞台为中心”的设计是为了保证最优的声学效果，进行此类有特殊要求的空间建设时，必须在

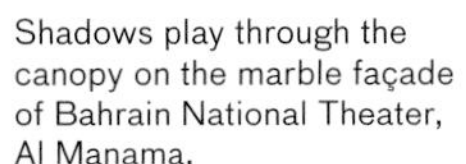

Shadows play through the canopy on the marble façade of Bahrain National Theater, Al Manama.

位于麦纳麦的巴林国家大剧院的大理石立面上，映着由遮篷透过的光影。

Curved rooms paneled in elm, with staff and terrazzo, Bahrain National Theater, Al Manama.

巴林国家大剧院内，由榆木曲线墙面和水磨石地面构成的空间。

from the acoustic requirements. Even beyond acoustics, evoking the rhythm of music itself, certain elements are repeated and, in the final analysis, there is an impression that the entire hall has been carved from the same material – the notion of massiveness.

Another tension or opposition that informs the materials and the appearance of the buildings of Architecture-Studio is expressed in the difference between interior and exterior. In the foyer of the Theatre in Bahrain (National Theatre of Bahrain, 2012), there is an intentional contrast with the simple form of the building itself. The foyer space, between the skin and the actual auditorium, develops what might be called a more "fantastic" image of the architecture. The partners worked with solid wood in the image of what was used to build boats in the region. "In warm countries itmight be said that the condition to obtain transparency is to create opacity," says Roueïda Ayache. The roof filters out 80 percent of the incoming sunlight. So, too, the canopy of the Theatre in Bahrain in woven aluminum recalls the local roof-making tradition in a modern way. The inner area brings to mind the human ear, or perhaps seashells. One steps from the real, outside world, into the fictional world of the theatre. These spaces have an organic element as opposed to the highly abstract nature of the exterior architecture.

Most frequently, there is a layering of meanings in the materials and surfaces of the work of Architecture-Studio. There is also an effort to evoke the poetry of the experience as well – an emotion.

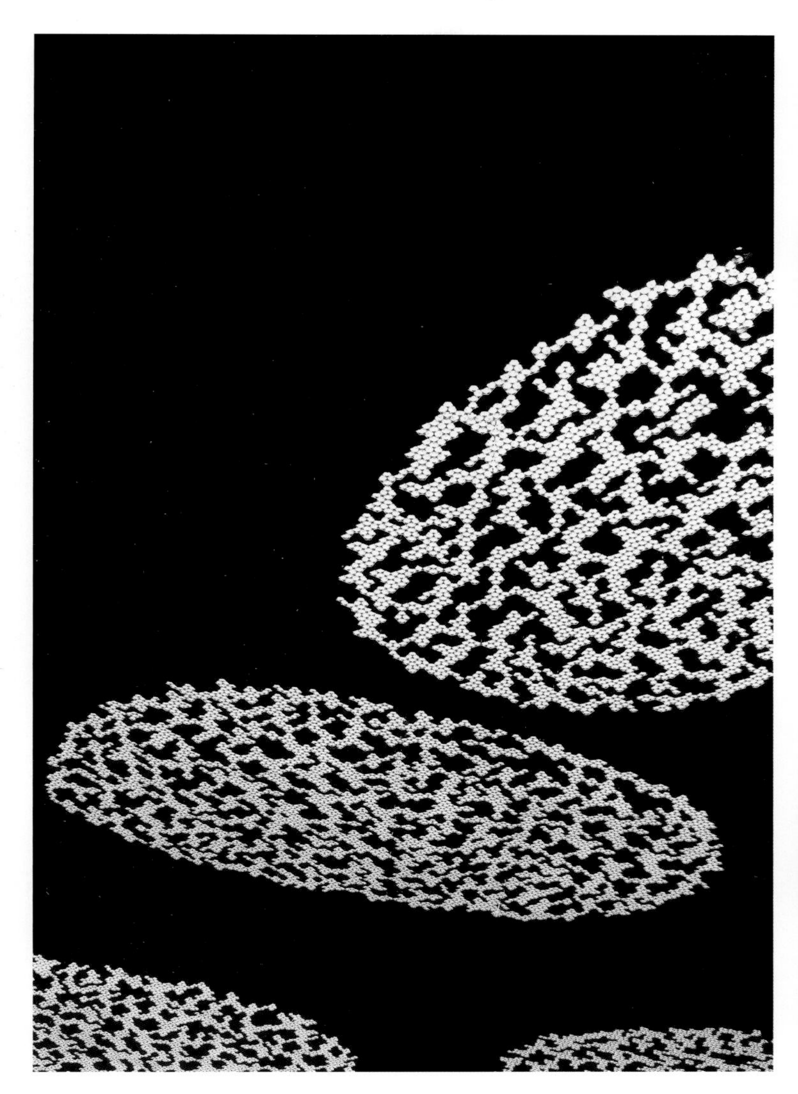

The ceiling lights of Daguan Theatre made of LED honeycomb mesh, Himalayas Center, Shanghai.

上海喜玛拉雅中心大观舞台的蜂窝网格状LED顶灯。

The metal balconies on the façade of the *Nouvelle Ligne* housing building, Montpellier.

蒙彼利埃Nouvelle Ligne住宅建筑外墙上的金属阳台。

建筑设计时优先考虑音响效果，独特的美感因此从音效需求中产生。在音乐厅内连续使用同种材料，不仅在音效上唤醒了音乐韵律的本身，也为大音乐厅带来视觉上的整体感，整个空间就像是由一个完整木块雕刻而来。

材料选择和建筑外观的对比和对立，也体现在法国AS建筑工作室所设计建筑的内部与外观的差异上。巴林国家大剧院（巴林麦纳麦，2012年）的内部设计与为其庇护的建筑外壳几何体有着鲜明的对比。梦幻般的剧院内部被包裹在金属贝壳状空间里，建筑师特意在建筑外立面与该巨型贝壳之间设置了留白。和当地传统的出海捕珍珠的渔船一样，剧院内部由全实木打造。“在炎热的国家，要获得透明度首先需要创建不透明度”，罗伊达 · 阿亚斯说道。事实上，由铝带编制而成的剧院华盖可滤过80%的太阳光，为室内带去阴凉，这是以当代的方式致敬当地铝制屋顶的传统。而剧院内部的抽象设计形似人的耳朵，又似许多的贝壳，和建筑方正的外立面对比显著。观众从真实的外部世界进入剧院内的虚构世界仅一步之遥。

The bronze aluminum anodized façade of Dor[a] Mar building, Montpellier.

蒙彼利埃Dor[a]Mar建筑的铜铝氧化膜立面结构。

The spruce arches and stained glass window of Créteil's Cathedral, artwork created by Udo Zembok.

克雷泰伊大教堂的云杉穹顶和由Udo Zembok设计的彩色玻璃窗。

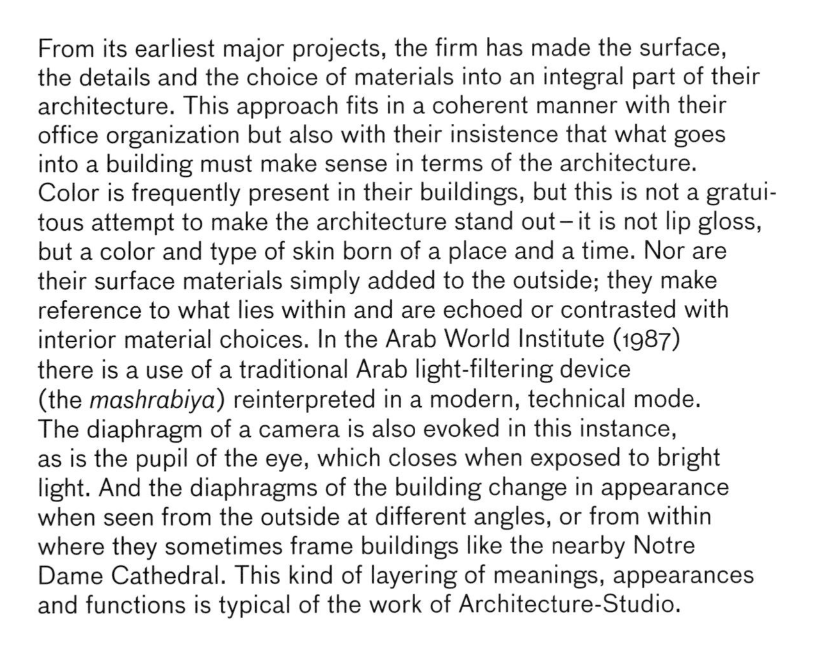

From its earliest major projects, the firm has made the surface, the details and the choice of materials into an integral part of their architecture. This approach fits in a coherent manner with their office organization but also with their insistence that what goes into a building must make sense in terms of the architecture. Color is frequently present in their buildings, but this is not a gratuitous attempt to make the architecture stand out – it is not lip gloss, but a color and type of skin born of a place and a time. Nor are their surface materials simply added to the outside; they make reference to what lies within and are echoed or contrasted with interior material choices. In the Arab World Institute (1987) there is a use of a traditional Arab light-filtering device (the *mashrabiya*) reinterpreted in a modern, technical mode. The diaphragm of a camera is also evoked in this instance, as is the pupil of the eye, which closes when exposed to bright light. And the diaphragms of the building change in appearance when seen from the outside at different angles, or from within where they sometimes frame buildings like the nearby Notre Dame Cathedral. This kind of layering of meanings, appearances and functions is typical of the work of Architecture-Studio.

At night, the red shell of the theatre of the Jinan Cultural Center.

夜幕降临山东省会文化艺术中心外立面透出的包裹着剧院的红色木质外壳。

法国AS建筑工作室为建筑外立面运用的材料和表面细节的处理赋予了含义，因此不难从中体会到诗意与情感。从最早期的重要项目开始，该工作室便将建筑表面，乃至建筑整体的材料选择和细节处理视为建筑的组成部分。设计方法与工作室一贯的坚持一致，他们坚持所有融入建筑的东西必须要赋予建筑以意义。色彩，是工作室经常运用到建筑中的工具，但这并不单纯是为了视觉上的凸显——而是需要让建筑表面与周边环境及时代背景产生共鸣。而建筑外表面上的材料也不是单纯的叠加，而应融入周围的环境与背景，同时与建筑内部的美学形成呼应或对照。阿拉伯世界研究中心（法国巴黎，1987年）的外立面设计源于阿拉伯传统用于遮光的雕窗，但是用更现代更科技的方法进行了重新诠释。其原理就像是照相机的光圈，又或者人类的瞳孔，在光亮的条件下便会自然的闭合。灵动的外立面随着视角的不同而不同，随着光线的变化而改变。从建筑内部往外看，大大小小的镂空处将巴黎的市景随机地截取，为建筑内部使用者带去惊喜。感官、功能和诗意等不同层面的特色，造就了法国AS建筑工作室的杰作。

INDEX + TIMELINE

索引 ＋ 时间线

ABB1
Hospital Centre
2003
Abbeville, France

ABD1
Opera
1977
Abu Dhabi, United Arab Emirates

ABD2
Urban Planning of Sheikh Zayed Mosque
Abu Dhabi, United Arab Emirates

ABD3
Al Reem Towers
2005
Abu Dhabi, United Arab Emirates

ABD4
Mixed Urban Development
2005
Abu Dhabi, United Arab Emirates

ABD5
Al Khalidiya Plaza Tower
2006
Abu Dhabi, United Arab Emirates

ABD6
Marina Towers
2006
Abu Dhabi, United Arab Emirates

ABH1
King Khalid University Campus
2002
Abha, Saudi Arabia

ABJ1
New Orange Headquarters
2015
Abidjan, Republic of Côte d'Ivoire

ABJ2
La Pyramide Building
2015
Abidjan, Republic of Côte d'Ivoire

ABJ3
Planning of "La Balade du Monde" Lagoon Area
2015
Abidjan, Republic of Côte d'Ivoire

ABJ4
Marcory's Eco-District
2016
Marcory, Republic of Côte d'Ivoire

ABM1
University Hospital Centre of Pointe-à-Pitre/Les Abymes
2011–2019
Pointe-à-Pitre, Guadeloupe

AGR1
Stadium for the 2006 Football World Cup
1999
Agadir, Morocco

AGS1
"Le Quai" Theatre
2003–2006
Angers, France

AIN1
Hotel
2009
Al Ain, United Arab Emirates

AIN2
Jebel Hafeet Mountain Hotel
2010
Al Ain, United Arab Emirates

AIX1
Aix Luynes II Penitentiary Centre
2013
Aix-en-Provence, France

AIX2
Voyage Privé Campus
2014
Aix-en-Provence, France

AKT1
Caspian Pearl
2007
Aktau, Kazakhstan

ALB1
Engineering School of Mines
1992–1995
Albi, France

ALB2
Engineering School of Mines – Student Housing
1994–1996
Albi, France

ALC1
UGC Ciné Cité
1998
Alicante, Spain

ALG1
Ministry of Foreign Affairs
2002
Algiers, Algeria

ALG2
BNP Paribas Headquarters
2007
Algiers, Algeria

ALG3
Grand Mosque
2007
Algiers, Algeria

ALH1
Exhibition Building At Abdulliyah Preserved Area
2014
Al Ahmadi, Kuwait

ALN1
Urban Planning Project
1996
Aulnay-Sous-Bois, France

ALX1
San Stefano Tourist Resort
1998
Alexandria, Egypt

AMN1
Amman Rotana Hotel
2005–2016
Amman, Jordan

AMN2
Abdali Gate East Tower
2006–2017
Amman, Jordan

AMN3
Abdali Gate West Tower
2006–2020
Amman, Jordan

AMN4
Abdali Complex
2007
Amman, Jordan

AMS2
Housing – Boulevard Faidherbe
1986
Amiens, France

AMS3
Restructuring of the Saint-Victor Hospital
1988
Amiens, France

AMS4
Urban Design of Amiens City Centre
1990
Amiens, France

AMS5
Sonacotra Residence
1993
Amiens, France

AMS6
Housing
2007
Amiens, France

AMS7
Facing Faces Institute
2015
Amiens, France

ANJ1
Anji Club Med
2014
Anji, China

ANT1
Theatre
2006
Antibes, France

AQB1
Hotel
2006
Aqaba, Jordan

AQB2
Beach Club
2006
Aqaba, Jordan

AQB3
Villa Sea View
2006
Aqaba, Jordan

AQB4
Residence Hill
2008
Aqaba, Jordan

ARC1
Urban Design of City Centre
1986
Arcachon, France

ARS1
Aloïse Corbaz Clinic
2001–2004
Arras, France

ARS2
Hospital Centre
2002
Arras, France

ASC1
Escolaco Borda Private House
1978–1979
Ascain, France

ATM1
C.N.R.A Nord Extension
1992
Athis-Mons, France

ATN1
Athens Olympic Village
2000–2004
Athens, Greece

ATN2
Onassis Cultural Centre
2002–2010
Athens, Greece

AVI1
Highschool
1988
Avignon, France

AXB1
Chevalley Thermal Baths
1993
Aix-les-Bains, France

AXN1
Rovaltain Research Centre
2010
Alixan, France

BAK1
Office Building
1996
Baku, Azerbaijan

BAK2
Apsheron Hotel
1996
Baku, Azerbaijan

BAK3
Oncological Hospital
2000
Baku, Azerbaijan

BAR1
Leroy Merlin Store
1997–2006
Bois d'Arcy, France

BAV1
Urban Design of City Centre
1980
Bavilliers, France

BBG1
Hospital Centre
2000–2005
Bobigny, France

BCH1
Urban Study
2006–2011
Bouchemaine, France

BDP1
Restructuring ot the Paris Department Store
2001
Budapest, Hungary

BDP2
Ethnography Museum
2016
Budapest, Hungary

BDX1
Urban Design for the Bastide District
1986
Bordeaux, France

BDX2
Technopolis
1987
Bordeaux, France

BDX3
Courthouse
1988
Bordeaux, France

BDX4
Urban Planning of Bordeaux South Ring Road
1995–2001
Bordeaux, France

BDX5
Regional Council
2001
Bordeaux, France

BDX6
Banque Populaire Sud Ouest Headquarters
2001
Bordeaux, France

BDX7
Hotel
2002
Bordeaux, France

BDX8
Administrative Office
2010
Bordeaux, France

BDX9
Talence, Pessac and Gradignan Campus - Eastern Sector
2011–2015
Bordeaux, France

BDX10
Caisse d'Epargne Headquarters of Aquitaine Poitou-Charentes
2012–2016
Bordeaux, France

BEY1
Urban Study
1993
Beirut, Lebanon

BEY2
Restructuring of Souks
1994
Beirut, Lebanon

BEY3
Theatre
2003
Beirut, Lebanon

BEY4
Marfaa Residence
2004
Beirut, Lebanon

BEY5
Hotel Rotana
2005
Beirut, Lebanon

BEY6
Wadi Hills Residence
2005
Beirut, Lebanon

BEY7
Bank Med Headquarters
2011
Beirut, Lebanon

BEY8
Franco-Lebanese Highschool
2012
Beirut, Lebanon

BEY9
Al Nahr Mixed-Use Building
2012–2018
Beirut, Lebanon

BEY10
Residential Project
2014–2018
Beirut, Lebanon

BF1
Urban Design of the Bougenel District
1976
Belfort, France

BF2
Restructuring of Housing
1985
Belfort, France

BF3
Convention Centre
1990
Belfort, France

BGB1
Departmental Council
1990
Bourg-en-Bresse, France

BGL1
A3 Highway
2000
Bagnolet, France

BHR1
National Theatre
2003–2012
Manama, Bahrain

BHR6
Extension of the Ritz Carlton
2011
Manama, Bahrain

BLB1
Edouard Vaillant Highschool
1989
Boulogne-Billancourt, France

BLB2
Renault Communication Centre
2000
Boulogne-Billancourt, France

BLB3
Sustainable Development and Urban Planning Project
2002–2003
Boulogne-Billancourt, France

BLB4
Housing
2004
Boulogne-Billancourt, France

BLB5
Muslim Cultural Centre
2006
Boulogne-Billancourt, France

BLC1
Airbus Customers Facilities
1999
Blagnac, France

BLG1
Bologna University
1998
Bologna, Italy

BLG2
Belgrade Waterfront
2014
Belgrade, Serbia

BNB1
Le Chesnois Highschool
2009
Bains-les-Bains, France

BNG1
Wanda Plaza
2013
Bengbu, China

BRL1
"Denkmal Oder Denkmodell" Project
1988
Berlin, Germany

BRL3
Reichstag Information Centre
2016
Berlin, Germany

BRS1
House of the Culture
2014
Bourges, France

BRT1
Multipurpose Sports Centre
2004
Biarritz, France

BRU1
Departmental Sports Equipment
2006
Bruz, France

BRV1
Sanatorium
2008
Barvikha, Russia

BRX1
Renovation of the Central Tower
1997
Brussels, Belgium

BRX2
New NATO Headquarters
2001
Brussels, Belgium

BRX3
Residence Palace
2005
Brussels, Belgium

BRX4
Northern Strategic Centre
2009
Brussels, Belgium

BRZ1
Urban Design of Brazzaville City Centre
1998
Brazzaville, Congo

BS1
"La City" Business District
1988 – 2000
Besançon, France

BS2
"La City" – Parking
1988 – 1990
Besançon, France

BS3
The C.L.A. – Centre de Linguistique Appliquée
1989 – 1992
Besançon, France

BS4
Caisse d'Epargne Franche-Comté Headquarters
1991 – 1994
Besançon, France

BS5
Students Housing
1993 – 1996
Besançon, France

BS6
"La City" Hotel
1993 – 1997
Besançon, France

BS7
Sonacotra Residence
1993
Besançon, France

BS8
Offices
1997 – 2000
Besançon, France

BSC1
Housing
2015
Bessancourt, France

BSG1
Leroy Merlin Commercial Mall
2006
Bussy-Saint-Geoges, France

BSR1
Stadium and Sports Village
2008
Basra, Iraq

BST1
Upper Corsica Prefecture
1994
Bastia, France

BTH1
Futura Business Centre
1988
Béthune, France

BUE1
Soler 5724 Office Building
2015 – 2017
Buenos Aires, Argentina

BV2
Sonacotra Residence
1993
Beauvais, France

BZR1
Media Library
2003
Béziers, France

CAL
Psychiatric Hospital
1985
Calais, France

CAN1
Courthouse
1993 – 1996
Caen, France

CAN2
Training and Research University
1994
Caen, France

CAN3
University Hospital Centre
2004 – 2009
Caen, France

CAR1
"New Cairo" Urban Design
1998
Cairo, Egypt

CAR2
Cairo Airport Building Complex
1999
Cairo, Egypt

CAR3
French Highschool
2001
Cairo, Egypt

CAR4
Westown Urban Design
2008
Cairo, Egypt

CAY1
Airport
1989 – 1994
Cayenne, France

CAY2
Women's and Children's Hospital
2003
Cayenne, France

CCH1
ESTP University Campus
2013 – 2018
Cachan, France

CDP1
Héraud Private House
1981 – 1983
Chasseneuil, France

CDX1
UGC Ciné Cité
2002
Cádiz, Spain

CFD1
Planning & Urban Design of the Science & Education City
2009
Caofeidian, China

CGC
Hospital Automated Storage Building
1983
Cognac, France

CGN1
La Villette District
2016
Cagnes-sur-Mer, France

CGV1
Technical College
1988
Cran-Gevrier, France

CH1
Guy Gasnier Junior School
1979 – 1980
Chelles, France

CHA1
S.F.E.N.A. Research Centre
1982
Châtellerault, France

CHB1
Hospital Centre
2009
Chambéry, France

CHB2
Private House
2010 – 2014
Chambéry, France

CHQ1
Development of Yuzhong Peninsula
2002
Chongqing, China

CHQ4
Chongqing Science & Technology Museum
2004 – 2009
Chongqing, China

CHR1
Chamrousse Olympic Ski Resort
2015
Chamrousse, France

CHS1
Hunan Provincial Museum
2010
Changsha, China

CHT1
Urban design for the north-east Plateau
2010 – 2030
Chartres, France

CL1
Retirement Residence
1972 – 1975
Clamart, France

CLF1
Renovation of the Saint-Pierre Place
1983
Clermont-Ferrand, France

CLF2
Court House
1985
Clermont-Ferrand, France

CLF3
French Institute of Advanced Mechanical Engineering
1989
Clermont-Ferrand, France

CLF4
Auvergne Great Hall
1997
Clermont-Ferrand, France

CLF5
School of Fine Arts
2001 – 2005
Clermont-Ferrand, France

CLF6
Regional Council of Auvergne
2006
Clermont-Ferrand, France

CLF7
Blaise Pascal School
2009
Clermont-Ferrand, France

CLM3
Paul Guiraud Hospital
2006 – 2012
Clamart, France

CLR1
Restructuring of the Jacques Brel Technical College
1999 – 2005
Choisy-le-Roi, France

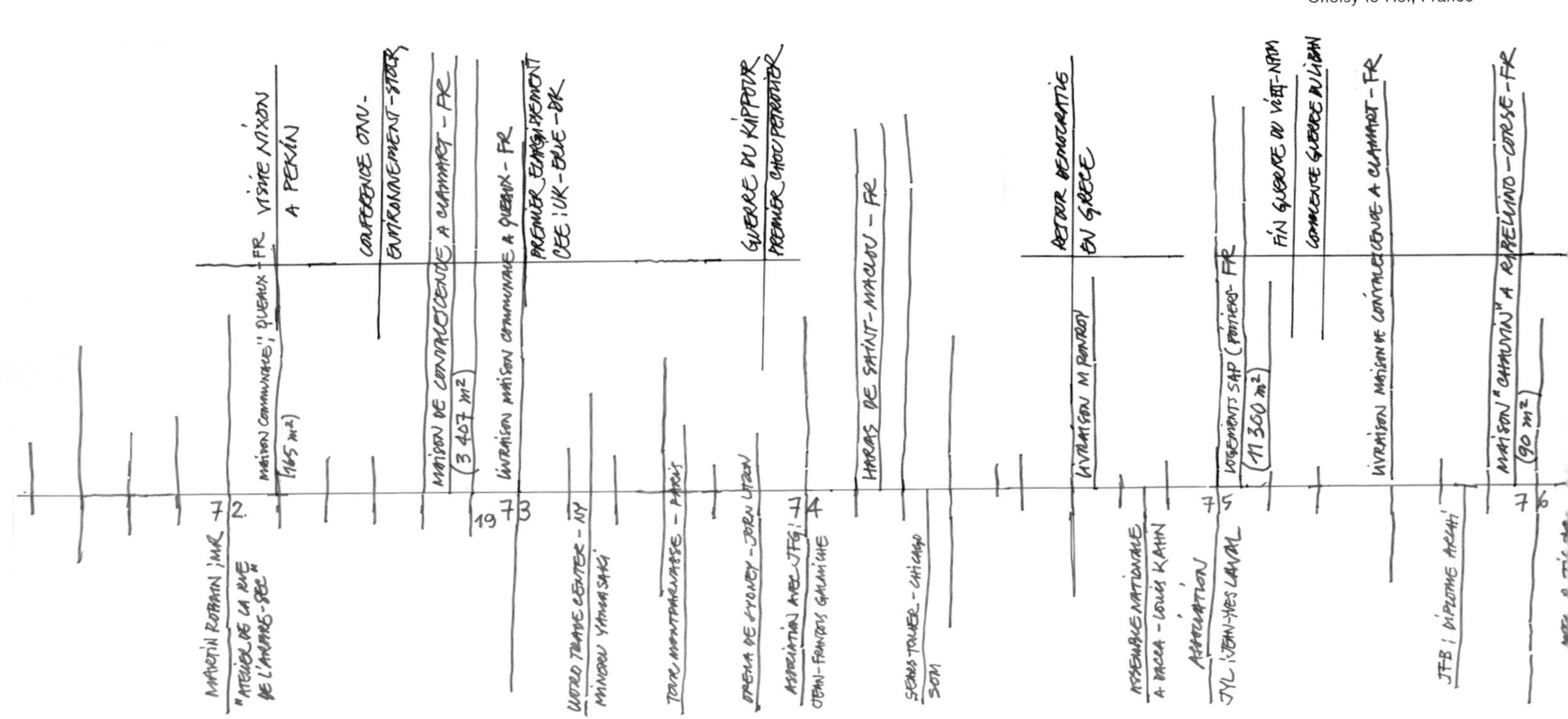

CLV1
Utah Beach American Cemetery
2004
Colleville, France

CLY1
Bechet Headquarters
1993 – 1996
Clichy, France

CLY3
Office Building
2006
Clichy, France

CLY4
Office Building
2011
Clichy-la-Garenne, France

CM1
Sonacotra Residence
1993
Colmar, France

CMT1
SDIS Training Centre
2010
Chaumont, France

CNN1
Hospital Complex
1994
Cannes, France

CNN3
"Palais des Festivals" Extension
2016
Cannes, France

CNS
Megat
1985
Cesson-Sevigné, France

CNT1
Hospital Complex
2006
Annemasse-Bonneville, France

CQHE1
Tikopia
2007

CRB1
Housing
2011
Corbeil-Essonnes, France

CRB2
Housing
2012
Corbeil-Essonnes, France

CRB3
Restructuring of Genopole Complex
2015
Corbeil-Essonnes, France

CRL1
Restructuring of an Industrial Wasteland
2000
Creil, France

CRT1
Pneumology Department – Créteil Hospital
2003
Créteil, France

CRT2
Créteil Cathedral
2008 – 2015
Créteil, France

CRT3
Urban Study for the Triangle de l'Échat District
2013 – 2014
Créteil and Maisons-Alfort, France

CRT4
Housing and Office Building
2014
Créteil, France

CRV1
Convention Centre
1998
Kraków, Poland

CRV2
Forum Orbis Hotel
1999
Kraków, Poland

CRV3
Concert Hall
2000
Kraków, Poland

CSB1
Urban Planning of the Avenue Royale
1999
Casablanca, Morocco

CSB2
Cheikh Khalifa Hospital
2008
Casablanca, Morocco

CST1
Restructuring of Hôtel Dieu
1987
Castres, France

CTN1
International Business Centre
1997
Kutonu, Benin

CY1
Urban Planning of Future District
1989
Cergy-le-Haut, France

CY2
Jules Verne Highschool
1991 – 1993
Cergy-le-Haut, France

CYR1
Martial Art Club
2005
Ceyrat, France

DJA1
Cultural Centre of Tihama
1981
Jeddah, Saudi Arabia

DJA2
Prince Abdullah Al Faisal Palace
1982
Jeddah, Saudi Arabia

DJA3
Creation of a new neighborhood
2007
Jeddah, Saudi Arabia

DJA4
Office Complex
2015
Jeddah, Saudi Arabia

DJA5
King Abdulaziz University
2015 - 2020
Jeddah, Saudi Arabia

DJN1
Urban Development of La République Area
2009
Dijon, France

DJN5
Sonacotra Residence
1992 – 1995
Dijon, France

DJN6
Building Complex
2003
Dijon, France

DJN7
Office Building
2007 – 2017
Dijon, France

DKR1
La Palmeraie Village Diamniadio Residence
2014
Dakar, Senegal

DKR2
Lac Rose District
2015
Dakar, Senegal

DLN1
Urban Planning of the Jimmalu District
2002
Dalian, China

DLN2
Tour de l'Humanité Project
2002
Dalian, China

DLN3
Convention Centre of Bangchuidao
2002
Dalian, China

DLN4
Urban Planning of the Town Hall Plaza
2003
Dalian, China

DLN5
Plaza Zhong Yin
2003
Dalian, China

DLN6
Plaza Dongrun
2003
Dalian, China

DLN7
Housing Tower
2003
Dalian, China

DLN8
Plaza Hai Wan
2003
Dalian, China

DLN9
Façade of the Jardin Jiadi Residence
2003
Dalian, China

DLN11
Housing Tower
2003
Dalian, China

DLN20
Science & Technology Museum
2011
Dalian, China

DLN21
Library of Dalian New City
2011
Dalian, China

DLN22
Museum of the New Puwan District
2012
Dalian, China

DLS1
Sofitel Hotel
1999
Dallas, United States of America

DMS1
French Highschool
2001
Damas, Syria

DMS2
Urban Planning
2004
Damas, Syria

DMS3
Souria Towers
2008
Damas, Syria

DNG1
Dongxing Free Trade Zone
2013
Dongxing, China

DNG2
Dongxing Free Trade Zone
2014
Dongxing, China

DNG3
Shaoguan Free Trade Zone
2014
ShaoGuan, China

DNZ3
Yangpu Economic Development Zone
2015
DanZhou, China

DOH1
Boulevard Commercial District
2006
Doha, Qatar

DOH2
Lusail Development Public Realm Design of Doha
2008 – 2022
Doha, Qatar

DOH3
Pier Exhibition Centre
2009
Doha, Qatar

DOH4
House
2011
Doha, Qatar

DOH5
South Marina Yacht Club House
2012
Doha, Qatar

DOH6
Restaurant
2012
Doha, Qatar

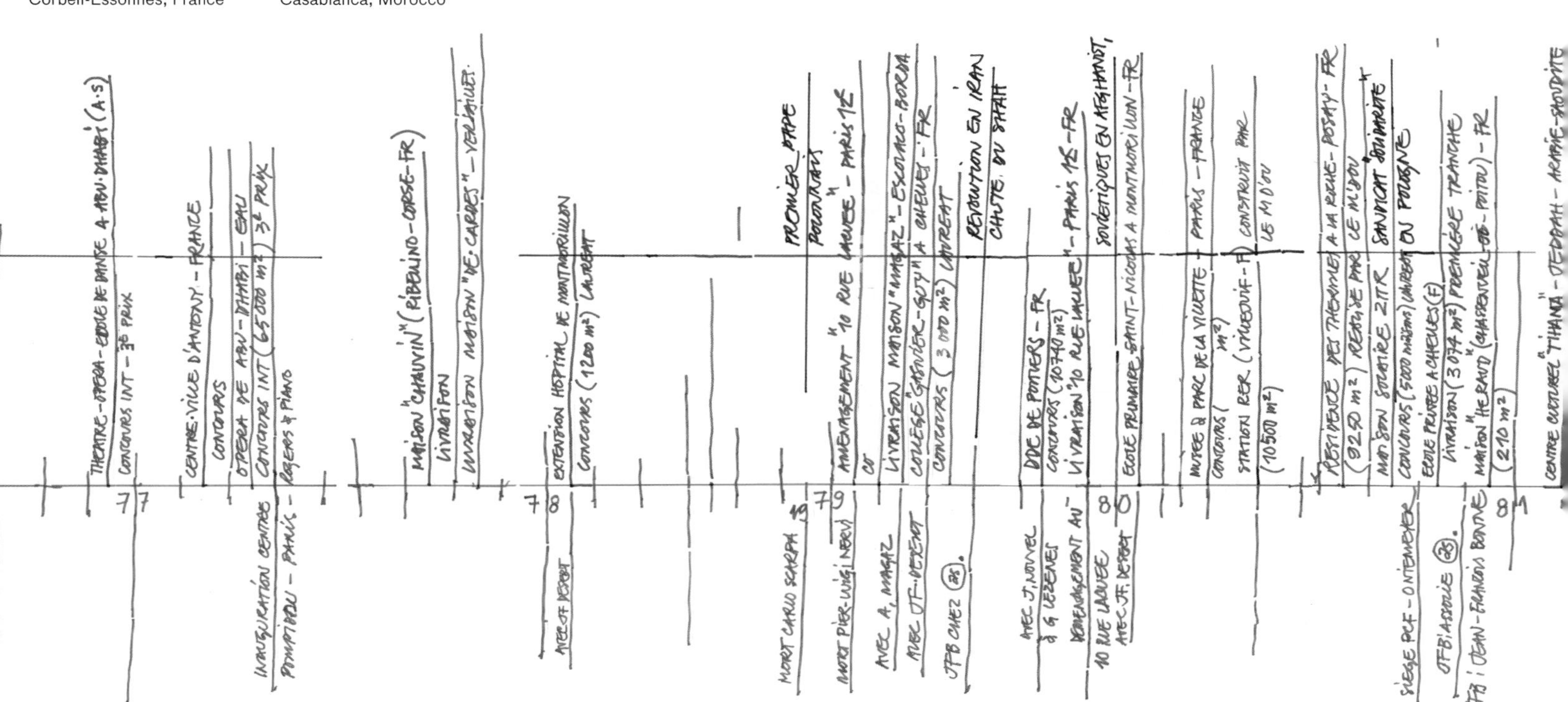

DOH7
Lusail Marina East Restaurant
2012
Doha, Qatar

DOH8
Wadi Park Agricultural Farm
2015
Doha, Qatar

DOH20
Al Waab City Development
2014
Doha, Qatar

DOH21
BeIN SPORT Tower
2015
Doha, Qatar

DOU1
Mesuroscope
1997
Douai, France

DQ1
Citadelle University
1987 – 1990
Dunkerque, France

DQ3
Citadelle University – Restaurant
1990 – 1993
Dunkerque, France

DQ4
Expansion Citadelle University
1990 – 1995
Dunkerque, France

DQ5
Expansion of the Citadelle University
1995 – 1998
Dunkerque, France

DQ6
Benjamin Morel Highschool
2003
Dunkerque, France

DRC1
Media Library
2003
Drancy, France

DRG1
Penitentiary Centre
2011
Draguignan, France

DRN1
Republican Security Companies Station
1986
Darnétal, France

DSL1
Masterplan of Saint-Martinus Hospital
2003
Düsseldorf, Germany

DSL2
Shopping Mall
2003
Düsseldorf, Germany

DTN1
Shopping Mall and Hotel
2010
Datong, China

DTN2
Office Building
2010
Datong, China

DUB1
Meydan Residential Tower
2008 – 2011
Dubai, United Arab Emirates

DUB2
Villas Umm Suqueim Second
2008 – 2010
Dubai, United Arab Emirates

DUB3
Villa Maktoum
2008
Dubai, United Arab Emirates

DUB4
Housing Tower
2008
Dubai, United Arab Emirates

DUB5
Seddiqi Headquarters
2009 – 2016
Dubai, United Arab Emirates

DUB6
Private Villa
2009
Dubai, United Arab Emirates

DUB7
Palm Villa
2013 – 2015
Dubai, United Arab Emirates

DUB8
KTL Residences
2013
Dubai, United Arab Emirates

DUB9
Villas Meraas Island
2014
Dubai, United Arab Emirates

DUB10
Form Hotel
2014
Dubai, United Arab Emirates

DUB11
Theatre
2015
Dubai, United Arab Emirates

DVL1
Medical and Social Establishment
2003 – 2010
Dainville, France

DVN1
AHNAC Polyclinic
2003 – 2006
Divion, France

DVN2
AHNAC Polyclinic Extension
2007 – 2008
Divion, France

DZY1
Retirement Home
1986 – 1991
Donzy, France

ECH1
Hospital Extension
2006
Echirolles, France

ECU
Residence
1985
Ecully, France

ELC1
UGC Ciné Cité
1998
Elche, Spain

ELC2
UGC Ciné Cité
1999
Elche, Spain

EPN1
Lorraine Regional Audit Office
1984
Epinal, France

EPN2
General Council
2008
Epinal, France

ERL1
Shopping Mall
2004
Erlangen, Germany

ERV1
Urban Planning of the Kond District
2008 – 2018
Erevan, Armenia

ESC1
House of Sciences
2009
Esch-sur-Alzette, Luxembourg

EVN1
Cultural and Conference Centre
2003
Evian-les-Bains, France

EVR1
Tax Administration Building
1995
Evry, France

EVR2
Conurbation Authority
2004
Evry, France

EVX1
Saint-Michel District
2003
Evreux, France

EVX2
Eure-Seine Intercommunal Hospital Center
2005 – 2010
Evreux, France

EVX3
Logistics Centre
2006
Evreux, France

EVX4
Sports Centre
2012
Evreux, France

FCV1
UGC Ciné Cité
2000
Franconville, France

FES1
Campus of the Euro-Mediterranean University
2014
Fez, Morocco

FLR1
Shopping Mall
2008
Fleury sur Orne, France

FNJ1
Sainte-Marie de Prouilhe Monastery
2004
Fanjeaux, France

FRB1
Sonacotra Residence
1993
Forbach, France

FRK1
Saint-Georges Church
1989
Frankfurt am Main, Germany

FRK2
Renault Booth for the 1999 International Motor Show
1999
Frankfurt am Main, Germany

FSG1
Automated Storage Building
1986 – 1986
Foussignac, France

FSH1
Urban Planning of the Foshan Axis
2013
Foshan, China

FZH1
Sciences Museum
2010
Fuzhou, China

GAZ1
Courthouse
2002
Gaza, Palestine

GEN1
Fire Station
1988 – 1995
Gennevilliers, France

GNG1
Administrative Office
2005
Gongju, South Korea

GNJ1
Cultural Centre of Tihama
2005
Gwangju, South Korea

GNV1
"La Nouvelle Comédie" Theatre
2009
Geneva, Switzerland

GNV2
Les Dardelles Penitentiary
2015
Geneva, Switzerland

GPA1
Grand Paris Express Railway Stations
2012 – 2014
Paris, France

GPA2
Western Undergound Line 15
2015
Saint-Cloud et Rueil-Malmaison, France

GPA3
The Green Line
2015
Orly, Massy, Palaiseau et Versailles, France

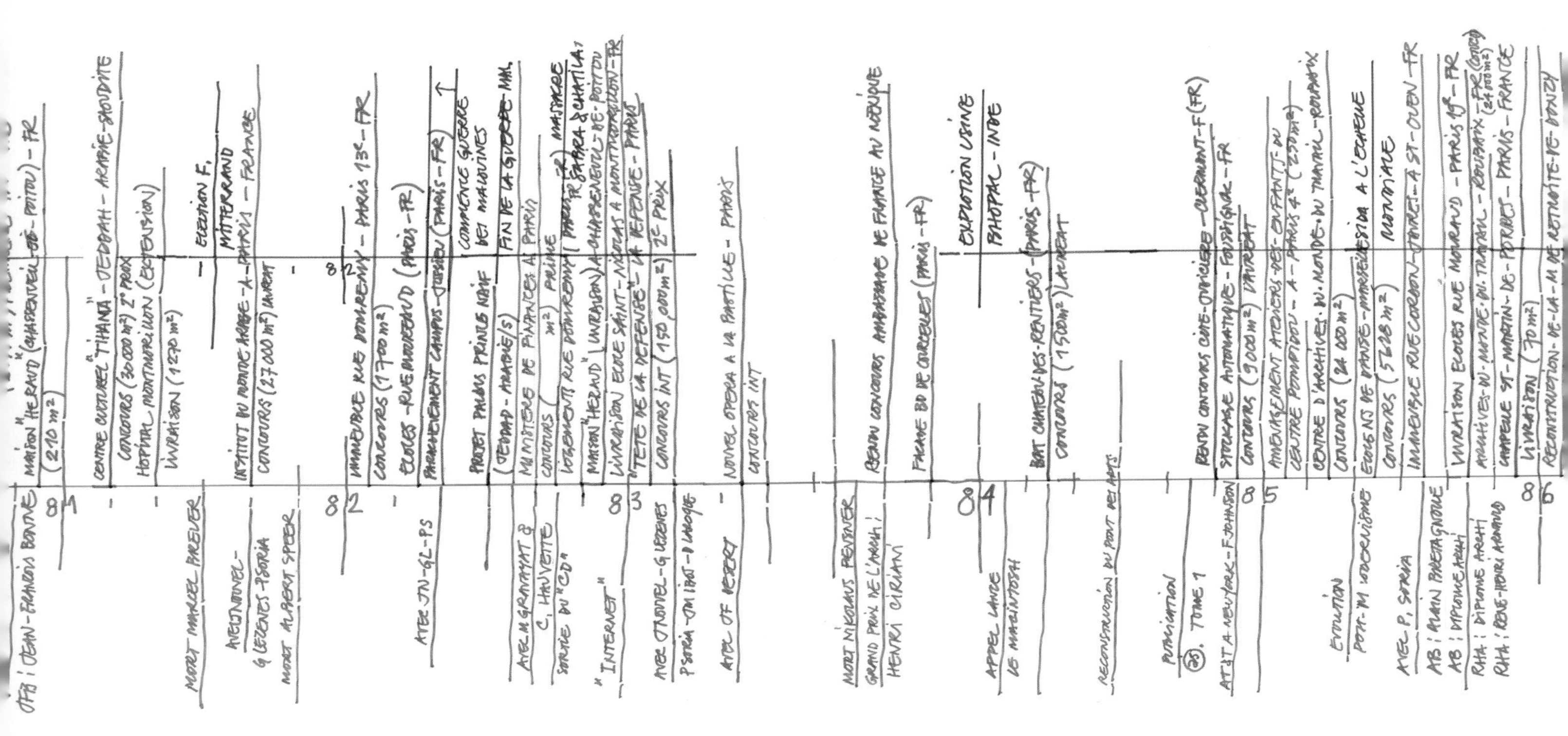

GRN1
GreEn-ER Centre
2011
Grenoble, France

GRN2
CEA Photonic Platform
2011
Grenoble, France

GRY
Community Centre
1986
Giromagny, France

GU/RVC
Renovation of a Spa Resort
2000
Ravine Chaude, Guadeloupe

GUY1
Renault Technological Centre
1990
Guyancourt, France

GVR1
Offices, Housing and Shopping Centre
2003 – 2007
Govorovo, Russia

GYC2
Saint-Quentin Eastern Sector
2014
Guyancourt, France

GZU1
Guangdong Olympic Stadium
1998
Guangzhou, China

GZU2
Guangzhou Opera
1999
Guangzhou, China

GZU3
Science City
2001
Guangzhou, China

GZU4
Television Tower
2004
Guangzhou, China

GZU5
International Conference Centre
2005
Guangzhou, China

GZU6
Dayi Villas
2007 – 2009
Guangzhou, China

GZU7
Dayi Villas
2007
Guangzhou, China

GZU8
Trendy Groupe Headquarters
2011
Guangzhou, China

GZU9
Arts and Science Museum
2013
Guangzhou, China

HCM1
Diamond Island District
2006
Ho Chi Minh City, Vietnam

HGN1
Restructuring of the Foch District
2008
Haguenau, France

HGP1
Cultural Centre
2015
Huangpi, China

HGS1
Medical Research Centre
2011
Huangshi, China

HHT1
Trading Centre
2014
Hohhot, China

HKG1
Hong Kong Parliament
2002
Hong Kong, China

HK01
Hainan Suntex International Tourist Centre
1997
Haikou, China

HK02
Urban Development of Haikou East Coast
2004
Haikou, China

HLR1
Transport Infrastructure
2014
L'Hay les Roses, France

HMB1
University Residence
1998
Hamburg, Germany

HMB2
Reception Pavilion of the University Residence
1996
Hamburg, Germany

HMB3
Business Centre
2001
Hamburg, Germany

HMB4
Media City Port
2001
Hamburg, Germany

HMB5
HASPA Headquarters
2001
Hamburg, Germany

HMB6
Office Building
2001
Hamburg, Germany

HMB7
Business Centre of Hafencity
2001
Hamburg, Germany

HMM1
Hotel Properties
2009
Hammamet, Tunisia

HMT1
UGC Ciné Cité
2002
Haumont, France

HNB1
Hospital Complex
2006
Hennebont, France

HN01
University of Science and Technology
2014 - 2021
Hanoi, Vietnam

HNZ1
Masterplan of the Hanzhong XinYuan Lake Zone
2011
Hanzhong, China

HNZ2
Museum
2012
Hanzhong, China

HRB1
Xin Ma Te Square
2006
Harbin, China

HRB2
Shidai Tower
2009
Harbin, China

HRB3
Jingwei Youyi Village
2010
Harbin, China

HRB4
Urban Planning of HaPing Road
2013
Harbin, China

HRN1
Penitentiary
2012 – 2017
Haren, Belgium

HYN1
Hakkas Cultural Centre
2009
Heyuan, China

HYR1
City Hall
2001
Hyères, France

HZH1
Housing and Commercial Complex
2006 – 2013
Hangzhou, China

HZH2
Appartements
2007 – 2013
Hangzhou, China

HZH3
Residential Buildings
2007
Hangzhou, China

HZH4
BeiTang River Commercial district
2013
Hangzhou, China

ILK1
University Centre
1990
Illkirch, France

ILK2
Commercial Complex
2013
Illkirch, France

ISP1
Research Lab
2006
Ispra, Italy

IST1
Multipurpose Sports Centre
1998
Istanbul, Turkey

IST2
Cultural Centre and Mixed-Use Building
2006
Istanbul, Turkey

ISY2
Corentin Celton Psychiatric Hospital
1983
Issy-les-Moulineaux, France

ISY3
Issy-les-Moulineaux Fort Masterplan
2000 – 2011
Issy-les-Moulineaux, France

ISY4
Office and Company Restaurant
2004
Issy-les-Moulineaux, France

ISY5
National Police Force Headquarters
2006
Issy-les-Moulineaux, France

ISY6
National Police Force Headquarters
2007
Issy-les-Moulineaux, France

ISY7
South-Eastern Bastion
2008 – 2013
Issy-les-Moulineaux, France

ISY8
Villa n°5
2007 – 2013
Issy-les-Moulineaux, France

ISY9
Villa n°1 et Belvedere
2007 – 2013
Issy-les-Moulineaux, France

ISY10
Office Tower
2011
Issy-les-Moulineaux, France

ISY11
Office Building
2011
Issy-les-Moulineaux, France

ISY12
Colas Headquarters
2012
Paris, France

ISY13
Corentin Celton Hospital – Housing
2013
Issy-les-Moulineaux, France

ISY14
Housing
2014 - 2021
Issy-les-Moulineaux, France

JC1
Highschool of the Future
1986 – 1987
Jaunay-Clan, France

JC3
University Residence
1988 – 1989
Jaunay-Clan, France

JLT1
Highschool
1988
Jouée-les-Tours, France

JNH1
Tourist Resort
2013
Jinhua, China

JNH2
Technological and Cultural Centre
2015
Jinhua, China

JNN1
Shandong Hotel Building
1999
Jinan, China

JNN2
Cultural Complex
2009
Jinan, China

JNN3
Cultural Centre
2010 – 2013
Jinan, China

JNN4
ShanDong Art Museum
2011
Jinan, China

JNN5
Office of the Western Railway Station
2012
Jinan, China

JNN6
Tower and Shopping Mall
2012
Jinan, China

JNN7
Jinan Valley Project
2012
Jinan, China

JNN8
Central Axis in Jinan Western City
2013
Jinan, China

JNN9
JiLu Gate
2013
Jinan, China

JNN10
Zhongtie Bauhinia Plaza
2013
Jinan, China

JNN11
Shandong Guard Station
2014
Jinan, China

JVY
Technical College
1987
Juvisy-sur-Orge, France

KB1
Retirement Extended Care Centre
1986
Le Kremlin-Bicêtre, France

KBL1
Conceptual Design for the New City of Deh Sabz
2007 – 2008
Kabul, Afghanistan

KLM1
Conceptual Planning of a Complex in the Mines Wonderland
2014
Kuala Lumpur, Malaysia

KLN1
Mushkino Leisure Park
2014
Kaliningrad, Russia

KMT1
Kama Hotel
2014
San-Pédro, Republic of Côte d'Ivoire

KRJ1
Alborz Science Park Masterplan
2012 – 2013
Karaj, Iran

KRJ2
Alborz Science Park
2014
Karaj, Iran

KRJ3
Biosun Pharmed Co. R&D
2015 – 2017
Karaj, Iran

KRM1
Karamay Campus
2014
Karamay, China

KTN1
Hospital and Retirement Home
2015
Kuitun Xinjiang, China

LBG1
Air Transport Union Headquarters
1985
Le Bourget, France

LBG2
Hospital Complex
1998
Le Bourget, France

LBL1
Sarthe et Loir Health Cluster
2002
Le Bailleul, France

LBN2
Shopping Mall
1997
Lisbon, Portugal

LBR1
Hotel and Commercial Complex
2007
La Brède, France

LDF1
"Le Cratère"
2006
La Défense, France

LDF2
Sogecap Tower
2007
La Défense, France

LDF3
"Tour Signal" Skyscraper
2008
La Défense, France

LDF4
"Les Saisons" – Landscape Design of Public Spaces
2012 – 2020
La Défense, France

LGG
Conference Room
1980
Ligugé, France

LGN1
Doctor's Surgery
1990 – 1991
Lagny-sur-Marne, France

LGN2
Hospital Complex
2005
Lagny-sur-Marne, France

LHR1
Workers' Hostel
2001
L'Hay des Roses, France

LHS1
Tibet Natural Science Museum
2009 – 2016
Lhasa, China

LHV1
Jules Lecesne Highschool
2006
Le Havre, France

LHY1
European Patents Office
1989
The Hague, Netherlands

LIS1
Technological Park
Lisses-Bois Chaland, France

LLL1
Haute Deûle Residence
2006
Lille, France

LLL2
Office Building
2008
Lille, France

LLL3
LFB R&D Centre
2010
Lille, France

LLV1
Housing
2004
La Louvière, Belgium

LLV2
Housing
2004
La Louvière, Belgium

LMB
Emile Roux Hospital
1985
Limeil-Brévannes, France

LMB1
Cultural Centre
2012
Limeil-Brévannes, France

LMC1
King Abdul Aziz Road Masterplan
2002
Makkah, Saudi Arabia

LMC2
Urban Development
2003
Makkah, Saudi Arabia

LMC3
King Abdallah Mosque
2004
Makkah, Saudi Arabia

LMC4
Mussallah
2005
Makkah, Saudi Arabia

LMC5
Darb Al-Khalil Urban Development
2005
Makkah, Saudi Arabia

LMC6
Haram Extension
2008
Makkah, Saudi Arabia

LMC7
King Abdallah Mosque
2010
Makkah, Saudi Arabia

LMC8
King Abdallah Mosque
2012
Makkah, Saudi Arabia

LMC9
Ejabah
2013
Makkah, Saudi Arabia

LMG2
Finance Office
1984
Limoges, France

LMG3
University Hospital – Women and Children's Unit
2000
Limoges, France

LMS1
Office Building
2010 – 2013
Limonest, France

LMS2
Office Building
2014 – 2017
Limonest, France

LNS1
Fitness Centre for Professional Athlete
2004
Lens, France

LNS2
Psychiatric Hospital
2005
Lens, France

LOM1
Organisation of African Union
2003
Lomé, Togo

LRN1
Vauban Residence
1998
La Réunion, France

LRN2
School of Fine Arts & Architecture
1999 – 2002
La Réunion, France

LRN3
Western Port Masterplan
1999 – 2001
La Réunion, France

LRN4
Penitentiary
2001
La Réunion, France

LRN5
Penitentiary
2004 – 2008
La Réunion, France

LRN6
Eastern Port Masterplan
2006 – 2010
La Réunion, France

LRN7
Restructuring of the Juliette Dodu Penitentiary
2014
La Réunion, France

LRP
"Les Thermes" Residence
1980
La Roche-Posay, France

LSN1
Children's Hospital
2013
Lausanne, Switzerland

LSN2
Health Campus
2016
Lausanne, Switzerland

LUS1
Molewa Ruichang New City
2015
Lushan Ruicheng, China

LVN1
Regional Covered Stadium
2004 – 2009
Liévin, France

LVR1
Psychiatric Hospital
2014
La Verrière, France

LYN1
International School Complex
1989
Lyon, France

LYN2
Tower of the Frankfurt Place
1992
Lyon, France

LYN3
"La Croix Rousse" Obstetrics and Gynecology Hospital
2006
Lyon, France

LYN4
Office Building
2006
Lyon, France

LYN5
Athletics Hall
2007
Lyon, France

LYN6
Gerland District
2011
Lyon, France

LYN7
Mixed-Use Building
2013
Lyon, France

LYN8
Restructuring of the Scientific Units of Lyon Tech La Doua Campus
2014
Lyon, France

LZH1
LiuDong Cultural Centre
2013
LiuZhou, China

MBL1
Touristic Resort
2007
Moulay Bousselham, Morocco

MCS1
French National Rugby Centre
2000
Marcoussis, France

MDB1
Technical University
1994
Magdeburg, Germany

MDN1
Retirement Home
1987
Meudon, France

MDN2
Renault Site
2000
Meudon, France

MEX1
French Embassy
1983
Mexico City, United Mexican States

MGY1
Danone Research Centre
2000
Magny-les-Hameaux, France

MKN1
Stadium for the 2006 Football World Cup
1999
Meknes, Morocco

MLB1
School Complex
1986
Montigny-le-Bretonneux, France

MLB2
Cycling Track
2007
Montigny-le-Bretonneux, France

MLB3
Cycling Track
2008
Montigny-le-Bretonneux, France

MLC1
Cycling Track
2001
Mouilleron-le-Captif, France

MLH
Sonacotra Residence
1993
Mulhouse, France

MLJ1
Innovative Secondary School
2016
Mantes-la-Jolie, France

MLK1
Office Building
2001
Malakoff, France

MLM1
Hospital Complex
2003
Meulan-les-Mureaux, France

MLN1
Urban planning of the Bovisa District
1997
Milan, Italy

MLN2
Contemporary Art Museum
1998
Milan, Italy

MLN3
AEM Office – Bovisa
1999
Milan, Italy

MLR1
Activity Centre
1987
Mandres-les-Roses, France

MLV1
Science and Technical Hub
2008
Marne-la-Vallée, France

MLV2
Restructuring of Copernic University
2015 – 2019
Marne-la-Vallée, France

MMD1
Kaweni Technical College
2003
Mamoudzou, France

MMZ1
The "Cité du Bois"
2005
Mimizan, France

MNB1
Montbéliard Urban District Headquarters
1993
Montbéliard, France

MNC1
Landscape Design of the Port Hercule
2002
Monaco, Monaco

MNC2
Cultural Centre
2007
Monaco, Monaco

MNH1
Commercial Complex
2007
Munich, Germany

MNS1
Domaine des Grands Prés
2004
Mons, Belgium

MNT1
Hérault Culture Sport
2002
Montpellier, France

MNT2
Urban Design of the New Saint Roch District
2002
Montpellier, France

MNT3
New City Hall
2003
Montpellier, France

MNT4
Parc Marianne Urban Development Area
2003 – 2017
Montpellier, France

MNT5
Millenium Avenue Building
2005 – 2010
Montpellier, France

MNT6
Mixed-Use Building
2006
Montpellier, France

MNT7
O'Zone Office
2006 – 2013
Montpellier, France

MNT8
Extension of the Parc Marianne Area
2003 – 2018
Montpellier, France

MNT9
Raymond Dugrand Avenue
2006 – 2020
Montpellier, France

MNT10
Park Avenue Residence
2007 – 2011
Montpellier, France

MNT11
Multipurpose Hall for the Fairground
2008
Montpellier, France

MNT12
Redevelopment of Nina Simone and Joan Miro Boulevards
2008 – 2011
Montpellier, France

MNT13
Nouvelle Ligne Residence
2010 – 2013
Montpellier, France

MNT14
Doramar Residence
2011 – 2014
Montpellier, France

MNT15
The "Mutuelle des Motards" Office
2012
Montpellier, France

MNT16
La Folie Richter Residence
2013 – 2014
Montpellier, France

MNT17
Housing
2013 – 2015
Montpellier, France

MNT18
Housing
2015
Montpellier, France

MOP1
Orbis Motels
1999
Warsaw, Poland

MOS1
Moscow Institute of Physics and Technology
2013
Moscow, Russia

MOS2
Circus
2013
Moscow, Russia

MRC1
Guy Dolmaire Secondary School
1999 – 2004
Mirecourt, France

MRG1
Maurice Genevoix Highschool
1992
Montrouge, France

MRG2
Crédit Agricole Headquarters
2011
Montrouge, France

MRK1
Stadium for the 2006 Football World Cup
1999
Marrakesh, Morocco

MRK2
Douja Golf Resort
2008
Marrakesh, Morocco

MRN1
Extension of the Hospital
1978 – 1981
Montmorillon, France

MRN2
Saint-Nicolas Primary School
1980 – 1982
Montmorillon, France

MRN3
Senior Citizen's Home
1982 – 1984
Montmorillon, France

MRS1
Roland Petit National Ballet School
1985
Marseille, France

MRS2
Extension of the City Stadium
1995
Marseille, France

MRS3
Regional City Library
1997
Marseille, France

MRS4
Carrefour Shopping Mall
2000
Marseille, France

MRS6
Sainte-Marguerite Hospital – Psychiatry Unit
2004 – 2007
Marseille, France

MRS7
Sainte-Marguerite Hospital – Senior's Unit
2006
Marseille, France

MRS8
Euromed Housing Complex
2013
Marseille, France

MSC1
French Embassy
1986 – 1989
Muscat, Sultanate of Oman

MSC2
Cultural Center
2008
Muscat, Sultanate of Oman

MSC3
Al Hail Sea Plaza
2009
Muscat, Sultanate of Oman

MSC4
Sultan Qaboos Sports Academy
2012
Muscat, Sultanate of Oman

MSH1
Wakilabad Masterplan
2010
Mashhad, Iran

MSM1
Retirement Home
1994
Campagne-les-Hesdin, France

MSY1
Restructuring of University Residence
1990
Massy, France

MSY2
Gustave Eiffel Highschool
2003
Massy, France

MTG1
Durzy Highschool
1991
Montargis, France

MTR1
Technical College
2001
Matoury, Guyana

MTZ1
National Engineering School
2005 – 2016
Metz, France

MTZ2
Hospital
2009
Metz, France

NCE1
Cancer Institut and University
2004
Nice, France

NCE2
Palazzo Meridia Office
2016 – 2020
Nice, France

NCH1
Office Building
2004
Nanchang, China

NCY1
Sonacotra Residence
1993
Nancy, France

NDJ1
CEBEVIRHA Headquarters
1998
N'Djamena, Chad

NDJ2
Pilgrim Village
2013
N'Djamena, Chad

NDR1
Stadium for the 2006 Football World Cup
1999
Nador, Morocco

NMA1
Grand Nouméa Highschool
1994
Nouméa, New Caledonia

NMA2
Koutio University Hospital
2007
Nouméa, New Caledonia

NMS1
Shopping Mall
2002
Nîmes, France

NMS2
Urban Development of the Central Station Area
2002
Nîmes, France

NMS5
Conference Centre
2002
Nîmes, France

NMS6
Housing and Office Building
2006 – 2010
Nîmes, France

NMS7
Commercial Complex
2007
Nîmes, France

NMS8
Cultural Centre
2007
Nîmes, France

NMS9
Urban Development of the Central Station Area
2008 – 2012
Nîmes, France

NMS10
Housing
2010
Nîmes, France

NNG2
Urban Development of the City Centre
2008
Ningbo, China

NNG3
Office Towers
2010 – 2016
Ningbo, China

NNG4
Ningbo University
2015
Ningbo, China

NNJ1
Memorial
2005
Nanjing, China

NNJ2
Conference and Exhibition Centre
2006
Nanjing, China

NNJ3
Hexi Olympic Village
2010
Nanjing, China

NNJ4
Hospital Complex
2012
Nanjing, China

NNJ5
Nanjing Baixia Cultural Centre
2013 – 2015
Nanjing, China

NNJ6
Urban Development of JinLuanXia District
2013
Nanjing, China

NNJ7
Urban Development of ZhongHua Men District
2013
Nanjing, China

NNJ8
Pukou Great Theatre
2014
Nanjing, China

NNJ9
Kirin Innovation Park
2014
Nanjing, China

NOY
Sonacotra Residence
1993
Noyon, France

NSH1
Pearl River Development Plaza
2015
Nansha, China

NTE1
Saint-Jacques Hospital
2000
Nantes, France

NTE2
Housing and Office Building
2007
Nantes, France

NTE3
Housing Complex
2014
Nantes, France

NTG1
Wison Factory
2004
Nantong, China

NTR1
Office
2006
Nanterre, France

NTR2
Masterplan of Paris Ouest University
2009 – 2010
Nanterre, France

NTR3
Arena Stadium
2010
Nanterre, France

NTR4
Nanterre la Folie Railway Station
2015 – 2021
Nanterre, France

OLF1
Arena
2011
Ouled Fayet, Algeria

OLF2
Arena
2013 – 2020
Ouled Fayet, Algeria

ORD1
Hotel and Conference Room
2011
Ordos, China

ORD2
Mixed-use Building
2011
Ordos, China

ORL1
Southern Highschool
1995
Orléans, France

ORL2
Masterplan of "Rue des Halles"
2004
Orléans, France

ORN1
Bab Es Salam Towers
2011 – 2012
Oran, Algeria

ORY1
Masterplan of "Coeur d'Orly" District
2006
Orly, France

OYX1
Haut Bugey Hospital
2004
Oyonnax, France

PAARN1
Apartment
2002
Paris, France

PAARS1
Scenography for the "Press Review" Exhibition – Pavillion de l'Arsenal
1994
Paris, France

PABAS1
Bastille Opera House
1983
Paris, France

PABAS2
Agence Architecture-Studio – boulevard Bastille
2005
Paris, France

PABDA1
Renovation of Apartment Building
1989 – 1991
Paris, France

PABEL
Housing
1988
Paris, France

PABER1
Ministry of Finance
1982
Paris, France

PABLY1
International Conference Centre
1989
Paris, France

PABNS1
Renovation of the Sarrailh University Centre
1996 – 1999
Paris, France

PABRT1
Housing
2011
Paris, France

PABSS1
Housing
2011
Paris, France

PACAB1
Restructuring of the Sainte-Anne Hospital Psychiatry Unit
1982
Paris, France

PACAB2
Restructuring of the Ferrus and Joffroy Pavilions
1997 – 2001
Paris, France

PACAB3
Masterplan of Sainte-Anne Hospital
2001 – 2002
Paris, France

PACAB4
Brain Disease Clinic
2004 – 2012
Paris, France

PACAP1
Société Générale Communication Department
1987 – 1988
Paris, France

PACAU1
Housing
1990 – 1997
Paris, France

PACAV1
Office
2008
Paris, France

PACHL1
Chalon Plaza
1991
Paris, France

PACHL1
Renault Area
1998
Paris, France

PACHP1
Post Office and Sorting Office
1988 – 1994
Paris, France

PACHR1
Apartment Building
1984 – 1986
Paris, France

PACHR3
Restructuring of an Apartment Building
2013 – 2014
Paris, France

PACHT1
National Judo Institute
1988 – 2004
Paris, France

PACIT1
Canal+ Headquarters
1988
Paris, France

PACL
Sonacotra Residence
1989
Paris, France

PACLY1
New Courthouse
2010
Paris, France

PACMD1
Hotel and Commercial Complex
2002
Paris, France

PACOT1
Saint Jean-Baptiste de Lasalle School
2015
Paris, France

PACOU1
Restructuring of Façade
1981 – 1984
Paris, France

PACR1
University Residence
1991 – 1996
Paris, France

PACRL1
19th Arrondissement Municipal School of Music
1985
Paris, France

PACRN1
Scenography for the "Paris and the Daguerrotype" Exhibition – Carnavelet Museum
1989 – 1989
Paris, France

PACRN2
Scenography for the *Paris in 3D* Exhibition – Carnavelet Museum
1998 - 2000
Paris, France

PACUR
Apartment Building – rue Curnonsky
Paris, France

PADAU1
Restructuring of an Office Building
2013
Paris, France

PADMN1
Urban Planning of the Daumesnil Viaduct
1987
Paris, France

PADOM1
Housing – rue Domrémy
1982 – 1984
Paris, France

PADX1
Apartment Building and Bilingual School
1990 – 1994
Paris, France

PAFRC1
Office Building
2001
Paris, France

PAFRK1
Extension of the Franklin Secondary School
2006 – 2016
Paris, France

PAFRN1
Alain Fournier School
2004
Paris, France

PAFRS1
Restaurant
2012
Paris, France

PAHCH
Restructuring of the Japanese Embassy
1995
Paris, France

PAHLP1
Paris Heliport
2005
Paris, France

PAINA1
Conference Centre of the World Bank
2001 – 2002
Paris, France

PAINA2
Sultanate of Oman Embassy
2004
Paris, France

PAJIB1
Saint-Martin-de-Porrès Chapel
1985 – 1985
Paris, France

PAJML1
Pin Up Studio
1999
Paris, France

PAJUS1
Rehabilitation of the Jussieu Campus
1982
Paris, France

PAJUS2
Jussieu Campus Library
1992
Paris, France

PAJUS3
Restructuring of the Jussieu Campus
1994
Paris, France

PAJUS4
Rehabilitation of the Central Tower of Jussieu Campus
2005
Paris, France

PAJUS5
Restructuring of the Eastern Sector of Jussieu Campus
2008 – 2015
Paris, France

PAKLB1
Office Building – Avenue Kléber
2015 – 2018
Paris, France

PALAC1
Architecture-Studio
1979 – 1980
Paris, France

PALDF
Tête de la Défense
1983
Paris-la-Défense, France

PALDF1
Restructuring of La Défense District
1989
Paris-la-Défense, France

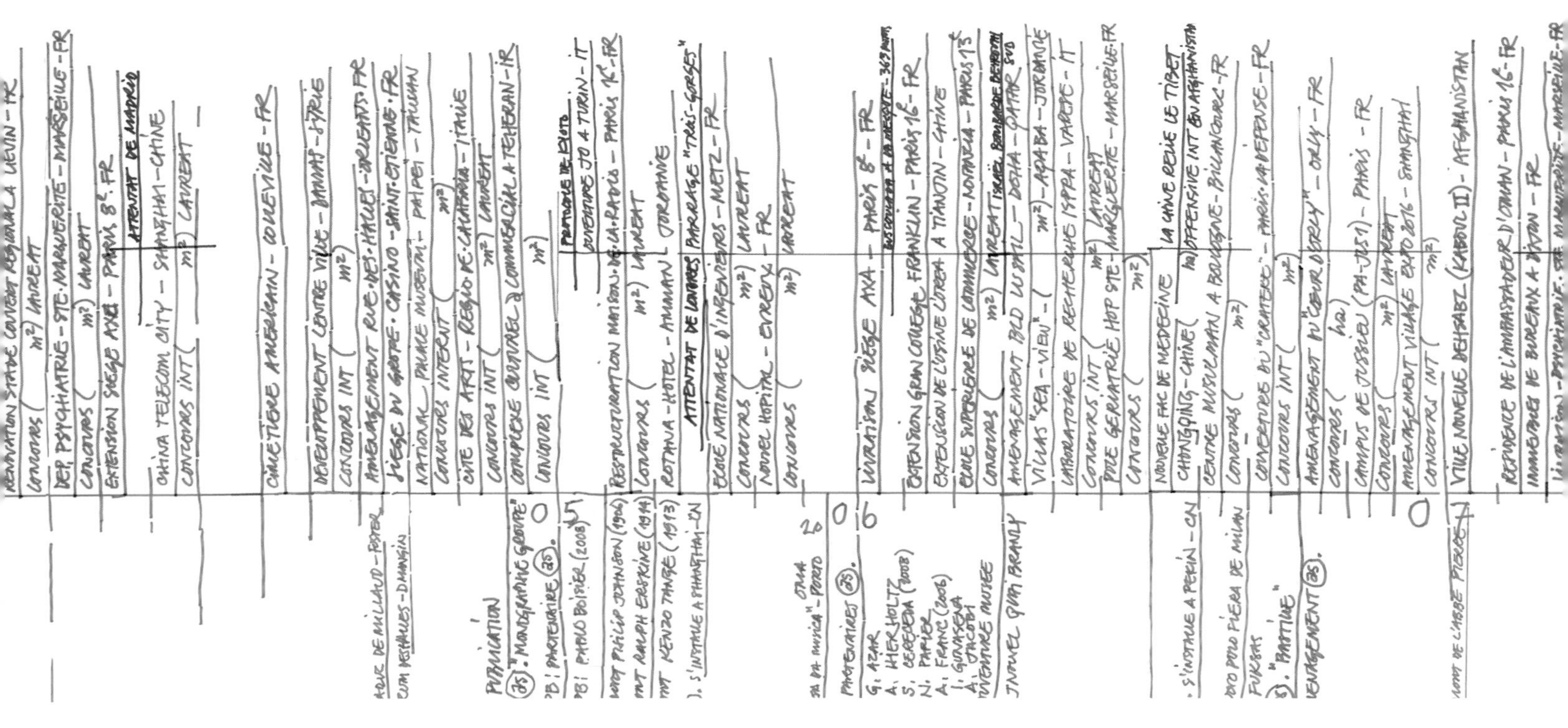

PALDF2
Notre-Dame de la Pentecôte Church
1989
Paris-la-Défense, France

PALES
Housing
1987
Paris, France

PAMAR1
Restructuring of a Factory
1982–1983
Paris, France

PAMAS1
Office Building
1988
Paris, France

PAMAT
Restructuring of an Apartment Building
1994
Paris, France

PAMIL1
Cognacq-Jay Hospital
1999
Paris, France

PAMMN1
The Ambassador of the Sultanate of Oman Residence
2007–2012
Paris, France

PAMOI1
Novancia Business School
2006–2010
Paris, France

PAMOI2
Novancia Business School – Offices
2011–2012
Paris, France

PAMON
Restructuring of an Apartment Building
1994
Paris, France

PAMOU1
Elementary and Primary School
1982–1985
Paris, France

PAORI1
Retirement Home
1990–1997
Paris, France

PAP1
Penitentiary
2011–2016
Papeari, French Polynesia

PAPAX1
China Eastern Airlines Head Office
1998–1999
Paris, France

PAPDA1
Project for the Commemoration of the Bicentinary of the Revolution
1988
Paris, France

PAPEL1
Housing
1994–2003
Paris, France

PAPER1
Pershing and Ternes-Villiers Area
2015
Paris, France

PAPLC1
Père Lachaise Crematorium
2013
Paris, France

PAPLG1
Floating Swimming Pool
2003
Paris, France

PAPNY
Restructuring of an Office Building – rue de Prony
2014
Paris, France

PAPOM1
Children's Workshop – Centre Pompidou
1986–1986
Paris, France

PAPRK1
Maison de la Radio – President's Office
2002
Paris, France

PAPRK2
Restructuring of the Maison de la Radio
2005–2019
Paris, France

PAPRO1
Notre-Dame de l'Arche d'Alliance Church
1986–1998
Paris, France

PAPRS1
Restructuring of an Apartment
2015–2017
Paris, France

PAPSY1
Passy Park Development
1988
Paris, France

PAPVR1
Renault Stand
2000
Paris, France

PARBR
Office Building
1980–1990
Paris, France

PARMR
Restructuring of an Apartment Building
1994
Paris, France

PAROB1
Private House
1992–1997
Paris, France

PARYL1
Office Building
1997–1999
Paris, France

PASAR1
Elementary School
2000–2003
Paris, France

PASTA1
Urban Planning of the Saint-Ange Islet
1987
Paris, France

PASTB1
Arab World Institute
1981–1987
Paris, France

PASTB2
Mohamed V Plaza
2001–2002
Paris, France

PASTB5
La Médina
2007
Paris, France

PASTB6
Restructuring of the Arab World Institute – Terrace
2007–2008
Paris, France

PASTB7
Temporary Exhibition Room
2008
Paris, France

PASTB8
Restructuring of the Arab World Institute – Museum
2009–2012
Paris, France

PASTB9
Arab World Institute – Square
2009–2010
Paris, France

PATEV1
Éole-Évangile District
2015
Paris, France

PATRB1
Apartment
2009
Paris, France

PATRH1
Business Centre
1990
Paris, France

PATRS
Façade for an Hotel
1986–1988
Paris, France

PATVS1
Fimat Banque Office – Hôtel de Bony
1993–1994
Paris, France

PAUVN2
National Assembly – Chaban Delmas Building
2005
Paris, France

PAVAR1
Rodin Museum
1987
Paris, France

PAVDM1
AXA Private Equity Headquarters
2004–2006
Paris, France

PAVLT1
Science Museum and La Villette Park Development
1980
Paris, France

PAVLT2
Paris School of Fine Arts
1992
Paris, France

PAVTA1
Pitié-Salpêtrière Hospital
1988
Paris, France

PAVTR1
Voltaire Substation Cinema
2015
Paris, France

PAWLM
Housing
1990
Paris, France

PCY1
Hospital's Logistic Centre
1999
Pacy-sur-Eure, France

PGM1
Secondary School
2008
Pégomas, France

PKN2
National Grand Theatre
1998
Beijing, China

PKN3
Continent Hypermarket
1998–1999
Beijing, China

PKN6
Urban Planning of San Li Tun District
Beijing, China

PKN7
Beijing Yao Hui International Building
2002
Beijing, China

PKN8
International Exhibition Centre
2004
Beijing, China

PKN9
Sino-European Exhibition Centre
2006
Beijing, China

PKN10
Museum, Housing
2007
Beijing, China

PKN11
CNPC Technological Park
2008
Beijing, China

PKN12
"China Life" Research Centre
2008
Beijing, China

PKN13
Beijing River Creative Zone
2009
Beijing, China

PKN14
Urban Design of China Mobile International Information Harbor
2009
Beijing, China

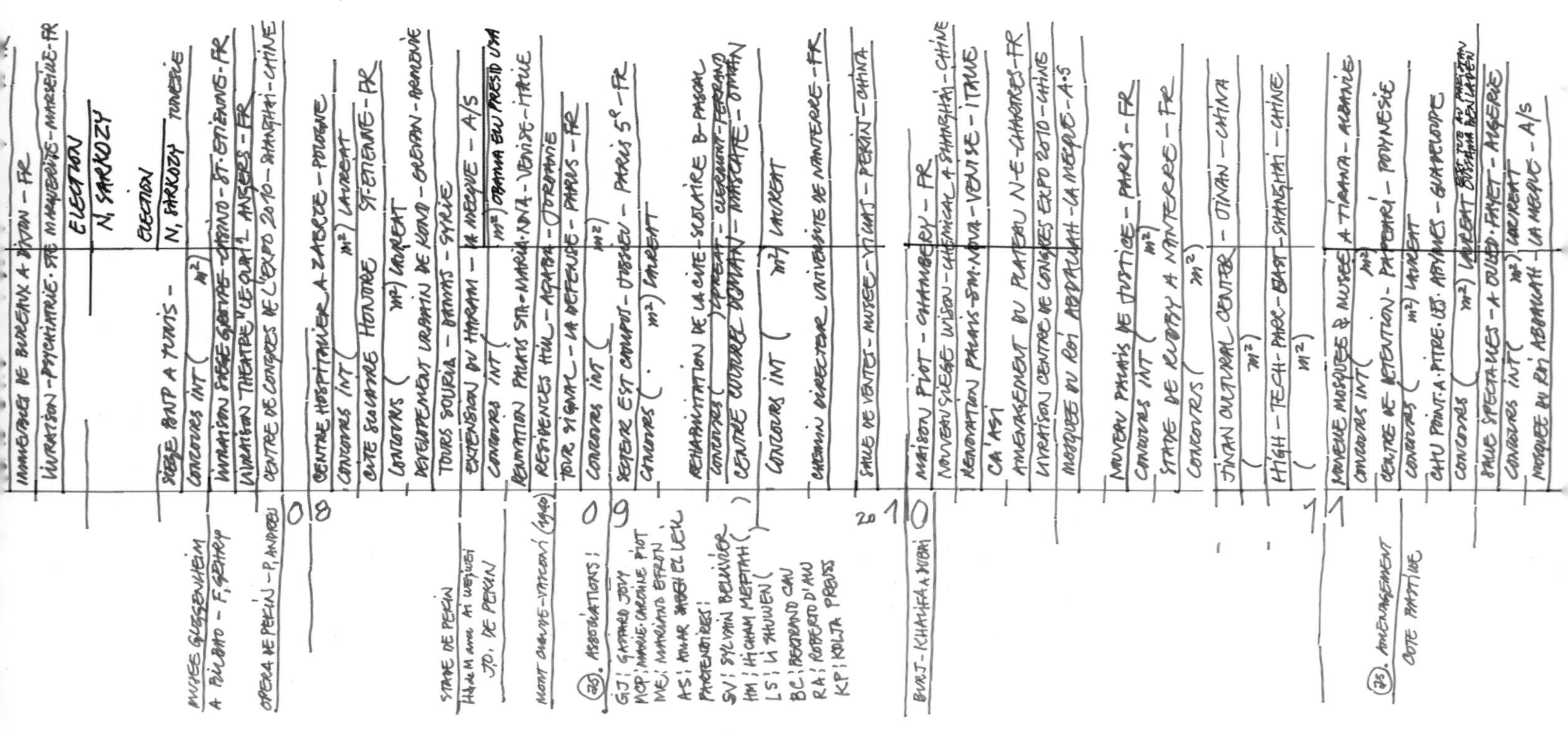

PKN15
Urban Design of Daxing Administrative Centre
2010
Beijing, China

PKN16
Conference Centre and Hotel
2010
Beijing, China

PKN17
3 Presidential Villas
2010
Beijing, China

PKN19
"Cité de la Mode" Masterplan
2011
Beijing, China

PKN20
Urban Design of Jinhai Lake Tourist Area
2012
Beijing, China

PKN21
Youth Training Centre
2012 – 2013
Beijing, China

PKN23
Restaurant
2013
Beijing, China

PKN24
Wangfujing West Club
2013
Beijing, China

PKN25
Qinglong Lake Aquatic Complex
2014
Beijing, China

PKN26
Jing XI
2014
Beijing, China

PKN27
Sheshenya Observatory
2015
Beijing, China

PKN28
Badaling Chadao East Project
2015
Beijing, China

PKN29
Motor Factory
2015
Beijing, China

PLN1
Lafarge Factory
1996
Port-la-Nouvelle, France

PLR1
Office Building
2001
Le Plessis-Robinson, France

PLS1
Danone Vitapole
2000 – 2002
Palaiseau, France

PLS2
Danone Research Centre
2002
Palaiseau, France

PLS3
Danone Research Centre – Conference Room
2004
Palaiseau, France

PLS4
National School of Advanced Technologies
2008
Palaiseau, France

PLS5
Restructuring of Danone Vitapole R&D Centre
2009 – 2009
Palaiseau, France

PLS7
Pernod Ricard R&D Centre
2014
Palaiseau, France

PLS8
AgroParis Tech and INRA Campus
2015
Palaiseau, France

PNA1
Tahiti Mahana Beach Masterplan
2014
Punaaula, French Polynesia

PNT1
Judiciary Police Office
2008
Pontoise, France

P02
"Shared Activity Areas" Housing Complex
1975 – 1980
Poitiers, France

P03
Equipment Departmental Council
1979
Poitiers, France

P04
Science Museum
1985
Poitiers, France

PRP1
TGV Railway Station
2005
Perpignan, France

PRS1
Seysses, Le Pontet and Sequedin Penitentiaries
1999
France

PRS2
La Farlède, Liancourt and Chauconin-Neufmontiers Penitentiaries
1999 – 2004
France

PRS3
Penitentiary
2004
France

PRS4
Ain, Ille-et-Vilaine and Landes Penitentiaries
2004
France

PRS5
4 Penitentiaries
2004
France

PRS6
Drôme, Puy-de-Dôme and Haut-Rhin Penitentiaries
2011
France

PSM1
UGC Ciné Cité
1999
El Puerto de Santa María, Spain

PSY1
Hospital Complex – Emergency Unit
2000
Poissy, France

PTL1
Caudan Mixed-use Building
2008
Port-Louis, France

PTX1
Gallery and Conference Centre
Puteaux, France

PTX2
Urban Planning of Les Bergères Area
2009
Puteaux, France

PZH1
Urban Planning Museum
2006
Pengzhou, China

PZH2
Masterplan and Mixed-use Building
2006
Pengzhou, China

QGD1
Haiyun Hotel
2008
Qingdao, China

QGD2
HaiYun Towers
2011
Qingdao, China

QGD3
Qingdao Eye Hospital
2015
Qingdao, China

QGD5
Masterplan of Pingdu Hospital
2015
Qingdao, China

QGD6
Urban Planning of Licang District
2016
Qingdao,Shandong, China

QHD1
Club Med Hotel
2013
QingHuangDao, China

QSW1
Design of a Club Med Village
2015
Wenchang, China

RATP1
Design of a Virtual Subway Station
2009
Paris, France

RB1
Archive Centre
1985
Roubaix, France

RB2
"Fosse aux Chênes" Project
1986
Roubaix, France

RB3
Euroteleport – Masterplanning
1988 – 1995
Roubaix, France

RB4
Euroteleport – Parking
1989 – 1990
Roubaix, France

RB5
Euroteleport – ATRIA Project
1989
Roubaix, France

RB6
Euroteleport Office Building
1990 – 1991
Roubaix, France

RB7
Sonacotra Residence
1993 – 1995
Roubaix, France

RB8
Canal de Roubaix – Urban planning
1995
Roubaix, France

RB9
Cycling Track for Lille 2004
1996
Roubaix, France

RB10
Union Area
2008
Roubaix, France

RB11
Hospital
2011
Roubaix, France

RBT1
New Headquarters
2005
Rabat, Morocco

RBT2
Ibn Sina Hospital
2015
Rabat, Morocco

RBT3
Ibn Sina Hospital
2016 – 2021
Rabat, Morocco

RDS1
Academic Centre
2007
Ruda l ska, Poland

RGC1
Art Centre
2004
Reggio Calabria, Italy

RLM1
André Malraux Theatre
1998
Rueil-Malmaison, France

RLM2
Urban Planning of the Bergasoli Stadium
2001
Rueil-Malmaison, France

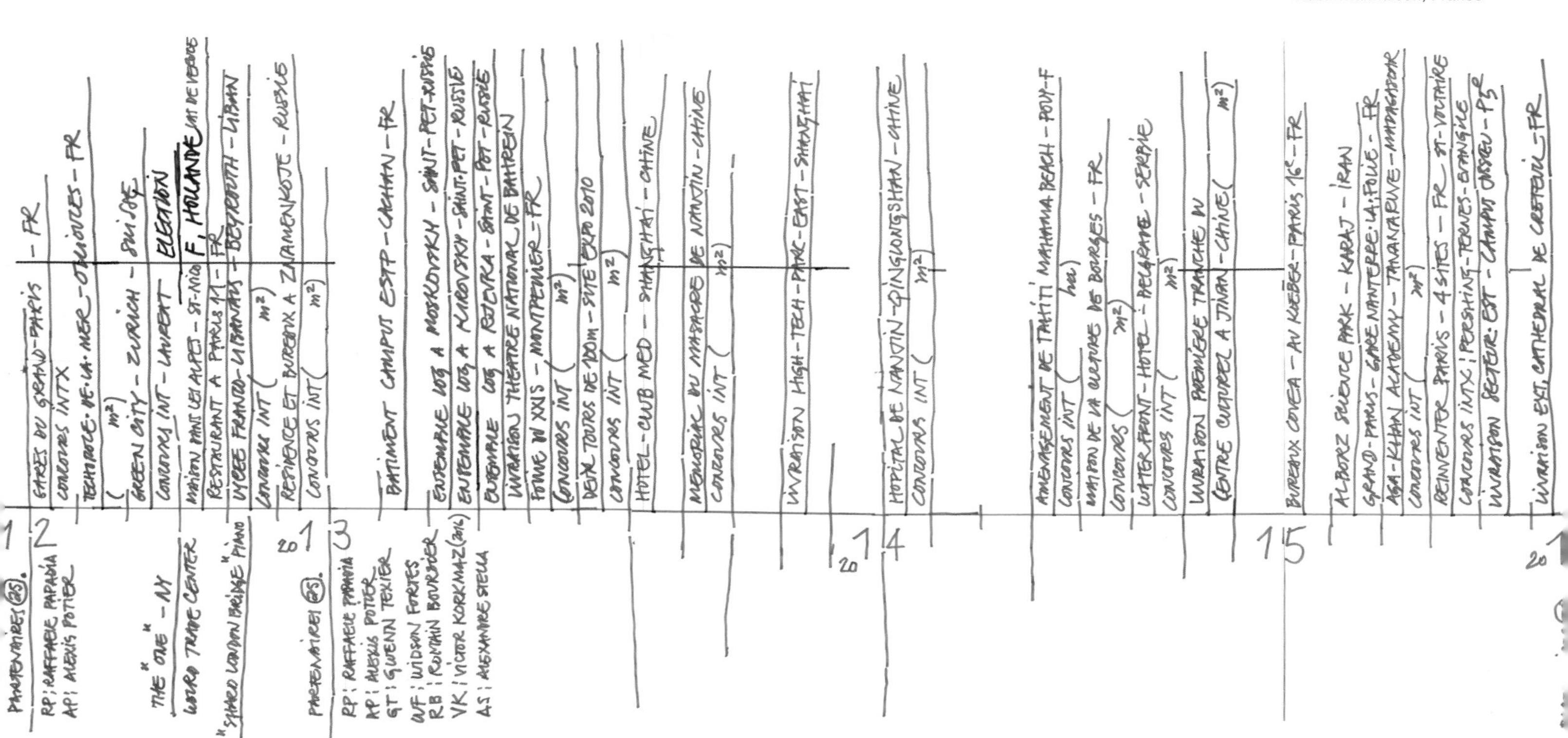

RM1
Sébastopol Hospital
Reims, France

RM2
Sonacotra Residence
1993
Reims, France

RML1
Courthouse
2002
Ramallah, Palestine

RMS3
Champagne P&C Heidsieck Pavilion
2006
Reims, France

RNE1
Leisure Centre
2003
Roanne, France

RNN2
University Centre
2005
Rennes, France

ROM1
Church
1994
Rome, Italy

ROM2
ISTAT Headquarters
2007
Rome, Italy

RON1
Natural History Museum
1993
Rouen, France

RSY1
Air France Headquarters
1992
Roissy-en-France, France

RSY2
Roissy 2 Area – Offices
2001
Roissy-en-France, France

RYD1
Administrative City
2007
Riyad, Saudi Arabia

SAS1
Urban Design and Development of Courbevoie, La Garenne-Colombes, Nanterre, Rueil-Malmaison and Suresnes
2011 – 2013
Nanterre, France

SCH1
Sonacotra Residence
1993
Schaeffersheim, France

SCH2
Louis Pasteur University Institute of Technology
1993
Schiltigheim, France

SCL1
HEC Campus
2007
Saclay, France

SDF1
Tourist Complex
2007
Sidi-Fredj, Algeria

SEN1
Theatre
2006
Sens, France

SER1
Engineering School
1993
Saint-Etienne-du-Rouvray, France

SET1
Hotel
2011 – 2015
Sète, France

SEV1
Universal Exposition "92 of Seville – French Pavilion
1989
Seville, Spain

SFN1
Trade Area
2004
Suifenhe, China

SGP1
Shopping Mall
2012
Saint-Genis-de-Pouilly, France

SHBTN1
Towers
2012
Shanghai, China

SHDNG1
Financial District
2013
Shanghai, China

SHG1
Cultural Centre
2013
ShaoGuan, China

SHG2
Shaoguan Train Station Front Plaza
2014
ShaoGuan, China

SHHNG1
Urban Design of Shanghai Art Centre
2013
Shanghai, China

SHN1
Art Centre
2001
Shanghai, China

SHN2
Expansion of Chongming City
2002
Shanghai, China

SHN3
World Expo 2010
2001
Shanghai, China

SHN4
Wison Chemical Headquarters and Laboratories
2002 – 2003
Shanghai, China

SHN5
North District
2002
Shanghai, China

SHN6
School of Visual Arts
2002
Shanghai, China

SHN7
Pudong Archive Centre
2002
Shanghai, China

SHN8
Conceptual Design of Nanhui District
2002
Shanghai, China

SHN9
Housing
2003
Shanghai, China

SHN10
Puxi Bund
2003
Shanghai, China

SHN11
China Telecom City
2003
Shanghai, China

SHN12
Cultural Centre
2003
Shanghai, China

SHN13
Housing
2003
Shanghai, China

SHN15
Jinqiao District
2004
Shanghai, China

SHN16
Hotel
2004 – 2013
Shanghai, China

SHN17
Technological Park
2004 – 2008
Shanghai, China

SHN18
Shi Liu Pu District
2004
Shanghai, China

SHN19
SGM Office
2004
Shanghai, China

SHN20
Agbar Tower
2005
Shanghai, China

SHN21
Zhiyin Mixed-use Building
2005
Shanghai, China

SHN22
CaoLu New Town Masterplan
2005
Shanghai, China

SHN23
Restructuring of the Gaoyang Road Warehouse
2005 – 2008
Shanghai, China

SHN24
Aquarium Expansion
2005
Shanghai, China

SHN25
The Bund Restaurant
2005
Shanghai, China

SHN26
Fazhan Tower Interior Design
2005
Shanghai, China

SHN27
Okaidi Shop
2005
Shanghai, China

SHNNR1
Mixed Use Buildings
2013
Shanghai, China

SHPDG1
BinHai Hotel
2007
Shanghai, China

SHPDG1
HuaMu Conceptual Design
2006
Shanghai, China

SHPDG2
Study for the World Expo 2010 Conference Centre
2006
Shanghai, China

SHPDG4
Shanghai World Exhibition Village
2006
Shanghai, China

SHPDG5
Shanghai World Exhibition Public Events Centre
2006
Shanghai, China

SHPDG6
Shanghai World Exhibition Art Centre
2006
Shanghai, China

SHPDG7
Office Building
2006 – 2009
Shanghai, China

SHPDG8
Restructuring of a Factory
2006
Shanghai, China

SHPDG9
High Tech Park
2007
Shanghai, China

SHPG12
Shanghai World Exhibition
2008 – 2010
Shanghai, China

SHPG14
Zhangjiang Mixed-Use Buildings
2008
Shanghai, China

SHPG15
Jinqiao Modern Industry Service Park
2009 – 2015
Shanghai, China

SHPDG19
Office Park for Wison Headquarters
2010 – 2013
Shanghai, China

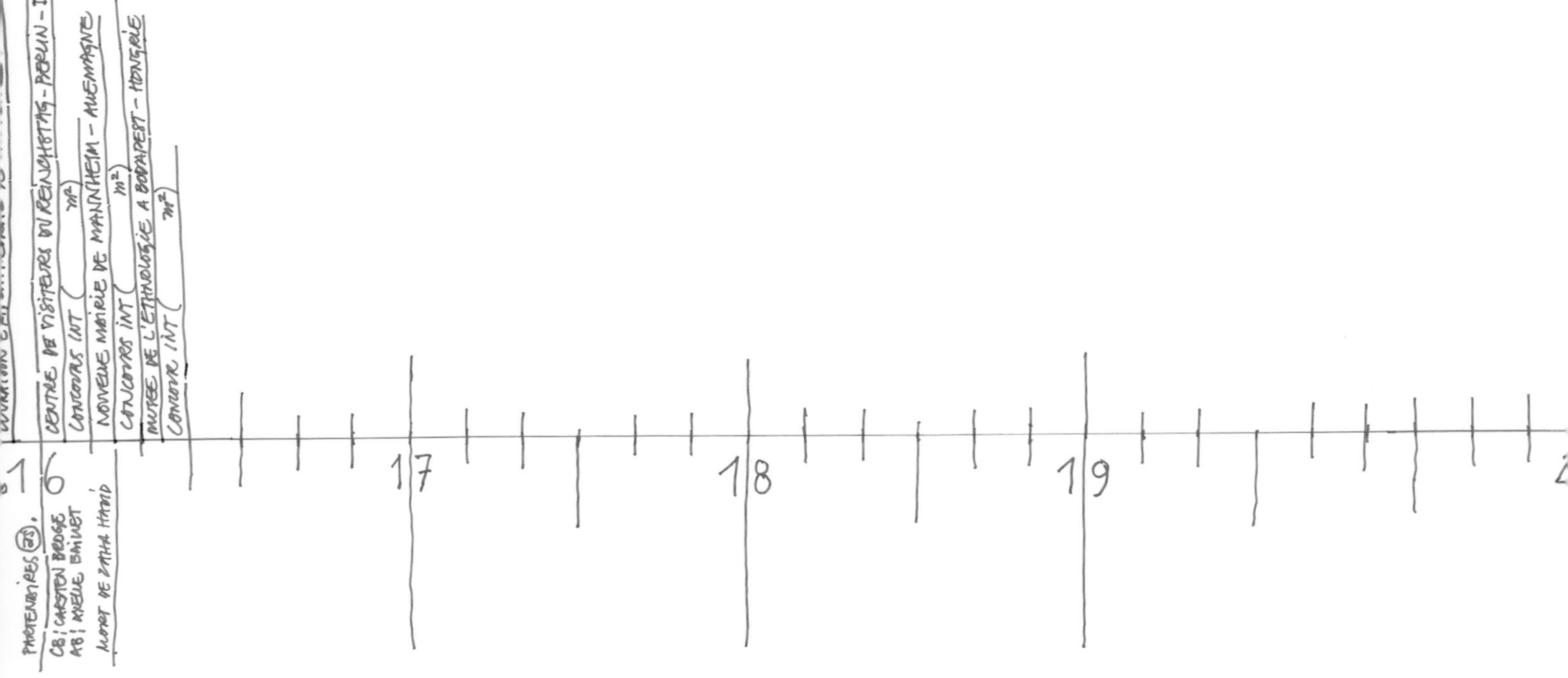

SHPG22
Interior Design of Zegna Office and Showroom
2009
Shanghai, China

SHPG23
Zhangjiang New Area
2010
Shanghai, China

SHPG24
Residential Hotel
2010 – 2013
Shanghai, China

SHPG25
Daguan Theatre
2010 – 2013
Shanghai, China

SHPG26
TV Tower in the Pudong Lingang District
2010
Shanghai, China

SHPG27
Wu Changshuo Memorial
2010
Shanghai, China

SHPG28
Urban Development of the ZhangJiang ZhongQu Area
2010
Shanghai, China

SHPG29
Housing
2010
Shanghai, China

SHPG30
Xie Zhiliu and Chen Peiqiu Art Gallery
2012 – 2013
Shanghai, China

SHPG31
Residential Hotel
2010
Shanghai, China

SHPG32
Interior Design of a Suite
2011
Shanghai, China

SHPG33
Expansion of Zegna Office
2011
Shanghai, China

SHPG35
Datang Office Park
2012
Shanghai, China

SHPG36
Tobacco Pudong Technological Park
2012
Shanghai, China

SHPG37
Zendai Villa
2012
Shanghai, China

SHPGR1
University of Performing Arts Campus
2013
Shanghai, China

SHPUX1
Xinhongqiao Business Centre
2007
Shanghai, China

SHPUX2
JianTe Biotechnology Office Park
2007
Shanghai, China

SHPUX3
Tobacco Factory
2008 – 2011
Shanghai, China

SHPUX4
Shanghai World Exhibition – French Pavilion
2008
Shanghai, China

SHPUX5
Restructuring of an Office Building
2008
Shanghai, China

SHPUX6
Urban Planning of Songjiang City
2009
Shanghai, China

SHPX7
Restructuring of the Tobacco Tower
2009
Shanghai, China

SHPX9
Zhangjiang Conference Centre
2009
Shanghai, China

SHPX10
Rolex Showroom and Exhibition Area
2010 – 2011
Shanghai, China

SHPX11
Serviced Apartments and Commercial Area
2011
Shanghai, China

SHPX12
Shanghai World Exhibition – Museum
2011
Shanghai, China

SHPX14
Interior Design of an Office Building
2016
Shanghai, China

SHQHY1
Mixed-use Buildings
2006
Shanghai, China

SHSHG1
Lingang DianLi University Campus
2013
Shanghai, China

SHXNJ1
3 Subway Stations
2013
Shanghai, China

SHY1
Conference and Exhibition Centre
2005
Shenyang, China

SHY2
Olympic Sports Training Centre
2005
Shenyang, China

SHY3
Hua Fu Tian Di Mixed-Use Towers
2006
Shenyang, China

SHY4
Museum, Conference and Exhibition Centre, Hotel and Courthouse
2009
Shenyang, China

SHY5
New Centre of Shenyang Hunnan New City
2010
Shenyang, China

SHY6
Qi Pan Mountain – National Game
2010
Shenyang, China

SHY7
Urban Planning Museum
2011
Shenyang, China

SHY8
Masterplan of the Shenyang Salon
2011
Shenyang, China

SHY9
Liaoning Art Centre
2012
Shenyang, China

SHY10
Youth Hostel
2014
Shenyang, China

SHZ1
Urban Planning of the Xichong Tourist Resort
2005
Shenzhen, China

SHZ2
Yidong Dasha
2009
Shenzhen, China

SHZ3
Founder South Headquarter
2013
Shenzhen, China

SHZ5
Merchant Retirement Home
2015
Shenzhen, China

SHZ6
Longhua Art Museum and Library
2015
Shenzhen, China

SHZ7
Nanfang University
2015
Shenzhen, China

SHZ8
Nanshan District
2015
Shenzhen, China

SHZHG1
Cement Factory Office Park
2014
Shanghai, China

SIG
Coca Cola Factory
1988
Signes, France

SIN1
Hospital of the Valais
2015
Sion, Switzerland

SJB1
Technological Park
2003 – 2009
Saint-Jean-Bonnefonds, France

SLC1
Hospital Centre
2001
Sallanches, France

SML1
La Grande Passerelle Cultural Hub
2009 – 2014
Saint-Malo, France

SML2
La Grande Passerelle Cultural Hub Esplanade
2009 – 2014
Saint-Malo, France

SMN1
Hospital Centre
2004
Saint-Ménéhould, France

SNS1
Foch Hospital
1997
Suresnes, France

SNV1
Private House
2012 – 2014
Saint-Nicolas de Veroce, France

SNY1
Club Med Resort
2012
Sanya, China

SNY2
Villa
2015
Sanya, China

SNZ1
Theatre
2007
Saint-Nazaire, France

SOF1
Research Centre
1987
Sofia, Bulgaria

SPD1
ITER Outbuildings
2007
Saint-Paul-lès-Durance, France

SPH
Paul VI Church
Sophia-Antipolis, France

SPM1
Sonacotra Residence
1993
Saint-Pol-sur-Mer, France

SPR1
Real Estate Study
2015
Saint-Pierre, France

SRG1
Hospital Centre
2004
Sarreguemines, France

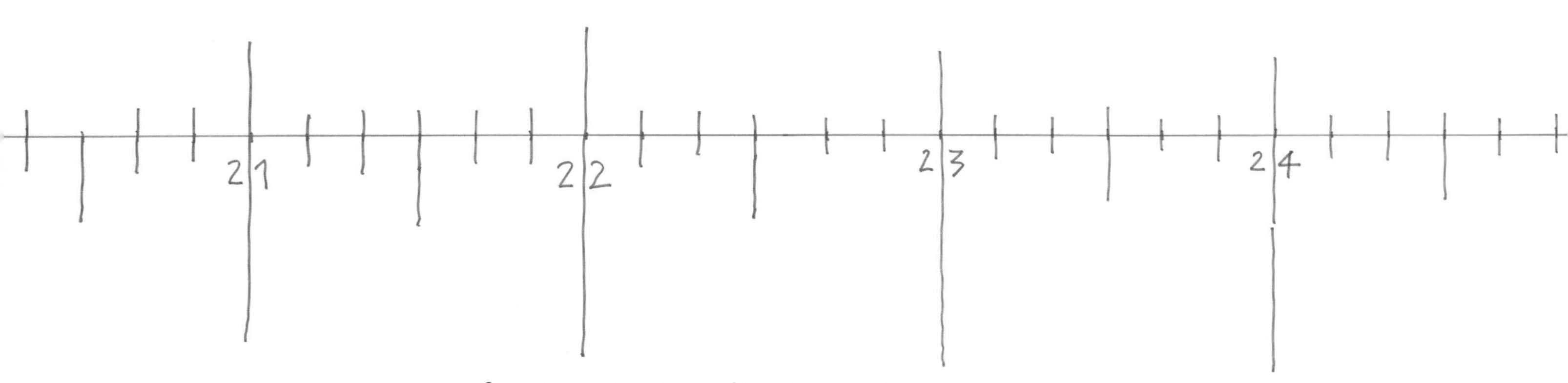

SSN1
Silesia Academy of Medicine
1999
Sosnowiec, Poland

STD1
Stade de France
1994
Saint-Denis, France

STD2
Office Building
2005
Saint-Denis, France

STD3
Office Building
2008
Saint-Denis, France

STD4
Saint-Denis Railway Station
2014
Saint-Denis, France

STE1
Extension of the Jean Monnet University
1993 – 1995
Saint-Étienne, France

STE2
Mutual Hospital Centre
1996
Saint-Étienne, France

STE3
Casino Group Headquarters
2004 – 2007
Saint-Étienne, France

STE4
Honoré d'Urfé School Complex
2008 – 2017
Saint-Étienne, France

STG2
Urban Planning of the Alameda Axis
2015
Santiago de Chile, Chile

STGO
3D simulator of the Sustainable City
2014
Santiago de Chile, Chile

STH1
Office Building
2013
Saint-Herblain, France

STJ1
Tourist Complex
2005
Saint-Julien-Mont-Denis, France

STL1
Development of City Centre
1993
Saint-Lô, France

STM1
Tourist Complex
2000
Saint Martin, France

STN1
Cordon Jaurès Apartment Building
1985
Saint-Ouen, France

STP1
Urban Development of Moskovsky District
2010
Saint Petersburg, Russia

STP2
Moskovsky District - Public Facilities
2010
Saint Petersburg, Russia

STP3
Rjevka Housing
2011
Saint Petersburg, Russia

STP4
Theatre
2011
Saint Petersburg, Russia

STP5
Office building
2014
Saint Petersburg, Russia

STP11
Moskovsky District - Housing
2013
Saint Petersburg, Russia

STP21
Kirovsky Housing
2013
Saint Petersburg, Russia

STP31
Rjevka Housing
2013 – 2015
Saint Petersburg, Russia

STP32
Urban Planning of Rjevka
2014
Saint Petersburg, Russia

STQ1
"La Cité du Risque" Museum
2013
Saint-Quentin, France

STR2
European Parliament
1991 – 1999
Strasbourg, France

STR5
Restructuring of the Stade de la Meinau
1997
Strasbourg, France

STR8
European Parliament – Entrance
2000
Strasbourg, France

STR10
European Parliament Extension
2003
Strasbourg, France

STV1
Restructuring of the Hospital
2004
Saint-Valéry-sur-Somme, France

SUL1
Opera
2005
Seoul, South Korea

SVT1
Archive Centre
1992
Savigny-le-Temple, France

SZH1
Danone New Factory
2003 – 2005
Suzhou, China

SZH2
Restructuring of the L'Oréal Factory
2004 – 2005
Suzhou, China

SZH3
Office Building
2012
Suzhou, China

TAP1
National Palace Museum
2004
Taipei, Taiwan

TAZ1
YiYaoChen Hotel
2014
Taizhou, China

TBF2
Hélène Bouchay Highschool
2000
Tremblay-en-France, France

TGL1
Tongliang City Planning
2015
Tongliang, China

THR1
Tourist Complex
2004
Tehran, Iran

THR2
Larak Garden
2011
Tehran, Iran

THV1
Sonacotra Residence
1993
Thionville, France

THV2
Sonacotra Residence
1994
Thionville, France

THV3
Sonacotra Residence
1995 – 2000
Thionville, France

THV4
Cultural Centre
2010
Thionville, France

TIR1
Millennium Building
2003 – 2012
Tirana, Albania

TIR2
Urban Design of Tirana City Centre
2003 – 2004
Tirana, Albania

TIR3
Vaso Pasha Building
2003 – 2012
Tirana, Albania

TIR4
Sulejman Pasha Tower
2004
Tirana, Albania

TIR5
Tirana International Hotel
2004
Tirana, Albania

TIR6
Scanderbeg Plaza
2008
Tirana, Albania

TIR7
New Mosque and Museum
2011
Tirana, Albania

TIR8
Diplomatic Mission Complex for the State of Kuwait
2014
Tirana, Albania

TJN1
Shopping Mall
2005
Tianjin, China

TJN2
Business Centre
2006
Tianjin, China

TJN3
Conference and Exhibition Centre
2008
Tianjin, China

TJN4
Cultural Centre
2008
Tianjin, China

TJN5
Venice Oriental Masterplan
2010
Tianjin, China

TJN6
Sunshine Masterplan
2010
Tianjin, China

TJN7
TETA HuiGu Service Centre
2013
Tianjin, China

TJN8
Employment Office
2013
Tianjin, China

TJN9
Restructuring of the 3rd Avenue
2014
Tianjin, China

TLN1
Mixed-use Buildings in La Loubière District
2010 – 2016
Toulon, France

TLN2
Technopole de la Mer
2012 – 2027
Ollioules, France

TLS1
Arena Highschool
1989 – 1991
Toulouse, France

TLS2
Maison de la Communication
1990
Toulouse, France

TLS3
Arena District
1990
Toulouse, France

TLS4
Arena Forum
1992 – 1994
Toulouse, France

TLS5
Escoulan District
1992
Toulouse, France

TLS6
COMIP District
1992
Toulouse, France

TLS7
French Civil Aviation Technical Unit
1992
Toulouse, France

TLS9
Sports Centre
2003
Toulouse, France

TLS10
Casino
2003
Toulouse, France

TLS11
Development of the Malepère District
2006 – 2012
Toulouse, France

TLS12
Urban Planning of the Malepère District
2014
Toulouse, France

TNG1
University Hospital of Tangier
2014 – 2021
Tangier, Morocco

TNN1
Hotel and Commercial Complex
2001
Antananarivo, Madagascar

TNN2
Restructuring of the Queen's Palace
2001
Antananarivo, Madagascar

TNN3
Organisation of African Unity
2001
Antananarivo, Madagascar

TNN4
Mixed-use Buildings
2005
Antananarivo, Madagascar

TNN5
Aga Khan Academy
2015
Antananarivo, Madagascar

TNS1
National Institute of Applied Science and Technology
1992
Tunis, Tunisia

TNS2
BNP Paribas/UBCI Headquarters
2007
Tunis, Tunisia

TRC1
National Customs Authority School
1999 – 2003
Tourcoing, France

TRN1
Regional Council
2000
Turin, Italy

TRS1
Conference Centre
1989
Tours, France

TRY1
University Residence
2001
Troyes, France

TRY2
"Stade de l'Aube" Official Gallery
2001
Troyes, France

TRY3
Expansion of the Technological University
2003
Troyes, France

TRY4
Departmental Council
2007
Troyes, France

TRZ1
City Complex and Administration Centre
2011
Tai Er Zhuang, China

TRZ2
Reception Centre
2011
Tai Er Zhuang, China

TYN2
Changfeng Tourist Area
2006
Taiyuan, China

TYN3
Shanxi Richesse International Plaza
2006
Taiyuan, China

TYN4
Administrative City
2007
Taiyuan, China

TYN5
Hospital Complex
2007
Taiyuan, China

TYN6
Commercial and Exhibition Centre
2007
Taiyuan, China

TYN8
Urban Planning of the Zhonggulou District
2009
Taiyuan, China

TYN9
Urban Planning of the Taiyuan Urban Xidajie District
2009
Taiyuan, China

TYN10
Urban Planning of the ZhongGu lou Plaza
2010
Taiyuan, China

TYN11
XiShan Exhibition Centre
2010
Taiyuan, China

TYN12
Urban Planning of the Zhonglou and Maoer Boulevards
2010
Taiyuan, China

TYN13
Urban Planning Museum
2010
Taiyuan, China

TYN14
Office Building
2010
Taiyuan, China

TYN15
Office Building
2010
Taiyuan, China

TYN16
Transport Centre
2011 – 2013
Taiyuan, China

TYN17
Podium Landscape
2012
Taiyuan, China

TYN18
New Shanxi-DongShan Campus
2014
Taiyuan, China

ULN1
Wulanchabu Theatre
2012
Ulanqab, China

VCN1
Paris Zoological Gardens
1990
Vincennes, France

VDA1
City Centre Masterplan
1990
Villeneuve-D'Ascq, France

VDR
Penitentiary Centre
2000
Val de Reuil, France

VFR1
Restructuring of the Hospital
2015
Villefranche-sur-Saône, France

VGL1
High-Mountain Refuge
2000
La Vogealle, France

VGT
Hotel
1999
Le Vigeant, France

VLC1
City Centre Masterplan
1999
Valenciennes, France

VLC2
Buildings for the SUPINFO Group
2009
Valenciennes, France

VLF1
Technological College
1993
Vitry-le-François, France

VLG
Sisley Port
1988
Villeneuve-la-Garenne, France

VLJ1
Gustave Roussy Institute
2013
Villejuif, France

VLN1
Hotel School
1988
Villers-lès-Nancy, France

VLP1
Technological College
1987
Vaux-le-Pénil, France

VLR1
Vlora Waterfront
2014
Vlorë, Albania

VLZ1
Housing Complex in the Louvois Area
2012
Vélizy-Villacoublay, France

VNK1
Business Centre
2007
Vnukovo, Russia

VNS1
Restructuring of the Palazzo Santa Maria Nova
2008 – 2009
Venice, Italy

VNS2
Palazzo Santa Maria Nova
2010
Venice, Italy

VPT
Georges Brassens Highschool
1994
Villepinte, France

VPT1
New Paris-Nord Villepinte Exhibition Park
1993
Villepinte, France

VPT2
Extension of Exhibition Park – Hall 7
1998 – 1999
Villepinte, France

VPT4
Extension of Exhibition Park – Hall 1
1999 – 2000
Villepinte, France

VPT5
Extension of Exhibition Park
1999 – 2000
Villepinte, France

VPT6
Reassembly of the Jean Prouvé Aluminium Pavilion
1998 – 1999
Villepinte, France

VRS1
Maison Decardes Private House
1976 – 1977
Versailles, France

VRS2
Office
1991
Versailles, France

VRS3
Office Park
1984
Versailles, France

VSL
Banlieues 89
1984
Vesoul, France

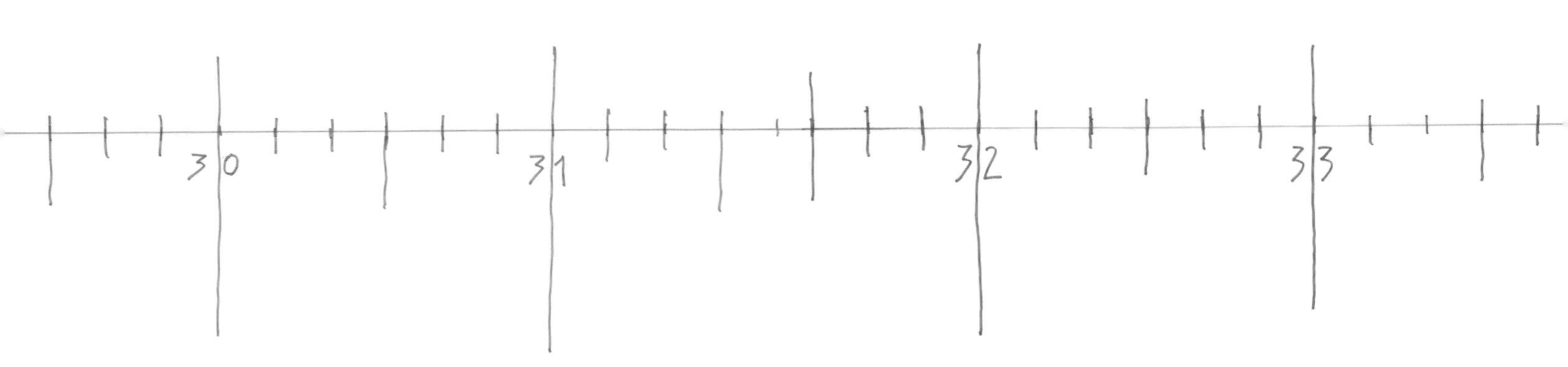

VSV1
Warsaw Clinic
1998
Warsaw, Poland

VSV2
Warsaw Hospital
1998
Warsaw, Poland

VSV3
Housing Tower
1999
Warsaw, Poland

VSV4
Mixed-use Building
1999
Warsaw, Poland

VSV5
Forum Orbis Hotel
1999
Warsaw, Poland

VSV6
Restructuring of the Koscielna Palace
2001
Warsaw, Poland

VTR1
"Grand Paris" Maintenance Site
2013
Vitry sur Seine, France

VVS1
Hachette Building
2001
Vanves, France

WFG1
Hospital Complex
2015
Weifang, China

WHN1
Cultural Complex
2010
Wuhan, China

WHN2
Wuhan Salon
2011
Wuhan, China

WHN3
Sino-French Hospital Complex
2012
Wuhan, China

WND1
3 Villas
2009
Wendeng, China

WQG1
Tiens International Campus
2013
Wuqing, China

WRC1
Orbis Motels
1999
Wrocław, Poland

WUH1
Masterplan of a Residential Zone
2009
Wuhu, China

WUH2
ChuangXin District
2010
Wuhu, China

WUH3
PanLongCheng Museum
2014
Wuhan, China

WUX1
Urban Planning of Taihu New City
2006
Wuxi, China

WUX2
Administrative Complex
2006
Wuxi, China

WUX3
Technological Park
2007
Wuxi, China

WUX4
Masterplan of Qianqiao
2007
Wuxi, China

WUX5
Art Centre
2008
Wuxi, China

WUX6
Urban Planning of the Wuxi Lihu Administrative Complex
2009
Wuxi, China

WUX7
Secondary School
2010
Wuxi, China

WUX8
Home-Mart Shopping Mall
2011 – 2014
Wuxi, China

WUX9
Wuxi Xidong Administrative Complex
2011
Wuxi, China

WUX10
Urban Planning of the Nanchang Tourist Area
2011
Wuxi, China

WUX11
Wuxi Historical Centre Canal
2012
Wuxi, China

WUX12
Masterplan of Lihu Lake Bay
2012
Wuxi, China

WUX13
Yunhe FengGuang Park
2013
Wuxi, China

WVR1
UGC Ciné Cité
1998
Wavre, Belgium

XAN1
Conceptual Design of Shan Xi Media Centre
2014
Xi'an, China

XMN1
Masterplan of Zhongzhou Residential Zone
2010
Xiamen, China

XNF1
Conference Centre
2010
Xiang Fan, China

XNG1
Club House
Xinghua, China

XNY1
XiangYang Library
2012
Xiang Yang, China

XNY2
Conceptual and Urban Planning of XiangYang City
2014
Xiang Yang, China

YCH1
Yidong Tourist Area
2010
Yichun, China

YCN1
Olympic Sports Training Centre
2013
Yichang, China

YCN2
Urban Planning of the Yichang New District
2013
Yichang, China

ZAB1
University Hospital Centre
1997 - 2003
Zabrze, Poland

ZHJ1
Urban Planning of the Gaohu Lake
2013
Zhuji, China

ZHN1
Museum
2015
Zheng Zhou, China

ZHN2
Archive Museum
2015
Zheng Zhou, China

ZHN3
Central China Securities New Office Building
2015
Zheng Zhou, China

ZHU1
Opera
2012
Zhuai, China

ZHU2
TCM Industrial Park
2014
Zhuhai, China

ZJK1
Archive Museum
2014
Zhang Jia Kou, China

ZJK2
Urban Design of Dahaoheshan Islet
2014
Zhang Jia Kou, China

ZNJ1
Cultural Centre
2014
Zhanjiang, China

ZNM1
Private Residence
2012 – 2013
Znamenskoyé, Russia

ZNM2
Private Residence
2013
Znamenskoyé, Russia

ZRH1
Green City Business Centre
2012 – 2021
Zurich, Switzerland

ZZH1
Administrative City
2011
ZaoZhuang, China

34 35 36 37 38

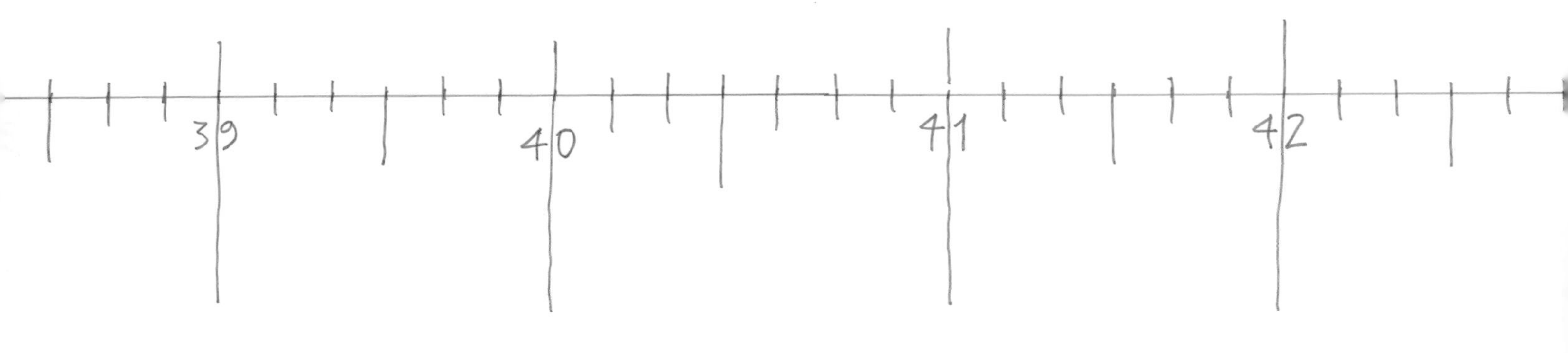

39
40
41
42

VOIE RAPIDE PERIPHERIQUE
VILLAGES EXISTANTS
ZONE AEROPORT + INDUSTRIES

BIOGRAPHIES 合伙人简介

Martin Robain
Born November 26, 1943
in Paris, France
Architect D.P.L.G.
Graduated from École nationale des Beaux-Arts of Paris in 1969
Co-founder of Architecture-Studio in 1973
马丁·罗班
1943年11月26日出生于法国巴黎
法国政府认可建筑设计师
D.P.L.G.
1969年毕业于巴黎国立高等艺术学院
1973年联合创立法国AS建筑工作室

Rodo Tisnado
Born June 18, 1940 in Hualgayoc, Peru
Graduated from Faculty of Architecture of the National University of Engineering of Lima in 1964
Partner architect at Architecture-Studio since 1976
罗多·蒂斯纳多
1940年6月18日出生于秘鲁瓦尔加约克
1964年毕业于利马国立工程大学建筑学院
1976年起成为法国AS建筑工作室合伙人

Jean-François Bonne
Born January 12, 1949
in Saint-Mandé, France
Architect D.P.L.G. and urban planner D.I.U.P.
Graduated from École nationale des Beaux-Arts of Paris in 1975
Graduated from Institut d'Urbanisme of Paris in 1976
Partner architect at Architecture-Studio since 1979
让-弗朗索瓦·博内
1949年1月12日出生于法国圣芒代
法国政府认可建筑设计师
D.P.L.G.及巴黎城市规划研究院认可规划师D.I.U.P.
1975年毕业于巴黎国立高等艺术学院
1976年毕业于巴黎城市规划研究院
1979年起成为法国AS建筑工作室合伙人

Alain Bretagnolle
Born January 17, 1961 in Vichy, France
Architect D.P.L.G.
Graduated from École d'Architecture Paris-Malaquais in 1985
Partner architect at Architecture-Studio since 1989
阿兰·布勒塔尼奥勒
1961年1月17日出生于法国维希
法国政府认可建筑设计师
D.P.L.G.
1985年毕业于巴黎-马拉盖国立高等建筑学院
1989年成为法国AS建筑工作室合伙人

René-Henri Arnaud
Born March 20, 1958
in Saint-Chamond, France
Architect D.P.L.G.
Graduated from École d'Architecture Paris-La Villette in 1985
Partner architect at Architecture-Studio since 1989
勒内-亨利·阿诺
1958年3月20日年出生于法国圣沙蒙
法国政府认可建筑设计师
D.P.L.G.
1985年毕业于巴黎-拉维特国立高等建筑学院
1989年成为法国AS建筑工作室合伙人

Laurent-Marc Fischer
Born November 10, 1964
in Paris, France
Architect D.P.L.G. and urban planner O.P.Q.U.
Graduated from École d'Architecture Paris-Belleville in 1989
Partner architect at Architecture-Studio since 1993
洛朗-马克·费希尔
1964年11月10日出生于法国巴黎
法国政府认可建筑设计师
D.P.L.G.以及法国专业认证规划师O.P.Q.U.
1989年毕业于巴黎-美丽城国立高等建筑学院
1993年成为法国AS建筑工作室合伙人

Marc Lehmann
Born April 13, 1962 in Rabat, Morocco
Architect D.P.L.G.
Graduated from École d'Architecture et d'Urbanisme de Bordeaux in 1989
Partner architect at Architecture-Studio since 1998
马克·莱曼
1962年4月13日出生于摩洛哥拉巴特
法国政府认可建筑设计师
D.P.L.G.
1989年毕业于波尔多国立高等建筑与景观规划学院
1998年成为法国AS建筑工作室合伙人

Roueïda Ayache
Born December 23, 1964
in Beirut, Lebanon
Architect D.E.S.A.
Graduated from École Spéciale d'Architecture of Paris in 1989
Master of advanced studies of Philosophy, Paris-Sorbonne University in 1999
Partner architect at Architecture-Studio since 2001
罗伊达·阿亚斯
1964年12月23日出生于黎巴嫩贝鲁特
建筑专业学院认证建筑设计师
D.E.S.A.
1989年毕业于巴黎建筑专业学院
1999年获得巴黎-索邦大学哲学高等深入研究文凭
2001年成为法国AS建筑工作室合伙人

Gaspard Joly
Born February 3, 1969
in Grenoble, France
Architect D.P.L.G.
Graduated from École d'Architecture de Bordeaux in 1993
Partner architect at Architecture-Studio since 2009
贾斯帕·朱利
1969年2月3日出生于法国格勒诺布尔
法国政府认可建筑设计师
D.P.L.G.
1993年毕业于波尔多国立高等建筑学院
2009年成为法国AS建筑工作室合伙人

Marie-Caroline Piot
Born October 26, 1978 in Paris, France
Architect D.E.S.A.
Graduated from École Spéciale d'Architecture of Paris in 2002
Partner architect at Architecture-Studio since 2009
玛丽卡·碧欧
1978年10月26日出生于法国巴黎
建筑专业学院认证建筑设计师
D.E.S.A.
2002年毕业于巴黎建筑专业学院
2009年成为法国AS建筑工作室合伙人

Mariano Efron
Born September 26, 1968
in Buenos Aires, Argentina
Graduated from National University of Buenos Aires in 1995
Partner architect at Architecture-Studio since 2009
马里亚诺·艾翁
1968年9月26日出生于阿根廷布宜诺斯艾利斯
1995年毕业于布宜诺斯艾利斯国立大学
2009年成为法国AS建筑工作室室合伙人

Amar Sabeh el Leil
Born May 8, 1973 in Damascus, Syria
Architect D.P.L.G.
Graduated from École d'Architecture de Paris Val de Marne in 2001
Partner architect at Architecture-Studio since 2009
艾马·萨布埃雷
1973年5月8日出生于叙利亚大马士革
法国政府认可建筑设计师
D.P.L.G.
2001年毕业于巴黎-马恩河谷国立高等建筑学院
2009年成为法国AS建筑工作室合伙人

EXHIBITIONS 展览

CA'ASI Exhibitions CA'ASI艺术展览馆策划展览

Mission Trans-Missions, Architecture-Studio & young awarded Chinese, Arab and African architects, CA'ASI, Venice, Italy, 05/26/2016 – 11/27/2016.

African Cities in Motion, CA'ASI, Venice, Italy, 05/09/2015 – 08/30/2015.

Young Architects in Africa, arc en rêve, Bordeaux, France, 09/24/2015 – 11/22/2015.

Young Architects in Africa, Franco-Namibian Cultural Centre, Windhoek, Namibia, 04/20/2015 – 05/05/2015.

Young Architects in Africa, CA'ASI, collateral events of the 14th Venice Biennale of Architecture, Venice, Italy, 06/06/2014 – 08/31/2014.

Construction Ahead, CA'ASI, Venice, Italy, 05/31/2013 – 08/31/2013.

Young Arab Architects, Institut du monde arabe, Paris, France, 10/15/2012 – 01/30/2013.

Young Arab Architects, CA'ASI, collateral events of the 13th Venice Biennale of Architecture, Venice, Italy, 08/28/2012 – 10/30/2012.

In the Crystal Palace, CA'ASI, with CNAP, Venice, Italy, 06/03/2011 – 07/31/2011.

New Chinese Architecture, Baiziwan Club, Beijing, China, 12/10/2010 – 12/19/2010.

Alex MacLean, Paris La Défense Seine Arche, CA'ASI, Venice, Italy, 09/01/2010 – 10/01/2010.

New Chinese Architecture, CA'ASI, collateral events of the 12th Venice Biennale of Architecture, Venice, Italy, 08/29/2010 – 11/21/2010.

Le Parlement Européen 1999 – 2009, CA'ASI, Venice, Italy, 11/2009.

Monographic Exhibitions 法国AS建筑工作室专场展览

Habiter Autrement le Monde : la ville écologique par Architecture-Studio, Museum of Architecture, Shenyang Architectural University, Shenyang, China, 12/05/2015 – 12/06/2015.

Habiter Autrement le Monde : la ville écologique par Architecture-Studio, Institut français, Beijing, China, 10/27/2014 – /09/2014.

Architecture-Studio Exhibition, Palazzo Dogano, Foggia, Italy, 01/03/2002 – 01/10/2002.

Architecture-Studio, Le Parlement européen, Galerie d'architecture, Paris, France, 03/10/2000 – 04/02/2000.

Beyond Contemporary Architecture : Architecture-Studio, NGMA, Mumbai, India, 1999.

Beyond Contemporary Architecture : Architecture-Studio, Kakatiya hotel, Hyderabad, India, 1999.

Beyond Contemporary Architecture: Architecture-Studio, India habitat center, Delhi, India, 1999.

Architecture d'Architecture-Studio, Galerie d'architecture, Perpignan, France, 1989.

Un urbanisme pour tous, Association Carcassonne 89, Carcassonne, France, 1989.

Architecture-Studio, Instituto Nazionale di Architettura, Rome, Italy, 1988.

Architecture-Studio, Fondation Claude-Nicolas Ledoux, Arc-et-Senans, France, 1988.

Architecture-Studio, Palais Attems, Vienna, Austria, 1987.

Architecture-Studio, Institut français, Lisbon, Portugal, 1987.

Architecture-Studio, arc en rêve, Bordeaux, France, 1986.

L'Institut du monde arabe, Micgoau, Athens, Greece, 1986.

L'Institut du monde arabe, Institut français, Stockholm, Sweden, 1985.

L'Institut du monde arabe, Institut français d'architecture, Paris, France, 1981.

Collective exhibitions 联合展览

In the End : Architecture, Journeys through Time 1959 – 2019, Architekturzentrum Wien, Vienna, Austria, 10/06/2016 – 03/20/2017.

Le foot, une affaire d'État, Archives nationales, Pierrefitte-sur-Seine, France, 05/27/2016 – 09/18/2016.

Réinventer Paris, Pavillon de l'Arsenal, Paris, France, 02/04/2016 – 05/08/2016.

15 ans déjà, Galerie d'architecture, Paris, France, 03/24/2015 – 04/09/2015.

Ailleurs/Outwards Exposition, Palmarès du Grand Prix AFEX 2014

Cité de l'architecture et du patrimoine, Paris, France, 01/09/2015 – 03/09/2015.

Ailleurs/Outwards Exposition, Palmarès du Grand Prix AFEX 2014

Palazzo Zorzi, Venice, Italy, 06/2014.

Habitat étudiant, Projets d'avenir, Pavillon de l'Arsenal, Paris, France, 06/25/2013 – 08/25/2013.

Kaboul 2050, French Cultural Center, Kabul, Afghanistan, 12/03/2012 – 12/04/2012.

Architecture Quatre Vingt, Pavillon de l'Arsenal, Paris, France, 05/02/2011.

Campus du 21e siècle, Cité de l'architecture et du patrimoine, Paris, France, 10/28/2010 – 12/06/2010.

Waterfront : New Trends in Urbanism and Architecture, Ajman University, Ajman, United Arab Emirates, 03/29/2010 – 03/31/2010.

Accords chromatiques, Pavillon de l'Arsenal, Paris, France, 07/10/2008 – 09/21/2008.

Architecture=Durable, Pavillon de l'Arsenal, Paris, France, 06/24/2008 – 10/19/2008.

Transitions Light on the Move, Touring exhibition in Europe, 2006 – 2008.

'Dehors Paris 2', Maison de l'architecture, Paris, France, 2008.

Lenexpo, Saint Petersburg, Russia, 2008.
Exo Architecture, Pavillon de l'Arsenal, Paris, France, 12/07/2007 – 05/04/2008.

Paris visite guidée,
Pavillon de l'Arsenal
(Permanent exhibit), Paris,
France, 2007.

Visions françaises,
Touring exhibition for the Year
of France in China, 2005.

Histoire de Paris et son
actualité,
Pavillon de l'Arsenal, Paris,
France, 12/2003.

Shanghaï by Architecture-Studio,
exhibition 'Alors, la Chine?',
musée national d'Art
moderne, Centre Georges
Pompidou, Paris, France,
06/03/2003 – 10/13/2003.

International Biennale
of Architecture, Mob Lab,
Las Palmas, Rotterdam,
Netherlands,
05/07/2013 – 07/07/2003.

Maquettes d'architectes,
Bordeaux, France,
07/02/2002 – 09/13/2002.

Le Goût du style,
Pavillon de l'Arsenal, Paris,
France, 02/2002.

Fragments français,
International Conference Center,
Beijing, China, 1999.

6 French Architects Exhibition,
GA Gallery, Tokyo, Japan, 1997.

European Architecture:
1984 – 1994,
Galeria del Ministerio de Orbas
Publicas, Madrid, Spain, 1996.

European Architecture:
1984 – 1994,
Fundacio Joan Miro, Barcelona,
Spain, 1996.

European Architecture:
1984 – 1994,
Galleria d'arte Moderna,
Bologna, Italy, 1996.

European Architecture:
1984 – 1994,
Nederlands Architectuur,
Instituut, Rotterdam,
Netherlands, 1996.

Arquitectura hoy,
Museo de arte, Lima, Peru,
1996.

European Architecture:
1984 – 1994,
Deutsche Architektur Museum,
Frankfurt, Germany, 1995.

Qualité, Label et Création,
Maison de l'architecture, Paris,
France, 1995.

Paris s'exporte,
Pavillon de l'Arsenal, Paris,
France, 1995.

Arquitectura francesa,
11 proyectos,
Junta de Andalucia, Seville,
Spain, 1993.

Les Écoles à Paris,
Pavillon de l'Arsenal, Paris,
France, 1993.

Le Dessin et l'Architecte,
Pavillon de l'Arsenal, Paris,
France, 1992.

Les Seuils de la ville-Paris,
des fortifs au périf,
Pavillon de l'Arsenal, Paris,
France, 1992.

Architettura e spazio sacro nella
modernità, Venice Biennale,
Venice, Italy, 1992.

Dix ans d'architecture
à Toulouse, Musée des
Augustins, Toulouse, France,
1992.

Beaubourg 90: commandes
publiques, musée national d'Art
moderne, Centre Georges
Pompidou, Paris, France, 1990.

Une excellente génération
d'architectes français,
French Cultural Center,
Belgrade, Yougoslavia, 1990.

French avant-garde architecture,
The Art Institute, Chicago,
United States, 1989.

Paris-Berlin: architecture
et utopie, Pavillon de l'Arsenal,
Paris, France, 1989.

Paris: grands projets,
1979 – 1989, US Custom House,
New-York, United States, 1988.

Profils en architecture & design,
Royal Swedish Academy of Arts,
Stockholm, Sweden, 1988.

Actualités architecturales 88,
Institut français d'architecture,
Paris, France, 1988.

Berlin: Denkmal oder
Denkmodel,
Galerie Aedes, Berlin, Germany,
1988.

Architecture française liée à l'art,
French Cultural Centre, Helsinki,
Finland, 1988.

Vienne – Architectures publiques,
Direction départementale
de l'équipement de la Vienne,
Poitiers, France, 1988.

Franco – Scandinavian
Architecture and Design
Exhibition, Charlottenborg
Museum, Copenhagen,
Denmark, 1988.

Arts et arts, actualités et histoire,
Venice Biennale, Venice, Italy,
1987.

Architecture et conception
hospitalière, Institut français
d'architecture, Paris, France,
1985.

Nouveaux plaisirs d'architecture,
musée national d'Art moderne,
Centre Georges Pompidou,
Paris, France, 1985.

Os grandes projectos
arquitectónicos em Paris,
Modern Art Centre, Lisbon,
Portugal, 1985.

Tendance de l'architecture
française, Academy of Fine Arts,
Denmark, 1984.

3 grands projets à Paris,
Institut français d'architecture,
Paris, France, 1983.

Décade d'architecture,
Batimat, Paris, France, 1983.

Extension d'un service
de médecine à Montmorillon,
Institut français d'architecture,
Paris, France, 1983.

Histoire d'y participer.
Voulez-vous y participer?,
Institut français d'architecture,
Paris, France, 1983.

30 French Architects Exposed,
2nd European Assembly
of Architecture Students,
Delft, Netherlands, 1982.

Première Biennale d'architecture
de Paris, musée national d'Art
moderne, Centre Georges
Pompidou, Paris, France, 1980.

Construire en quartier ancien,
Grand Palais, Paris, France,
1980.

AWARDS 荣誉和奖项

Chaptal Award: Construction and Fine Arts 2016
Architecture-Studio

Green Building Solution Award, Smart City Grand Price
Issy-les-Moulineaux Fort, France

Trophée France Bois Ile-de-France, Special Mention 2016
Créteil's Cathedral, France

USITT Architecture Honor Award, 2015
Bahrain National Theatre, Manama, Bahrain

Mount Lu Estate World Architecture Prize, 2015, UIA, Second
Architecture-Studio Practice

Best Landscape Silver Award 2014
Parc Marianne – Port Marianne, Montpellier, France

France China Best Entrepreneur Trophy 2014
Architecture-Studio Practice

Best Office Building in China Award 2014
Wison Headquarters, Shanghai, China

Silver Pyramid Award 2014
Dor[a]mar Mixed-use Building, Montpellier, France

Gold Award for Planning 2013
Urban Design of Tai Er Zhuang, China

Tianjin Gold Award for Planning, Sciences and Technology 2013
Industrial Center of TETA HuiGu, Tianjin, China

AFEX Award 2013, Palmares
Bahrain National Theatre, Manama, Bahrain

Gold Prize Interarch of the Ministry of Culture 2012
Onassis Cultural Centre, Athens, Greece

AFEX Award 2012, Palmares
Onassis Cultural Centre, Athens, Greece

Eco-neighborhood Award 2011
Parc Marianne Development, Montpellier, France
Issy-les-Moulineaux Fort, France

Honorary Credential Award, 5th China International Design and Art Fair, 2010
Architecture-Studio Practice

EDF Habitat Bleu Ciel Award 2009
Park Avenue, Montpellier, France

SIMI New Building Award 2007
Casino Headquarters, Saint-Étienne, France

Environmental Prize of Essonne's companies, Landscape Insertion, 2007
Danone Vitapole, Palaiseau, France

Wood Construction Laurels 2006
Guy Dolmaire High School, Mirecourt, France

Solar Construction Observ'ER, Award for Public Facilities 2006
Guy Dolmaire High School, Mirecourt, France

Unesco – International Festival of Art and Pedagogical Films 2004
"Architectural Heritage", mention for the documentary *Manama, Doha, Dubai, three cities from the Gulf*

Gold Steel Award 2003, Metal Architecture Category
Wison Chemicals Headquarters, Shanghai, China

AMO Award 1998
Bechet Company Headquarters, Paris, France

Architecture Palmares 1996
Our Lady of the Ark of the Convenant Church, Paris, France

Aga Khan Award 1989
Arab World Institute, Paris, France

Équerre d'Argent Award 1988
Arab World Institute, Paris, France

Losange d'Argent Prize 1985
Domrémy Residence Building, Paris, France

BIBLIOGRAPHY 出版物

CA'ASI Publications / Exhibitions CA'ASI艺术展览馆展览相关出版

African Cities in Motion, Architecture-Studio, 2015, 48p, Catalog of the exhibition

Young Architects in Africa, Architecture-Studio, 2014, 134p, Catalog of the exhibition

Construction Ahead, Architecture-Studio, 2013, 70p, Catalog of the exhibition

Young Arab Architects, Architecture-Studio, 2012, 152p, Catalog of the exhibition

Nel Palazzo Di Cristallo, Architecture-Studio & CNAP, 2011, 38p, Catalog of the exhibition

The New Chinese Architecture, Architecture-Studio, 2010, 123p, Catalog of the exhibition

Paris La Défense Seine Arche. Carte blanche à Alex MacLean, France, Dominique Carré Éditeur, 2010, 63p

Monograph 法国AS建筑工作室专著

Créteil, la cathédrale déployée, Marie-Pierre Etienney, France, Édition du Signe, 2015, 72p, English Version: *Créteil, the Expanded Cathedral*

Pôle culturel de Saint-Malo, tomes 1 & 2, France, Archibooks + Sautereau Éditeur, 2015, 111p and 103p

50 Projets, 50 Idées, Architecture-Studio, China, Tianjin University Press & MARK China, 2014, 191p, Catalog of the exhibition

OZ'One, ZAC Parc Marianne, Montpellier, Architecture-Studio, France, EiffageImmobilier&Pragma,2013, 58p

The Ecological City, Contribution for a Sustainable Architecture, Architecture-Studio, China, AntePrima & MARK China, 2012, 272p

Architecture-Studio, Pocket Book, France, 2012

Novancia, la grande école du « Business Development », Catherine Sabbah, France, Archibooks + Sautereau Éditeur, 2012, 80p

Parlement européen de Strasbourg. Au cœur des étoiles, France, Carré Blanc Éditions, 2010, 69p

Le Parlement européen, 1999-2009, Lionel Blaisse, France, AAM Éditions, Silvanaeditoriale & AntePrima, 2009, 134p

Danone Research, Centre Daniel Carasso, Architecture-Studio, France, 2009, 134p

La Ville écologique. Contribution pour une ville durable, Architecture-Studio, France, AntePrima & AAM Éditions, 2009, 270p

Centre pénitentiaire de Saint-Denis, La Réunion, Olivier Pioch, France, Sira Imprimeur, 2008, 31p

L'Institut du monde arabe, Hugo Lacroix, France, Éditions de la Différence, 2007, 128p

Architecture-Studio, Global Architecture, Italy, l'Arca Edizioni, 2006, 128p

Architecture-Studio, Mario Pisani, Italy, Edil Stampa editrice dell 'Ance, 2005, 127p

Architecture-Studio : Monographie d'un groupe, China, Dalian University of Technology Press, 2004, 391p

Le Collège Guy-Dolmaire à Mirecourt, Architecture-Studio, France, 2004, 19p

Contemporary Architecture, Korea, CA Press Co., Ltd., vol. 51, 2003, 239p

Architecture-Studio, Pocket Book, France, 2002

"Architecture-Studio", China, *World Architecture Review,* n°140, 2002, 88p

Architecture-Studio : Le Parlement européen à Strasbourg, Italy, Edizioni Charta Srl, 2001, 84p

Architecture-Studio. The Master Architect Millennium Series, Australia, Images Publishing Group, 2000, 256p

Le Parlement européen à Strasbourg, Pierre-François Mourier, France, Les Éditions de l'Imprimeur, 1999, 186p

Notre-Dame-de-l'Arche-d'Alliance à Paris, Cardinal Lustiger and Frédéric Boyer, France, Architecture-Studio Edition, 1999, 60p

Architecture-Studio. The Master Architect Series II, China, China Architecture & Building Press, 1999, 256p

"Architecture-Studio, 1989-1998", Italy, *Metamorfosi,* n°37, Quaderni Di Architettura, 1998, 72p

École des Mines Albi-Carmaux. L'Architecte et l'ingénieur, France, Éditions Odyssée, 1998, 100p

Architecture-Studio, Pocket Book, France, 1998

Architecture-Studio. The Master Architect Series II, Australia, Images Publishing Group, 1996, 256p

Architecture-Studio. Selected and Current Works, Japan, Meisei Publications, 1996, 256p

"Le Parlement européen. Chronique d'un grand chantier", *Techniques & Architecture,* Special Edition, France, Éditions Altédia, 1996, 32p

Architecture-Studio. Rites de passage, Mario Pisani, Italy, Carte Segrete & Wordsearch Publishing, 1995, 306p

Architecture-Studio, 3 tomes, France, Éditions Techniques & Architecture, 1991, 48p, 57p and 83p

Architecture-Studio, in Arch, Italy, Carte Segrete, 1988, 97p, Catalog of the exhibition

Architecture-Studio. Stimuli, France, Éditions Champ Vallon & Les Éditions du Demi-Cercle, 1987, 61p

Architecture-Studio, Architecture-Studio and Mario Pisani, Italy, Carte Segrete, 1986, 102p, Catalog of the exhibition

Architecture-Studio, 1984-1985, tome 2, France, Graphic Procédé, 1986, 45p

Architecture-Studio, 1972-1984, tome 1, France, Architecture-Studio, 1985, 143p

Publications 其他出版物

Ailleurs/Outwards, AFEX (Architectes français à l'export). 50 architectures françaises dans le monde, Frédéric Lenne, France, Dominique Carré Éditeur, 2016, 224p

La Rénovation du campus de Jussieu, Virginie Picon-Lefebvre, Jean-Philippe Hugron and Christian Hottin, France, Archibooks, 2016, 200p

Les 101 Mots de la lumière dans l'architecture à l'usage de tous, Collectif, France, Archibooks, 2016, 68p

Atlas of European Architecture, Markus Sebastian Braun and Chris van Uffelen, Switzerland, Braun, 2015, 512p

Bouygues Immobilier. L'innovation par nature, Virginie Luc, France, Zoko Production, 2015, 167p

City Guide Vuitton, Venice, Collectif, France, 2015, 320p
Ordo administratif 2016, France, Bayard Service Édition, 2015, 720p

Forum Bois Construction Épinal / Nancy 2015, Switzerland, Forum Holzbau, 2015

Journées nationales AMO, AMO Région Languedoc-Roussillon, France, 2015, 176p

Les Espaces de la musique. Architecture des salles de concert et des opéras, Antoine Pecqueur, France, coédition Parenthèses-Philharmonie de Paris, 2015, 288p

Ville durable, ville désirable, France, Bouygues Immobilier, 2015, 199p

50 Ans de coopération franco-chinoise, Agence Kas Éditions, 2014, 170p, p. 72

Clermont-Ferrand. Patrimoine et architecture, Arnaud Frich, France, Page centrale Éditions, 2014, 177p

Comprendre simplement l'acoustique des bâtiments, Loïc Hamayon, France, Le Moniteur, 2014, 248p

Facade Detail in Architecture, China, Liaoning Science & Technology Publishing House, 2014

Le Cadre conceptuel des projets connexes aux gares du Grand Paris, Société du Grand Paris, France, 2014, 61p

Proximité(s), France, special edition of AMC & Fondation d'entreprise Bouygues Immobilier, Paris, 2014, 74p, Catalog of the exhibition at Cité de l'architecture et du patrimoine

Science, Culture & Creative Park, China, CNS, 2014, 334p

Un tour de Babel. Les tribulations d'une agence d'architecture, Jean Vermeil, France, Archibooks + Sautereau Éditeur, 2014, 211p, p. 178-181

*Les Annales de l'*École de Paris du *management,* vol. 20, Association des amis de l'École de Paris du management, France, 2014, 477p, p. 297-303

Monumental Athens Urban, G.F. Zaimis, IDEAlab Editions, 2014, 129p

2012 Architecture Report, China, Rihan.cc, 2013, 333p

Anticiper, prendre un temps d'avance, France, Ernst & Young, 2013, 134p

Architekten Objekte Lösungen, Austria, Kaba GmbH, 2013, 55p

Dunkerque, l'Armateur et l'Architecte. La reconquête des espaces portuaires, Région Nord-Pas-de-Calais, France, 2013, 127p, p. 61

La Maison de la Radio fête ses 50 ans, 2 tomes, France, Radio France Édition & Somogy Éditions d'art, 2013, 60p, 58p

LEC Lyon, France, Storm, 2013, 93p

Les Cahiers de la métropole, n°3, Mairie de Paris, 2013, 115p

*Les Édifices religieux du XX*e *siècle en Île-de-France, 1905-2000*, France, Beaux-Arts Éditions, 2013, 159p, p. 154

Hommes et projets d'exception, France, Vinci Construction, 2013, 123p

Innovative High-Rise Building, China, Fujian Science & Technology Publishing House, 2013, 463p

Qu'est-ce que la lumière pour les architectes ?, France, Archibooks + Sautereau Éditeur, 2013, 167p

1000 x European Architecture, Chris van Uffelen, Switzerland, Braun, 2012, 1006p
Arena Nanterre-La Défense, France, Racing-Arena, 2012, 32p

Contract Division, Italy, Poltrona Frau, 2012, 138p

Grand Prix AFEX 2012 de l'architecture française dans le monde, France, 2012, 56p, Catalog of the exhibition

Global Architecture Today, Xu Xiaodong, China, Tianjin University Press, 2012, 311p

Le Savoir-faire dans la peau, Poltrona Frau, Mario Piazza, Italy, RCS Libri Spa, 2012, 261p

Light in Architecture, Chris van Uffelen, Switzerland, Braun, 2012, 139p

Hopsca, Hi-Design, China, Ifengspace, 2012, 351p

Penser la ville durable. L'approche française, France, AFEX (Architectes français à l'export), 2012, 70p

Reconnaître les styles de l'architecture, Christophe Renault, France, Éditions Jean-Paul Gisserot, 2012, 32p

Top International Residential Building, Chine, Design Vision International Publishing Co., Ltd., 2012, 320p

Wallpaper City Guide Strasbourg,* United Kingdom, Phaidon Press, 2012, 128p

100 x N Architectural Shape and Skin, vol. I & II, Yang Feng, Chine, Ifengspace & Fenghuang Publishing, 2011, 379p

Annuel optimiste d'architecture 2011, France, French Touch, 2011, 400p

A Theatre Project, Richard Pilbrow, United States, Plasa Media, 2011, 459p

Écoquartiers, Olivier Namias, France, Éditions PC, 2011, 95p

International Housing Design Yearbook 2011, Zhang Xianhui, China, Tianjin University Press, 2011, 320p

International Interior Design Yearbook 2011, vol. 1, 2 & 4, Zhang Xianhui, China, China Forestry Publishing House, 2011, 320p, 344p, 360p

Les Villes et les Formes. Sur l'urbanisme durable, Serge Salat with Françoise Labbé and Caroline Nowacki, France, Hermann & CSTB, 2011, 544p

Les Styles de l'architecture, Christophe Renault, France, Éditions Jean-Paul Gisserot, 2011

Quand la ville grandit, Société d'équipement de la région montpelliéraine, France, 2011, 163p

Tirana. Contemporaneity in Suspension, Andrea Bulleri, Italy, Quodlibet Studio, 2011, 204p

Architecture 100 Ideas, China, H.K. Rihan Int'L Culture Spread Limited, 2010, 276p

Architecture Zone II, Rihan.cc, China, Tianjin University Press, 2010, 253p

Architecture Zone III, Rihan.cc, China, Tianjin University Press, 2010, 250p

Contemporary Architects II, Lizhuang, China, Hua Zhong University of Science & Technology Press Co., Ltd., 2010

Contrats export. Négocier et bâtir en dix points, AFEX (Architectes français à l'export), France, Le Moniteur, 2010, 130p

Construire en Chine, France, AFEX (Architectes français à l'export), 2010, 117p

Écrire le futur, ENIM, France, Press'Etic, 2010, 147p
Das Europaviertel Strasburg, Germany, Verlag Schnell & Steiner Gmbh Regensburg, 2010, 31p

From Z to A, France, Umicore Building Products, France, 2010, 315p

Guide découverte Strasbourg, Gilbert Poinsot, France, ID L'Édition, 2010, 64p

Le Fort d'Issy. Un patrimoine en devenir, Henri Ortholan, France, Archibooks + Sautereau Éditeur, 2010, 126p

La Défense 2030, Michèle Leloup, France, Archibooks + Sautereau Éditeur, 2010, 76p

Masterpieces Performance Architecture and Design, Chris van Uffelen, Switzerland, Braun, 2010, 304p

Montpellier. La ville inventée, France, Parenthèses, 2010, 262p

Paris. Métropole sur Seine, France, Éditions Textuel, 2010, 159p

The Architecture Guide, Chris van Uffelen and Markus Golser, Switzerland, Braun, 2010, 568p

The Latest Office Space Design of the World, Shenzhen Chuangyang Culture, China, Fujian Science Publishing House, 2010, 327p

World Hotels Architecture, Hong Kong Polytechnic International Publishing, China, China Forestry Publishing House, 2010, p. 355-362

Construire pour un développement durable, France, AFEX (Architectes français à l'export), 2009, 160p

European Housing Concepts 1990-2010, Luisella Gelsomino and Ottorino Marinoni, Italy, Editrice Compositori, 2009, 437p

La Revue #10 ans, France, La Galerie d'architecture, 2009

M2 -360° Interior Desing II, China, Sandu Publishing Co., 2009, 400p

Montpellier. Chroniques de Port Marianne. Une histoire urbaine (1989-2009), France, Dominique Carré Éditeur, 2009, 192p

Rodolphe Huguet, France, Monografik Édition, 2009, 288p

100 x 400, vol. 1, China, Rihan. cc, 2008, 485p, p. 64-75

La Devise républicaine sur les bâtiments scolaires parisiens, France, Mairie de Paris, Direction du patrimoine et de l'architecture, 2008

Les Styles de l'architecture. De la préhistoire à nos jours, Christophe Renault, France, Éditions Jean-Paul Gisserot, 2007, 192p

Shanghai Expo, Huang Yaocheng, China, Thomson & Shanghai Century Publishing Co., Ltd., 2007, 170p

Construire un projet de ville. Saint-Étienne « In Progress », France, Le Moniteur, 2006, 111p

Het Andere Parijs, Krista Bracke, Belgium, Uitgeverij EPO, 2006

Scénographies d'architectes, France, Pavillon de l'Arsenal, 2006, 486p

Master Architects in the New Millennium, China, China Scientific & Technical Publishers, 2005, 197p

Plus Architecture & Interior Design 'Collective architecture', vol. 222, Korea, Plus Publishing Co., Ltd., 2005

1000 Architects, Robyn Beaver, Australia, Images Publishing Group, 2004, p. 56

Notre démocratie, Italy, Quattroemme Editore, 2004, 105p

Les Eaux blessées, Direction générale Information and Relations publiques du Parlement européen, 2002, 40p

Architecture, China, 2001, p. 334-335, 'Architecture-Studio: Science City of Guangzhou'

Cafés Designers & Design, Christina Montes, Spain, Silverback Books, 2001, 216p, p. 124-127

Guide d'architecture contemporaine, Bas-Rhin, 1980-2000, France, Strasbourg, CAUE du Bas-Rhin, 2001, 272p, p. 208-209

Les Actes du séminaire 'Ville et Port', France, Graphica, 2001, 101p

The House Book, United Kingdom, Phaidon Press, 2001, 512p, p. 24

Water Spaces of the World, vol. 3, Australia, Images Publishing Group, 2001, 224p, p. 205-207

Aventures architecturales à Paris, France, Éditions A. et J. Picard, 2000, 236p, p. 124-139

Earthquake Architecture: New Construction Techniques for Earthquake Prevention, Bélen Garcia, United States, Paco Asensio, 2000, 208p, p. 45-52

Les Années 90, Anne Bony and others, France, Éditions du Regard, 2000, 669p

Panorama de l'architecture contemporaine, Francisco Asencio Cerver, France, Könemann, 2000, 998p, p. 316-318, 330-331

99 Architectures en 99: France, Diana Chan Chieng, China, China Architecture & Construction Publishing House, 1999, 250p, p. 38-39

Beyond Contemporary Architecture, India, Indo-French Technical Association, 1999, 44p, Catalog of the exhibition

Construire un nouveau millénaire, Philip Jodidio, Germany, Taschen, 1999, 576p, p. 76-79

Dunkerque en projet: Neptune 1989-1999, France, Archives d'architecture du Nord, 1999, 128p, Catalog of the exhibition

European House Now: Contemporary Architectural Directions, Daralice Boles and Susan Doubilet, United States, Universe, 1999, 240p, p. 162-167

Proposals of the International Design Contest of Guangzhou Stadium, China, China Architecture & Building Press, 1999, p. 123-134

Residences for the Elderly, Arian Mostaedi, Spain, Instituto Monsa de Ediciones, 1999, 237p, p. 102-111

Un siècle de constructions, 1900-2000, Dominique Barjot and Frédéric Lenne, France, Le Moniteur, 1999, p. 87, 94, 298, 301

L'Architettura delle università, Paola Coppola Pignatelli and Domizia Mandolesi, Italy, 1998, p. 40-45

L'Élaboration des projets architecturaux et urbains en Europe, vol. 3: *Les Pratiques de l'architecture: Comparaisons européennes et grands enjeux,* France, CSTB, 1998, 133p

The Contemporary Architecture Guide, vol. 1, Masayuki Fuchigami, Japan, Toto, 1998, p. 32-33, 62, 78, 93, 103, 106, 126-127

Formes nouvelles. Architecture des années 90, Philip Jodidio, Germany, Taschen, 1997, p. 39-41

International Architecture Yearbook III, Australia, Images Publishing Group, 1997, p. 342-345

Le Stade de France, France, Le Moniteur, 1997, 197p, p. 57

Paris sous verre. La ville et ses reflets, Bernard Marrey, France, Éditions A. et J. Picard & Pavillon de l'Arsenal, 1997, 223p, p. 101, 103, 137, Catalog of the exhibition

Residential Architecture, Carles Broto, Spain, Books Nippan, 1997, 237p, p. 120-129

Arquitectura Hoy, Peru, Arkinka – Museo de Arte, 1996, Catalog of fourth Art Biennale

Architettura per lo spazio sacro, Fabrizio Ivan Apollonio, Italy, Umberto Allemandi, 1996, p. 54, Catalog of the exhibition

Contemporary European Architects, Philip Jodidio, Germany, Taschen, 1996, p. 62-67

Églises parisiennes du XXe siècle, Simon Texier, France, Action artistique de Paris, 1996, 246p, p. 144, 244-245

International Architecture Yearbook II, Australia, Images Publishing Group, 1996, 432p, p. 390-393

L'Architecture du Futuroscope, Odile Perrard, France, Futuroscope & Hachette, 1996, 64p, p. 26-27
Le Logement social, France, 1996, p. 67-69

Le Logement collectif, Françoise Arnold, France, Le Moniteur, 1996, p. 39

581 Architects in the World, Masayuki Fuchigami, Japan, Toto, 1995, 632p, p. 76

International Architecture Yearbook I, Australia, Images Publishing Group, 1995, 256p, p. 52-55

Les Années 80, Anne Bony and others, France, Éditions du Regard, 1995, 694p, p. 436, 438, 461

Contemporary European Architects, Wolfgang Amsoneit, Germany, Taschen, 1992, 160p, p. 32

Crosscurrents: Fifty-One World Architects (Contemporary Architects – Ideas and Works), Masayuki Fuchigami, Japan, Synetics Inc., 1995, p. 18-21

L'Architecture contemporaine à Paris, Jean-Michel Hoyet, France, Éditions Techniques & Architecture, 1994, 155p, p. 51-83

Arquitectura francesa, 11 proyectos, Spain, Junta de Andalucia, 1993, p. 34-43, 64-71, Catalog of the exhibition

Centrum: Jahrbuch Architektur und Stadt, Reinhart Wustlich and others, Germany, Birkhäuser Basel, 1993, 223p, p. 241-249

Architecture for a Changing World, Steele James, United Kingdom, Academy Editions, 1992, 212p

Architettura e spazio sacro nella modernità, Paola Gennaro, Italy, Abitare Segesta, 1992, 341p, p. 225, Catalog of Venice Biennale

Construire en coût global, France, 1992, p. 18-19

Des fortifs au périf, Jean-Louis Cohen and André Lortie, France, Éditions A. et J. Picard & Pavillon de l'Arsenal, 1992, 320p, p. 301, Catalog of the exhibition

Jahrbuch Für Licht und Architektur 1992, Germany, Ernst & Sohn, 1992

Paris, Die Grossen Projekte, Paul Hans Peters, Germany, Ernst & Sohn, 1992, 151p, p. 62-73

Paris, la ville et ses projets, Pavillon de l'Arsenal, France, Babylone, 1992, 331p, p. 55-61, 220, 259-266

World Survey of the Top Three Hundred Architecture Firms, World Architecture, United Kingdom, 1992, p. 120-142

Bétons, matière d'architecture, Jean-Michel Hoyet, France, Regirex, 1991, 157p, p. 30-31

Die Neue Französische Architektur, Wojciech Lesnikowski, Germany, Kohlhammer, 1991, 223p, p. 54-67

Guide de l'architecture moderne à Paris, Hervé Martin, France, 1991, 320p, p. 36, 97, 136-137, 153-173

Iritecna per l'Europa 1991, Italy, L'Arca Ediz, 1991, 109p, p. 80-81

Les Lycées du futur, Jacques Garnier, France, L'Harmattan, 1991, 103p, p. 29, 73-75

Architectures publiques, Belgium, Pierre Mardaga, 1990, 128p, p. 21-22

Regards sur l'architecture. Comprendre l'art et la technique des architectes, Borel Yolande et Véronique Girard, France, Éditions du Sorbier, 1990, 61p

The New French Architecture, Wojciech Lesnikowski, United States, Rizzoli Intl Pubns, 1990, 223p

Construire Nel Construito. Conversation en architettura moderna, Italy, Éditions Kappa, 1989

Paris Bygger, Sten Gromark, Sweden, Svensk Byggtjänst, 1989, 168p

Tendenze nell'architettura degli anni'90, Mario Pisani, Italy, Dedalo, 1989, 166p, p. 11-17

Université d'été Grands Projets de l'État, 1979-1989, France, Axium, 1989, 85p

Architecture contemporaine 88-89, Switzerland, Anthony Kraft Éditeur, 1988, p. 210-215

Concours d'architecture 1988, France, Mairie de Paris, Direction de l'architecture, 1988, 112p

Paris – Architecture et utopie. Projets d'urbanisme pour l'entrée dans le XXIe siècle, Kristin Feireiss, France, Pavillon de l'Arsenal and Wilhelm Ernst & Sohn, 1988, 238p, Catalog of the exhibition

10 Ans d'équipements publics, France, Ville de Paris, Direction de l'architecture, 1987, 95p

Horizon 90, France, 1987, p. 27-30

Paris. Nouvelle architecture, Michèle Behar and Manuelle Salama, France, Regirex-France, 1986, 175p

À travers l'architecture. 128 réalisations, 128 architectes, France, La Décade d'architecture, 1983, 187p

À la recherche de l'urbanité. Savoir faire la ville, savoir vivre la ville, France, Academy Editions, 1980, 176p, Catalog of the first Architecture Biennale of Paris

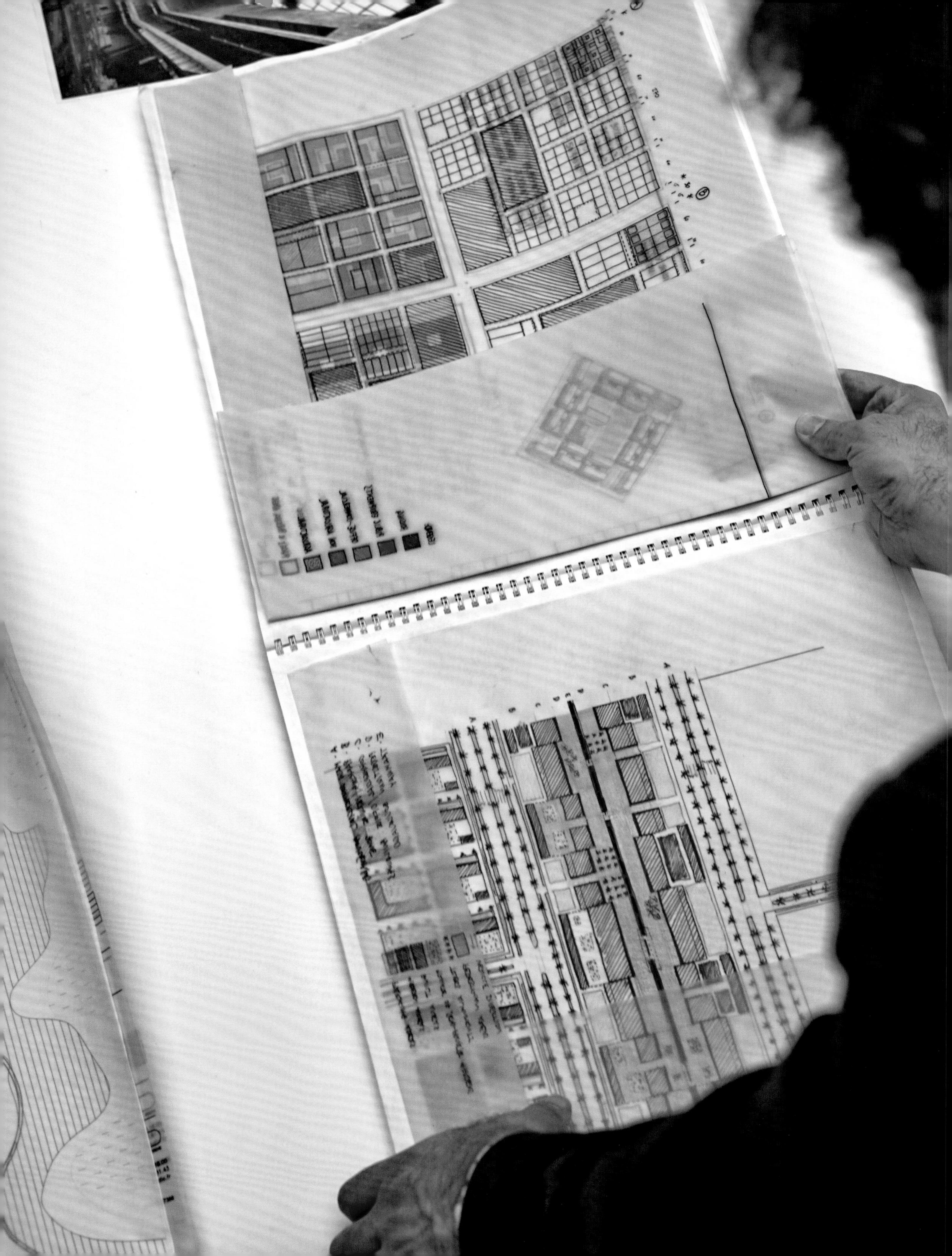

CREDITS 图片提供

All the plans of this book have been made by Architecture-Studio.
本书中所有的设计图由法国AS建筑工作室完成并提供

Architecture-Studio: 16, 17, 18, 19, 23, 27, 28, 29, 30, 31, 33, 34, 39, 40, 41, 48, 49, 51, 55, 59, 60, 61, 62, 65, 68, 71, 72, 73, 75, 76, 79, 91h, 101, 102, 103, 108, 109, 110, 111, 118, 119, 120, 122, 123, 124, 125, 126, 127, 149, 150, 151, 152, 155, 156, 157, 164, 168, 169, 170, 171, 172, 182, 183, 184g, 186, 187, 205, 229, 230, 231, 241, 243, 246, 248g, timeline
Architecture-Studio, Babel & L'Agence: 89, 90, 91b
Arc en rêve centre d'architecture: 53
Ayman Alakhras: 166, 167
Frédéric Baron: 175b, 177
Louidgi Beltrame: 50
Gaston Bergeret: 8, 15, 66, 67, 69, 153, 206, 247
Luc Boegly: 143, 144, 145, 176, 203, 204, 208, 212, 213, 217, 219, 221, 239g, 242, 250d
Nicolas Buisson: 64g, 185, 196, 198, 199, 201, 235, 248d
Change is good: 4, 12, 24, 42, 56, 104, 146, 178, 232, 252, 272, 275, 278, 284, 287
Stéphane Couturier: 54
Nikos Daniilidis: 139, 140, 141, 238
Hervé Douris: 63,
Matthieu Ducros: 85
Antoine Duhamel: 93, 94, 95, 96, 97, 98, 211, 214, 240h
Georges Fessy: 37, 45, 113, 116, 117, 189, 191, 192, 193, 236, 240b
Shu He: 107g, 132, 133, 134, 136, 137, 160, 237, 251
Valérie Jouve: 52
Yongjin Luo: 107d
Olivier Marceny: 22, 129, 131, 159, 162, 163, 244, 245, 249g
Mereglier-Coudrais: 175h
Anna Puig Rosado: 64dcd
Roger Rothan: 114
Claude O'Sughrue: 78
Takuji Shimmura: 194, 227, 239d
Julien Thomazo: 86, 250g
Patrick Tourneboeuf: 184d
Benoit Werhlé: 80, 82, 83, 84, 249d
Yiming Zhang: 223, 224
Christophe Bourgeois : 71, 72, 73

The editor would like to thank
the partners of Architecture-Studio
for their trust.

编者感谢法国AS建筑工作室合伙人的信任。

Thanks also go to Vanessa Clairet and
Tiphaine Riva, for their unqualified support.
Finally, thanks to Philip Jodidio for his
synthetic approach to the multi-input story
and to Hugo Lacroix for his analyses
of the design projects.

同时也要感谢Vanessa Clairet和Tiphaine Riva两人的无条件支持。最后，感谢菲利普·朱迪狄欧从多角度呈现工作室的历史和理念，以及雨果·拉克鲁瓦对每个项目的介绍与分析。

PLAN DE TOITURE
JUILLET 98 ECH 1/100

图书在版编目（C I P）数据

Architecture-Studio 法国 AS 建筑工作室 /（法）菲利普 · 朱迪狄欧（Philip Jodidio）编著 .. -- 天津 ：天津大学出版社，2018.3
ISBN 978-7-5618-5888-2

Ⅰ . ① A… Ⅱ . ① A… Ⅲ . ①建筑设计－作品集－世界
Ⅳ . ① TU206

中国版本图书馆 CIP 数据核字（2017）第 163196 号

ARCHITECTURE
—STUDIO

法国AS建筑工作室 /（法）
菲利普 · 朱迪狄欧编著

出版发行　天津大学出版社
地　　址　天津市卫津路 92 号天津大学内（邮编：300072）
电　　话　发行部 022-27403647
网　　址　publish.tju.edu.cn
印　　刷　北京华联印刷有限公司
经　　销　全国各地新华书店
开　　本　220mm × 292mm
印　　张　18
字　　数　450 千
版　　次　2018 年 03 月第 1 版
印　　次　2018 年 03 月第 1 次
定　　价　240.00 元